Agro-Techniques of
MEDICINAL PLANTS

Agro-Techniques of
MEDICINAL PLANTS

RAVINDRA SHARMA 'Nature's Friend
M.Sc., Ph.D., F.N.R.S.
Consultant & Technical Expert (Medicinal Plant Cultivation)
GREEN FOUNDATION
(For health care, social equity & biodiversity conservation)
DEHRADUN (Uttaranchal)

2013
DAYA PUBLISHING HOUSE
Delhi – 110 035

Published by	:	**Daya Publishing House®**
		A Division of
		Astral International Pvt. Ltd.
		– ISO 9001:2008 Certified Company –
		4760-61/23, Ansari Road, Darya Ganj
		New Delhi-110 002
		Ph. 011-43549197, 23278134
		E-mail: info@astralint.com
		Website: www.astralint.com
Laser Typesetting	:	**Classic Computer Services**, Delhi - 110 035
Printed at	:	**Thomson Press India Limited**

PRINTED IN INDIA

FOREWORD

Age old traditional sources of medicines *viz.* medicinal plants with their myriad curing abilities of many dreaded ailments with hardly any side-effects are getting prominence. Demand for medicinal plants is increasing at an exponential rate. This up-surging demand for medicinal plants has led to indiscriminate and unscientific exploitation of their natural habitat in forests. This has resulted in extinction or near extinction of many invaluable medicinal plants in their own habitat. Thus, an alarming situation is emerging wherein in near future many of these plants will be lost irrecoverably. Every effort should, therefore, be made immediately to reverse this trend before many of them are lost. Now, time has come to apply modern tools of science to improve the genetic make up of these medicinal plants, their agro-technique in agricultural lands away from their native sites and; to tailor agro-techniques to improve their yield potential. The expanding local and global market and immense export potential to Europe, America and other Western Countries has opened up a new prospective field for Indian agriculture. To fully make use of this emerging demands world-wide, appropriate knowledge regarding their improved cultivation, processing and marketability is of utmost important. However, there is very little systematic information on these aspects.

The book by Dr. Ravindra Sharma on *"Agro-Techniques of Medicinal Plants"* is a welcome step in fulfilling this need. It has covered the most relevant aspects pertaining to nursery technology, cultivation, agro-technique, trade and commerce in medicinal plants. I hope the techniques described in the book are well tested in the field. If so, this book will be valuable source of information to policy makers, researchers and progressive farmers interested in cultivation of medicinal plants.

I am delighted to write this foreword and congratulate to *Dr. Ravindra Sharma* for his painstaking efforts in producing this work, which I hope will go a long way and contribute significantly to our existing knowledge in this area. I wish to Dr. Sharma success in his future endeavor in this direction.

Padmshree (Prof.) A. N. Purohit (Retd.)
Ex. Vice Chancellor, H.N.B.G. University, Srinagar (UA)
Ex. Director, HAPPRC, Srinagar (UA)

PREFACE

With the increasing awareness about the usefulness of medicinal plants, more and more people are taking to their cultivation. Changing cropping pattern and higher returns have prompted many farmers to growing the medicinal plants commercially. Since these crops are new and uncommon to the growers and they often lack of knowledge of their scientific cultivation, there is hesitation even among the interested growers, to take up their cultivation. However, the information on progress made in improvement of medicinal plants, cultivation practices and marketing prospects are out of reach to many interested people. Most of the advancements made in this regard are either confined to research stations or lie scattered in the research journals. Not much information is available in bound and readily usable form at one place. This book entitle *"Agro-Techniques of Medicinal Plants"* make an earnest effort to present the progress made in crop improvement, cultivation technique and trade related matter pertaining to some important medicinal plants.

The book consists of eleven chapters *viz.* 1. Introduction, 2. Nursery Technology, 3. Biofertilizers and Biological Pest Control, 4. Organic Farming, 5. Agro-Techniques of Medicinal Plants, 6. Harvest Technology and Value-Addition of Medicinal Plants, 7. Adulteration and Substitution of Crude (Herbal) Drugs, 8. Market Potential (Export/Import) of Crude Drugs, 9. Standardization and Quality Control of Medicinal Plants, 10. Legislation and Policy of Medicinal Plants, 11. Sustainable Conservation and Development Strategies of Medicinal Plants.

The Chapter first 'Introduction' covers history, traditional system of medicine, market trade etc. Chapter two 'Nursery Technology' provides some important tips of nursery management. The chapter third ' Biofertilizers & Biological pest control' has covered the basic knowledge of biofertilizer and Integrated pest management. The fourth chapter 'Organic farming' an attempt has been made to compile available information and present the current status of organic farming, its strength and weaknesses, and research needs, for providing a basis for development of alternate farming system.

The chapter fifth ' Agro-Techniques of Medicinal Plants' consists latest agro-technique for realizing higher productivity in the 50 commercially important medicinal plants like *Aonla, Ashwagandha, Babchi, Basella, Belladonna, Bishop's weed, Brahmi, Buckwheat, Chebulic myrobalan, Cinchona, Coleus, Coptis, Costus, Cowhage, Danti, Datura, Digitalis, Dioscorea, Duboisia, Ergot of Rye, Glory Lily, Guggal, Henbane, Himalayan Ginseng, Himalayan Yew, Indian Aloe, Indian long Pepper, Indian tejpat, Ipecacuanha, Isabgol, Jatamansi, Kalmegh, Kuth, Liquorice, Musk mallow, Neem, Opium Poppy, Periwinkle, Podophyllum, Primrose, Pyrethrum, Quinghao, Red sorrel, Safed musli, Serpentine root, Senna, Shatawar, Silybum, Steroid- Solaum, Sweet flag.*

The sixth chapter 'Harvest technology and value addition' has covered collection practices, part used, active constituents and various value added (semi-processed) products. The seventh chapter 'Adulterants & substitution of crude drugs' included an important information on adulterants & substitution of crude drugs. The chapter eight 'Market potential (export/import) of crude drugs' covering trade and commerce, export and import in medicinal plants and quality aspects. The chapter ninth 'Standardization and Quality control of Plant drugs' an attempt the basic concept of standardization and quality control for exportable crude drugs. The chapter tenth 'Legislation and policy of medicinal plants' include about the national policy and National Medicinal Plants Board. The last chapter 'Sustainable Conservation and Development Strategies for medicinal plants' deals with sustainable conservation of medicinal plants in the face of indiscriminate exploitation of these plants.

It is hoped that this publication provide recent information covering major aspects related to the cultivation of medicinal plants will be highly helpful to the student, teachers, extension workers, researchers, traders, ayurvedic practitioners and grower of medicinal plants as a source book of information on the said topics.

As it is my first attempt in this direction, there may be many shortcomings. Any criticism and suggestions from the research workers and the enlightened farmers in this regard are welcome, so that in the future editions, such mistakes can be avoided.

It is my pleasure to acknowledge the help which I received from many publications for completing the book. While writing this book we were very fortunate in having exceptional and invaluable cooperation from a number of libraries of the Botanical Survey of India (Northern Circle), Forest Research Institute (FRI), Dehradun, Indira Gandhi National Forest Academy (IGNFA), Dehradun, Indian Agricultural Research Institute, New Delhi, Central Institute of Medicinal & Aromatic Plants (CIMAP), Lucknow, National Botanical Research Institute (NBRI), Lucknow, Central Council of Research in Ayurveda & Siddha (CCRAS), New Delhi. In particular, we would like to acknowledge the generous help to Dr. D. K. Sharma, M.D., Om Herbal Remedies & Research Centre, Haridwar (UA) and Ms. Hemlata Sharma, M.Sc. (Pharmaceutical Chemistry) S.B.S. (P.G.) College, Dehradun.

I am also grateful to numerous inhabitants who had been kind enough to provide local information about the medicinal plants. Sincere thanks are also due to my wife Smt. (Dr.) Ranjana Sharma who has been a source of inspiration to pursue the work in all the circumstances. Words fail to express sense of indebtedness to my parent Sri M. C. Sharma (Civil Engineer) and Smt. Vimla Devi, who imbibed in my mind the aesthetics of the Nature.

I am also thankful to Mr. Anil Mittal, The Publisher, Daya Publishing House, deserves special appreciations for his painstaking and enthusiastic approach to complete this strenuous task.

Dr. RAVINDRA SHARMA
MIG-149-B, Indirapuram, GMS Road, Dehradun, Uttaranchal
Telefax: 0135-2621932, Mobile: 9412381158
E-mail: gf_india@yahoo.com

CONTENTS

     6.1. Collection

     6.2. Harvesting

     6.3. Drying

         6.3.1. Natural Drying (Sun-drying)

         6.3.2. Artificial Drying

     6.4. Garbling (Dressing)

     6.5. Packing

     6.6. Storage

     6.7. Value-Addition or Processing

         6.7.1. Decoction

         6.7.2. Extraction

IMPORTANT TIPS FOR CULTIVATION OF MEDICINAL PLANTS

Indian is endowed with a rich wealth of medicinal plants, which ranked our country in the list of top producers of herbal medicines. This is alarming that the total forest area is reducing which has reduced the total cultivation/collection of medicinal plants in India. If this kind of damage continues, we may lose the wealth of medicinal plants, which will affect the growth of our economy. Therefore, it is essential to cultivate at large scale and to conserve heritage of medicinal plants.

It was an old interest of the farmers to cultivate medicinal plants, and unlimited research efforts have been made to improve the varieties of medicinal plants, however, we are still at the developing stage with regard to cultivation of medicinal plants because there is no standard organization/ agency for the certification of these plants in India. Therefore, in such situation it is essential to know the tips to follow them carefully in the cultivation of medicinal plants.

☞ Try to collect the available literature on medicinal plants which you can able to collect. Read it carefully for many times. The literature is available in the forms of books, magazines, research paper, films etc. You may develop habit to participate in seminars/conferences or training programme, etc.

☞ Try to visit the farms of those farmers who are cultivating medicinal plants successfully and listen their experiences with special reference to their techniques for growing these plants, storage and packing practices, procedure of their sale, etc.

- ☞ You have to evaluate your own capacity honestly in following ways:
    - ☞ You will cultivate the plants at your own.
    - ☞ The plants will be cultivated by others on your farm.
    - ☞ How much of land you owned?
    - ☞ How much time you will spare for cultivation?
    - ☞ The land value
    - ☞ It is ownership/rental land.
    - ☞ Type of land and soil quality
    - ☞ Irrigation facilities and climatic condition of your land.
- ☞ If you have few selected plants for cultivation, find out its Market potential (price, sale value, etc) by their market survey. You should known about the buyer/company, exporters. The selected medicinal plant should be discussed with these buyers very carefully. When you are satisfied with the capacity and requirement of buyer along with his term and condition, then proceed with the cultivation.
- ☞ You should know the part of Medicinal Plant has demand in the market *e.g.* root, fruit, leaves, flowers.
- ☞ You should know various parameters (active constituents) of the standard as required by buyer.
- ☞ You should know the climatic (rainfall, temperature, moisture etc.) edaphic (soil nature, pH value, mineral status, etc) and other requirement of the plant, which you are proposed to cultivate.
- ☞ After extensive survey of the medicinal crops, select only 3 or 5 medicinal plants for cultivation purpose.
- ☞ The growers should carefully purchase planting material (seeds/seedling) either from institution or farmer directly.
- ☞ Never use chemical fertilizers or minimize their use. As far as possible, use organic manure (cowdung, etc) only.
- ☞ If the selected medicinal plants are infected with diseases, try to cure them using indigenous practices or bio-pesticides (natural biological agents).
- ☞ The cultivation of medicinal plants may be a new field for you. Therefore, at initial stage, try to cultivate at small scale to know the characters/properties of the particular plants.
- ☞ When you have started cultivation of medicinal plants, do not advertise much among your fellow-farmers. As and when you get success and launch commercial cultivation, let you propagate your success to others, if you are failed, you may be depressed.
- ☞ After getting success in your venture, let other farmers should known your success and people will join you. This will result into reorganization of your village/place as a prime growing area of a particular medicinal plants.
- ☞ Your success and cultivated area should come to record of local/state Government, specially for those plants which can not be exported.

The above guidelines help you to become a successful grower of medicinal plants.

Chapter 1

INTRODUCTION

India enjoys the privilege of having time tested traditional systems of medicines based on the natural products. Plants based products have been in use for medicinal, therapeutic or other purposes right from the dawn of history. The traditional remedies of the ancient world were all based on natural products, *e.g.* morphine from opium for use as narcotic and analgesic, latex exuded from the poppy seed for gout, cocaine from cocoa-leaf as potential local anesthetic, ergot for obstetric use, castor oil, senna etc. as laxatives, turmeric as an antiseptic, garlic for blood and heart remedies to mention just a few. The traditional Indian system of medicine, namely Ayurveda which involves dispensing of herbal and plant products in various forms such as powders, extracts, decoctions etc., dates back to the vedic period, when the first mention of diseases and drugs is found in the Rigveda and Yajurveda (*i.e.* around 2000 B.C.) and the earliest comprehensive description of Ayurveda is available in the Atharvaveda (*i.e.* 1600-1000 B.C), which contains inter alia descriptions of human anatomy, rudiments of classifications of diseases and reference to herbal medicine.

Recent past has witnessed an upsurge in the popularity of the herbal medicines. In the developing countries, about 80% of the people depend upon the traditional system of medicine and 95% of the industrial need of this is met through indiscriminate collection from wild. According to an estimation, the over half a million tonnes of dry raw material is indiscriminately collection of plants has led to considerable genetic erosion and loss of biodiversity including plants of medicinal value.

Among endangered plant species, medicinal herbs account for almost 1/3 of the species mentioned in the red data book. In the absence of organized cultivation, 14 species have become endangered or rare and 35 species are vulnerable.

Medicinal plants as a group comprise approximately 8000 species and account for around 50% of all the higher flowering plant species of India. India possesses almost 8% of the estimated biodiversity of the world. India has 16 Agro-climatic zones, 45,000 different plant species and 15,000 medicinal plants that include 7,000 plants used in Ayurveda, 700 in Unani medicine, 600 in Siddha medicine, 450 in Homeopathy and 30 in modern medicines. India is one of the 12 mega biodiversity centers with 2 hot-spots of biodiversity in Western Ghat and North-Eastern Region. There are about 400 families in world of the flowering plants, atleast 315 are represented in India

In India, there are 7843 licensed pharmacies of Indian System of Medicine in addition to 857 of Homeopathy and a number of unlicensed small scale processing unit engaged in the manufacturing of the medicines to meet the requirement of 4.65 lakhs registered practitioners of ISM&H and other users. Although about 8000 species of plants are estimated to be used in human and animal health care and over 10,000 herbal drug formulations have been recorded in codified medical texts of Ayurveda, the pharmaceutical industries are largely based on about 400 plant species.

According to 1994 UNDP report, the annual value of medicinal plants derived from developing countries is approximately 32 billion US dollars. There are 47 major modern pharmaceutical plant-based drugs already in the world market and the predicted 328 drugs, yet to be discovered, have a market potential of 147 billion US dollars. The sale in the very first year of the anticancer drug, taxol from *Taxus* spp. has been more than 2000 billion US dollars. Phyto-remedies and health-food over-the-countries sales in USA is expected to touch 2 billion US dollar according to American Botanical Council.

Unsustainable ways of harvesting and unrestricted marketing have led to the reduction in population of some of the high demand medicinal plants leading to sudden escalation in prices of these crude drugs in the market.

Herbal medicine continues its unbridled growth. Sales have now reached a record breaking $ 1.5 billion annually in United States and a figure of $ 5 billion is predicted by the end of decade (Tyler, 1996). In Germany, annual sales of herbal drugs is at $ 2.5 billion and per capita spending of $ 37 on phytomedicine (Gruenwald, 1998). The value of global trade of the medicinal plant products has been put over US$ 75 billion per year and is growing @ 12.5% annually. Of the total value of trade, about 20 billion US dollar account for over the counter (OTC) drugs, US$ 25 billion for prescription drugs and remaining US$ 30 billion for the nutritional supplements.

Bulk of the raw material (90%) is produced in Asia, Africa and Latin America and some in Europe and USA (10%). About 60% of the total material is imported and processed in USA, Canada, UK, Australia, Germany, France, Italy, Switzerland and Japan and about 50% of that is used there and the rest is exported to the raw material producing countries to be sold at high rates. About 90% of marketed material is collected from the wild resources and there is some cultivated in China, India, USA, Germany, France, Italy and Eastern Europe.

The trade of medicinal plants in India is estimated to the tune of Rs 675 crores per year. Of India's total turnover of Rs. 3,100 crores of Ayurvedic and herbal products, major OTC products contribute around Rs. 1,700 crores, Ayurvedic ethical formulations constitute around Rs. 850 crores and Ayurvedic classical formulations constitute remaining Rs. 550 crores. Indian export has steadily grown at the rate of 65%, since 1991-92 and grew to Rs. 215 crores in 2000-01 from Rs.130 crores in 1991-92. It is estimated that the global market for herbal drugs is nearly Rs. 800 crores per year. The export of medicinal plant from India appears to be growing faster.

Conservative estimates put the economic value of medicinal plants related trade over US$ 90 billion. Demand and trade in medicinal plant species globally indicates a upward trend and World trade in medicinal plant and related products is expected to rise to US$ 5 trillion by 2050 A.D.

Chapter 2

NURSERY TECHNOLOGY

Nurseries are going to play a very important part in afforestation and reforestation programme in India. A few decades back, plantations used to be raised mainly in the reserved forest area which are generally moist and hospitable, and direct seed sowing technique in the plantation site could be successfully practiced. The method consumed excess seeds, but since the total planting program was small, the absolute seed requirement did not amount to much. The chance of damage to the small germinating seedlings by grazing existed, but it was much less than it is today, since protection by intense policing was practicable. The present program of plantations in the country also includes planting in private farm lands, in community grazing lands and in government wastelands. Environmental factors at these sites are often harsh, with one or more adverse conditions present, such as erratic rainfall, infertile soils, high alkalinity and salinity, water logging, excessive biotic pressure and so on. At such sites, plantations by direct seed sowing are likely to be unsuccessful. Nursery raised seedlings of appropriate quality to raise plantations are of primary importance.

In order to raise healthy seedling in the nursery, the important factor to be considered are site selection, preparation of beds and potting mixture, pricking out of seedling, nursery establishment, manuring and fertilizer, watering, plant protection, grading and storing. Genetic quantities of the mother plant is very important in seed collection as trail like better yield, resistance to pests and diseases, hardy nature to adverse environmental conditions like drought and flood are passed from one generation to the other. The quality of the seeds depends on number of plants required for actual planting, germination percentage of seeds, viability of seedlings and post planting mortality. Screening and selection of superior seedling in the nursery itself is very important.

For the development of ideal nursery for the propagation of germplasm of medicinal plants, few nursery management tips are given below:

2.1 Nursery Management Tips

2.1.1. Site Selection

☞ The site of the nursery should be as near the planting site as possible. The nursery should lie within a radius of 5 km from planting sites in the plains and 2 km in rugged hilly terrain.

☞ The area of the nursery should be about 0.4 ha (1 acre) for every 100,000 seedlings

2.1.2. Potting Mixture (Nursery-Soil)

☞ There is a tendency to collect the topsoil area to fill the polythene begs. The practice is not good as it contributes to soil erosion.

☞ The whole success of nursery depends on good soil mixture. It should be light in weight, well decomposed and well drained and free from insects as well as weed seeds.

☞ Always use compost in the nursery. It can easily be prepared by decomposing organic materials (green leaves). It increases the nutrients contents of the potting mixture and enhances the air spaces and water holding capacity of the mixture.

2.1.3. Seed Collection and Sowing

☞ Always collect seeds from plus mother plants or phenotypically superior plants.

☞ A chart should be prepared in each nursery indicating the seeds collection period of medicinal plants along with the location of the herbs.

☞ Seedling developed from poor or abnormal medicinal plants will never produce good plants.

☞ After collection seeds should be processed carefully otherwise they may get damaged and loose viability.

☞ Each medicinal plants required different processing after seed collection *i.e.* seeds with pulp (kadam) are to be processed in a different way than the pods (Mandhani), drupes (bansum and dhuna) and capsules (opium).

☞ In tropical herbs, the most of the seeds have short viability; therefore, sowing is to be done immediately after collection and processing the seeds *eg.* (Nagkesar).

☞ For the pretreatment of seeds, if needed (seed coat hard), it is always advisable to put the seeds in water for 24 hours and dry them in the shade before sowing.

☞ Sow large seeds (Amla) 2 cm to 3 cm deep in the soil.

☞ Sow small seeds (Kalmegh) about 0.5 to 1 cm deep or broadcast and cover them with a small amount of fine soil. The beds can also covered with thatch or plastic sheets till the germination of seeds.

☞ Before storing, the seeds should be mixed with prophylactics like gamaxene or neem leaves.

2.1.4. Watering

☞ Keep seeds beds moist not wet.

☞ Do not water at a fixed time each day. Water when the plants need it.

☞ Small seedling require less amount of water.

☞ Large seedling require more water more often.

☞ Seedling require more water more often on windly days and sunny days.

- Water less often when seedling are kept in shade.
- The moss and algae growth is an indication of excess watering.
- Always use clean water; dirty water may cause diseases problem.
- Over-watering results into weak plants and cause diseases such as root-rot and damping-off of seedling.

2.1.5 Transplanting

- Water mother beds thoroughly before Transplanting the seedling.
- Always use some tools (bamboo-sticks, etc.) to loosen the soil before pulling out seedling from the beds.
- Make a deep and wide hole in the polythene bag or container for Transplanting the seedling.
- Hold seedling at the base of the stem and pull it out gently from the mother beds.
- Never bend the roots and do not force the seedling to fit into the hole.
- Keep transplanted seedling under proper shade until they have recovered.

2.1.6. Nursery Hygiene and Disease Control

- Most of the micro-organism, insects and pests that cause disease in the nursery live in weeds, trash and puddles. Therefore, keeping nursery in neat and clean condition to reduce the chances of attack of common diseases. All trash, waste polythene bags and diseased plants should immediately be removed and burnt at a place far from the nursery.
- *Pesticides Recommended*: Diazinon, Melathion, Sevin.
- *Fungicides Recommended*: Bavisteen, Benlate, captan, Dithane M-45.
- *Dangerous Pesticides*: Aldicarb, Aldrin, DDT, Lindane, Nuvacron, Heptachlor etc. These are very dangerous. Never use them.
- *Natural insecticides*: Neem seeds,
- Proper doses should be prepared by carefully reading the label or guidelines.
- Always provide protective clothing, mask, gloves and goggle to the person spraying the pesticides.
- Never mix insecticides and fungicides together in the same sprayer.
- Never eat, drink or smoke while spraying.
- Extra pesticides should be disposed by digging a hole away from rivers and wells.

Chapter 3

BIOFERTILIZERS AND BIOLOGICAL PEST CONTROL

In recent times, the cropping pattern has undergone perceptible changes. During post-green revolution period, 1967-68 to 1992-93, growth rate in agricultural production was assessed at 2.84 per cent per annum. Production of food grains during this period increased from 950.5 lakh tonnes to 1795 lakh tonnes. Per capita net amiability of food grains increases to 468.5 g per day from 395 g per day in early fifties. Modern agriculture rely heavily on a wide range of synthetic chemicals which include different types of fertilizers and pesticides. Use of these chemicals has played a very vital role in increased agricultural production. During 1986-87, fertilizer consumption was 8.64 million tonnes, and it has increased to 13.83 million tonnes during 1994-95. Use of chemical fertilizers is a costly input and their manufacture depends on the dwindling resources of energy such as petroleum and coal. Productivity of the soil is undergoing strains as per unit crop yields are stagnating and higher doses of fertilizers are not giving commensurate increase in yields. Chemical fertilizers being industrial products are subjected to cost increases every year, and becoming unaffordable to a large segment of farmers. In order to withstand these constraints, a scheme "Balanced and Integrated Use of Fertilizers" is implemented to promote the use of chemical fertilizers in conjunction with enhanced use of organics, compost, green manuring and biofertilizers. Similarly, due to widespread use of pesticides the food consumed by man today is contaminated with chemicals. Rain wash down pesticides into rivers causing pollution of water. In recent years, Integrated Pest Management (IPM) was given high priority to enable farmers to adopt measures which are environment friendly and less costly. Use of biocontrol agents, pesticides of plant origin, use of plant protection chemicals with residues not exceeding toxic limits and employing newer methods for rodent, birds, nematodes and mite control are being promoted.

Under a given situation, the system of the farming, soil management and manuring practices, etc. influence the productivity of soils and crop yields obtained from them. It is estimated that the different agricultural crops in India remove about 4.27 million tonnes of nitrogen, 2.13 million tonnes if phosphorus, 7.42 million tonnes of potash and 4.88 million tonnes of lime per year. Obtaining larger yields through improved varieties of crops and intensive cultivation increases the depletion of nutrients still further. Moreover, erosion and leaching cause additional losses. It is thus obvious that the huge depletion of nutrients impoverish the soil unless these supplies are replenish by natural or artificial means. The principle methods of supplementing natural recuperation and for improving the productive capacity of the soil are:

℞ To add organic matter (manures) to the soil so that through decay it may furnish a more or less continuous supply of nutrients to crops, and

℞ To restore or increase the amount of deficient nutrients by the application of fertilizers.

3.1. Manures

Manures are the organic substance obtains by the decomposition of animal wastes and plant residues, which supplies essential element, and humus of the soil. They are added mainly to improve the physical condition of the soil, to replenish and keep us the humus status to maintain the optimum condition for the activities of soil micro organism and make good a small part of the plant nutrients removed by crops of otherwise lost through leaching and soil erosion .thus they supply practically all the element of fertility which crops require, though not in adequate proportions.

3.1.1. Farmyard Manure

Good-quality farmyard manure is perhaps the most valuable organic matter applied to a soil. It is the most commonly used organic manure in India. It consists of a decomposed mixture of cattle dung, the bedding used in the stable and any remnants of straw and plant stalks fed to cattle. Partially rotten farmyard manure should generally be applied to the soil about 3 to 4 weeks before the sowing of a crop. However, if the manure is already well rotted, it is advisable to apply it just before sowing to a crop. Farmyard manure can be applied to all crops including medicinal plants grown in the rainy season or grown under irrigation. However, the use of farmyard manure alone causes an imbalance in nutrition owing to its relatively low content of phosphates. Therefore, to keep the soil well supplied with all the essential elements of plants food in a readily available form. It is one of the most important agricultural by-products. Unfortunately more than 50 per cent of the cattle dung produced in India is burnt as fuel and is thus lost to agriculture.

It must be stressed that the value of farmyard manure in soil improvement is due to its content of macronutrients and its ability to *(i)* improve the soil tilth and aeration, *(ii)* increase the water-holding capacity of the soil, and *(iii)* stimulate the activity of micro-organisms that make the plant-food elements in the soil readily available to crops. The supply of organic matter, which is later convened into humus is a property of farmyard manure. One tonne cattle dung can supply only 2.95 kg of nitrogen, 1.59 kg of phosphoric acid and 2.95 kg of potash.

Table 1: Average Nutrient Content (in %) of Compost Manures

Manure Items	Nitrogen (N)	Phosphoric Acid (P_2O_2)	Potash (K_2O)
Cattle dung, fresh	0.3-0.4	0.1-0.2	0.1-0.3
Horse dung, fresh	0.4-0.5	0.3-0.4	0.3-0.4
Sheep dung, fresh	0.5-0.7	0.4-0.6	0.3-1.0
Poultry manure, fresh	1.0-1.8	1.4-1.8	0.8-0.9
Raw sewage, fresh	2.0-3.0	-	-
Sewage sludge, dry	2.0-3.5	1 .0-5.0	0.2-0.5
Cattle urine	0.9-1.2	tr.	0.5-1.0
Horse urine	1.2-1.5	tr.	1.3-1.5
Human urine	0.6-1.0	0.1-0.2	0.2-0.3
Sheep urine	1.5-1.7	tr.	1.8-2.0
Wood ashes Ash, coal	0.73	0.45	0.53
Ash, household	0.5-1.9	1.6-4.2	2.3-12.0
Ash, wood	0.1-0.2	0.8-5.9	1.5-36.0
Rural compost, dry	0.5-1.0	0.4-0.8	0.8-1.2
Urban compost, dry	0.7-2.0	0.9-3.0	1.0-2.0
Farmyard manure, dry	0.4-1.5	0.3-0.9	0.3-1.9
Filter-press cake	1.0-1.5	4.0-5.0	2.0-7.0

Table 2: Average Nutrient Content (in %) of Crop Residue, Straw and Stalks Manures

Manure Items	Nitrogen (N)	Phosphoric Acid (P_2O_2)	Potash (K_2O)
Rice hulls	0.3-0.5	0.2-0.5	0.3-0.5
Groundnut husks	1.6-1.8	0.3-0.5	1.1-1.7
Pearl millet	0.65	0.75	2.50
Banana, dry	0.61	0.12	1.00
Cotton	0.44	0.10	0.66
Sorghum	0.40	0.23	2.17
Maize	0.42	1 .57	1.65
Paddy	0.38	0.08	0.71
Tobacco	1.12	0.84	0.80
Pigeon pea	1.10	0.58	1.28
Wheat	0.53	0.10	1.10
Sugarcane trash	0.35	0.10	0.60
Neem cake	5.2	1.0	1.4
Mahuwa leaves	1.66	0.5	2.0
Castor cake	4.3	1.8	1.3
Tobacco dust	1.10	0.31	0.93

Table 3: Average Nutrient Content (in %) of Dry Leaves Manures

Manure Items	Nitrogen (N)	Phosphoric Acid (P$_2$O$_2$)	Potash (K$_2$O)
Calotropis gigantean	0.35	0.12	0.36
Careya arborea	1.67	0.40	2.20
Cassia actriculata	0.98	0.12	0.67
Dhilenia pentagyna	1.34	0.50	3.20
Madhuca indica	1.66	0.50	2.00
Pengamia pinnata	3.69	2.41	2.42
Pterocarpus marsupium	1.97	0.40	2.90
Terminalia chebula	1 .46	0.35	1.35
Terminalia paniculata	1.70	0.40	1.60
Terminalia tomntosa	1.39	0.40	1.80
Xylia dolabriformis	1.37	0.30	1.61

3.1.2. Green Manure

Green manuring is the principle supplementary means of adding organic matter to the soil. It is done by growing of a quick growing medicinal plants and ploughing it under to incorporate it into the soil. The green-manure crop supplies organic matter as well as additional nitrogen, particularly if it is a legume crop, which has the ability to fix atmospheric nitrogen with the help of its root nodule bacteria. A leguminous crop producing 8 to 25 tonnes of green matter per hectare adds about 60 to 90 kg of nitrogen when ploughed under. This amount equals an application of 3 to 10 tonnes of farmyard manure on the basis of organic matter and its nitrogen contribution. The green manure crops also exercise a protective action against erosion and leaching.

Table 4: Average Nutrient Content (in %) of Fresh Green Manures

Manure Items	Nitrogen (N)	Phosphoric Acid (P$_2$O$_2$)	Potash (K$_2$O)
Cowpea (Vigna catjang)	0.71	0.15	0.58
Dhancha (Sesbania aculeate)	0.62	-	-
Cluster-bean (Cyamopsis tetragonoloba)	0.34	-	-
Horse-gram (Dolichos biforus)	0.33	-	-
Mothbean (Vigna aconitilalla)	0.80	-	-
Greengram (Vigna radiata)	0.72	0.18	0.53
Sunhemp (Crotalaria juncea)	0.75	0.12	0.51
Blackgram (Vigna mungo)	0.85	0.18	0.53

The increase in yield after green-manuring is usually to the order of 30 to 50 per cent. The fertilizing value of the legume crop can be increased a great deal by manuring it with super-phosphate. This practice not only increases the phosphorus content of the green manure plants, but also encourages their growth on the whole, thus converting an inorganic fertilizer into an organic fertilizer.

3.2 Biofertilizers

The use of artificial riched soil cannot support microbial life and hence there is less humus and less nutrients and the soil thus becomes poor and eroded by wind and rain. Since, chemical fertilizers are made up of only a few minerals, they impede the uptake of other minerals and imbalance the whole minerals pattern of the plants body. To overcome these problems, the use of biofertilizers is being encouraged.

Biofertilizers are organisms which can bring about soil nutrients enrichment. They are biological active products or microbial inoculants of bacteria, algae and fungi separately, or in combination, which may help biological nitrogen fixation for the benefit of crop plants. They mostly include nitrogen fixing micro-organism. The following are some well known biofertilizers:

- ☞ Legume-Rhizobium symbiosis
- ☞ Azolla-anabaena symbiosis
- ☞ Loose association of nitrogen-fixing bacteria
- ☞ Diazotrophs (Azotobacter, Azospirillium)
- ☞ Cynobacteria, Mycorrhiza
- ☞ Phosphate solubilizing bacteria and fungi
- ☞ Frankia.

3.3. Biopesticides

The natural occurrence of diseases caused by microorganism is common in both insect and weed populations and is a major natural mortality factor in many situations. For many practical reasons however, use of micro-organisms for pest control involve their culture in artificial media and later introduction of comparatively large amounts of inoculum. into the field at an appropriate time and place. The technique is possible only with those microorganisms that can be readily cultured in artificial media and moreover than can be induced to produce spores or other suitable resting stages which permit storage and application. Many fungi and bacteria can be handled in this way but insect viruses have the limitation that they have to be raised in living insects. This is unfortunate as insect viruses probably have greater potential for pest control than any other group of microorganisms because of their virulence and selectivity.

As these biocontrol agents (microbial pathogens) are applied on the targeted pests in much the same way as chemical pesticides they are often termed as biopesticides or natural pesticides. Microbial phytotoxins, allelochemicals and other biocides of plant origin are also discussed here.

Bacillus thuringensis a bacterial pathogen infecting a wide range of insect pests is the most common microbial insecticide in use today. The disease agent is marketed by several companies and is registered for use against the insect caterpillar pests that attack a wide variety of vegetables, flower and ornamental crops. Unlike most chemical insecticides, it can be used on edible products upto the time of harvest. It is selective in action and does not harm parasites or predators of pests of any extent.

3.3.1. Natural Pesticides

Some useful weeds like Lantana, Notchi, Tulsi, Adathoda etc; act as natural repellent to many pests. The indigenous trees like pungam, wood apple, anona and their byproducts have excellent insecticidal value in controlling diamond black moth, heliothis, white files, leafhopper and aphid infestation. The fish oil resin soap (FOS) is a safe, non-posionous natural product used for the control of developed whitefiles, mealybug, wooly aphid etc. A wider range of bio extracts are proved to be more effective in controlling many insect Pests and Diseases.

Farmers in their traditional wisdom have identified and used a variety of plant products and extracts for pest control, especially in storage. As many as 2121 plant species are reported to possess pest management properties, 1005 species of plants exhibiting insecticidal properties, 384 with antifeedant properties, 297 with repellent properties, 27 with attractant properties and 31 with growth inhibiting properties have been identified. The most-commonly used botanicals are neem (*Azadirachta indica*), pongamia (*Pongamia glabra*) and manhua (*Madhuca indica*). Neem seed kernel extract (2 to 5%) has been found effective against several pests including rice cutworm, diamond backmoth, rice brown plant hopper, rice green leafhoppers, tobacco caterpillar, several species of aphids and mites. Mahua seed kernel extract (5%) is effective against sawfly (Athalia lugens proxima) and others.

The efficacy of vegetable oils in preventing infestation of stored product pests such as bruchids (*Callosorbuchus* spp.), rice and maize weevils (*Sitophitus* spp.) has been well documented. Root extracts of Tagetes or Asparagus work as a nematocide for plant parasitic nematodes. Similarly, leaf extracts of many higher plants can inhibit a number of fungal pathogens. A number of plants like *Chenopodium*, *Bougainvillea, etc.* have also been reported to be sources of antiviral principle. A number of herbal natural pesticides preparation which not only play a vital role to control the pathogen but also increase the growth as well as yield of crops, are given below:

1. Pongam (*Derris indica*)

Its extracts serve as natural pesticides. They can be used for controlling a variety of pests. Its extract is beneficial against the leaf eating caterpillars.

(a) Pongam Cake Extract

100 gm of Pongam cake is required for use in 1 litre of water. The Pongam cake is powdered well and tied in a muslin pouch and soaked overnight in water. The next morning the pouch is squeezed and the extract is taken out. The extract is then filtered and emulsifier is added at the rate of 1 ml for 1 litre of water. It is now ready for spraying. 10 kg of Pongam cake is required for preparation of 100 litres of spray, which can be used for one acre.

(b) Pongam-Neem Oil Extract

Pongam oil and Neem oil should be mixed in the ratio of 1:4. Mix 10 ml Pongam oil and 40 ml Neem oil with the emulsifier (1 ml/ 1 litre). It should be mixed well to enable the water to mix with the oil properly. To this add one litre of water. Shake it well and spray the mixture immediately.

(c) Pongam, Aloe and Neem Extract

1 kg of pounded Pongam cake, 1 kg of pounded Neem cake and 250 gms of pounded Poisonnut tree seeds are taken in a muslim pouch. This is soaked overnight in water. In the morning the pouch is squeezed and the extract is taken out. This is mixed with 1/2 litre of *Aloe* (*Aloe barbedensis*) leaf juice. To this 15 litres of water is added. This is again mixed with 2-3 litres of cow's urine. Before spraying 1 litre of this mixture is diluted with 10 litres of water. For an acre 60-100 litres of spray is used. This controls various pests effectively.

2. Tulsi (Ocimum spp)

Use of Tulsi for pest control has a long history. It has both repellent and herbicidal properties.

Various extracts can be prepared from dried and powdered leaves, flowers and entire plants but their effectiveness appears to vary. Whole plants used as an effective insect repellent.

(a) Tulsi Leaf Extract

100 gms of Tulsi leaves are needed to prepare 1 litre extract. The leaves should be soaked in water overnight. The next day the leaves are ground and the extract is filtered. To this 1 ml of khadi soap solution is added and stirred properly. This can be sprayed in the morning hours for effective control of pests.

3. Aloe (*Aloe barbedensis*)

Aloe has anti-fungal and anti-bacterial properties. It also has an attractant property when mixed with castor cake. Aloe in combination with other plant products also control a variety of pests efficiently. In this section we can see how various Aloe extracts can be prepared.

(a) Aloe-Nirgandi Extract

For preparation of 50 litres of this extract we require 2 litres of Aloe leaf juice. For extracting Aloe juice the outer part of the leaves are removed and the mucilaginous part is taken and ground. 5 kg of Nirgandi leaves (*Vitex negundo*) are taken and immersed in water. These leaves are boiled for half an hour and the extract is filtered. The Aloe juice and the Nirgandi extract are mixed together. This can be diluted in 50 litres of water and sprayed.

4. Melia (*Melia azadirachta*)

The leaves, bark and fruits are accredited with insect repellant properties. Leaves are placed inside books and folds of woollen garments to protect them from ancient times. The active principle is reported to be an alkaloid, soluble in hot water. A caretenoid, meliatin present in the aqueous extract of leaves acts as a repellant to grasshoppers. Few preparation of Melia extracts are given below;

(a) Leaf Extract

For 5 litres of water, 500 grams of green Melia leaf is required. Since the quantity of leaves required for preparation of this extract is quite high, (nearly 40 kilograms are required for I hectare) this can be used for nursery and kitchen gardens. The leaves are soaked overnight in water. The next day, the leaves are ground and the extract is filtered. The emulsifier helps the extract to stick well to the leaf surface.

(b) Cake Extract

50 gms of Melia cake is required for 1 litre of water. The Melia cake is put in a muslin pouch and soaked in water. It is soaked overnight before use in the morning. It is then filtered and emulsifier is added at the rate of I ml for 1 litre of water. It is now ready for spraying.

5. Custard apple (*Annona* spp)

The seeds of custard apple possess insecticidal and anti-feedant properties. The powdered seeds are used to destroy worms. Seeds are reported to be contact poison to flies, aphids and several beetles. The combined seed extract of neem and custard apple is toxic to housefly. It is also used as a nematicide. Different alkaloids namely annonine, municine, artabotrine etc. are reported to be found in custard apple.

(a) Leaf Extract

Take 500 grams of custard apple leaves and boil it in 1-2 litres of water. Allow to boil till it becomes 1/4th of its original volume. Mix it with 10- 15 litres of water and spray over the crop. For an acre, 2-3 kg of leaves are required.

(b) Mixed Extract (Custard Apple, Calotropis & Tobacco)

Take 500 grams of custard apple leaves and boil it in 1-2 litres of water. Allow to boil till it becomes thick. Filter the solution to get the decoction. Take 250 to 300 ml of calotropis extract. Take 500 grams of tobacco leaves and boil it in 1 to 2 litres of water for 45 minutes. Then filter the extract. Take 250 ml of biogas waste (whitish fluid which deposits in the biogas digester) and 100 grams of copper sulphate. Mix the above ingredients with 60 litres of water and spray it over the crop. The above quantity is recommended for an acre.

6. Nirgandi (Vitex negundo)

Its leaves possess insecticide properties. It is used in storage, godowns to ward off insects. Extracts of the leaves and twigs show anti-bacterial activities. Nirgandi in combination with other plant extracts (viz. Aloe, Pongam, castor, calotropis) proves to be successful in controlling certain pests.

7. Sweet Flag (Acorus calamus)

The powdered rhizome is used as an insecticide for the destruction of fleas, bed bugs, moths, lice etc. It is effective in killing insect pests of stored grains and is considered to be better than chemicals which pose the problem of residual toxicity. The ether extract of the rhizome exhibits ovicidal and insecticidal properties with no residual toxicity. The alcoholic extract inhibits the growth of certain fungi.

Its rhizome powder and cow's urine are used for the pre-treatment of seeds before sowing in the fields and protect them against pathogens and pests. One litre of cow's urine and 50 gm of rhizome powder is required for treating 1 kg of seed.

8. Garlic (Allium sativum)

Garlic is a safe pesticide. It causes no adverse effects on man, animals and environment. The effectiveness of garlic as a pesticide is due to an acrid volatile oil. Various extracts of garlic can be used for pest control. The whole plant, bulbs, leaves and flowers may be used either fresh or dried. Preparation of different Garlic extracts are given in this section. The preparations/extracts should be used as soon as it is prepared due to the volatile nature of the oil. It remains effective for 4-13 days after application.

(a) Bulb Extract

Add 85 grams of chopped garlic to 50 ml of mineral oil (Kerosene). Allow this to stand for 24 hours. Then mix it with 450 ml of water to which 10 ml of soap solution has been added. Shake this mixture well and filter it through a fine cloth. Store this in a container. For application, dilute one part of the emulsion with nineteen parts of clear water. For example, 50 ml of emulsion is mixed With 950 ml of water. The best time to spray it is in the morning. Shake well before application.

(b) Garlic-Neem Spray

Take 1/2 a handful of Neem seeds, 1 kg of garlic, 10 kg neem leaf (tender branches included) and 10.5 kg stem and root bark of Neem. Crush these and reduce it to pulp. To this, add 12 litres of water. Boil this in copper or clay vessel on slow fire till it is reduced to half. If Neem seeds are not available, add 50 g of Neem oil to the finished preparation or boil the mixture to increase viscosity of preparation. After boiling, allow the mixture to cool. Pass this preparation through a sieve and collect the filtered liquid. The material left over in sieve is fibrous. Grind this well and mix it with a light application of Neem oil and a small quantity of turmeric boiled in water. Pack it in cellophane bags. This can be spread on ground before cultivation and on plants. This will prevent disease bearing insects and flies from resting on plants. It can be used in addition to the liquid spray, if necessary. Allow the boiled liquid preparation that has been filtered to settle for 4-5 weeks in a sealed jar. At the end of the period, filter it again so that it can be used in a sprayer. For maximum effect, use it at regular intervals of 3 days, and increase it from 3 to 7 days as plants grow. In wet weather, reduce the intervals and apply the undiluted preparation.

(c) Mixed Extracts in Cow's Urine

The extracts of Garlic, Chilli, Ginger, Neem oil, Tobacco, Asafoetida are dissolved in 72 hours old Cow's urine and dissolved. Before spraying, an emulsifier is added. For spraying on cotton 30-40 ml of this spray is used with 15ml of water and 70 ml per 15 ml water is used for spraying in pigeonpea.

(d) Garlic-Pepper Extract

Take two finely grated Garlic bulbs and two teaspoon pepper. Stir into four litres of hot water. Add a nut sized piece of soap and spray. This is very effective in controlling caterpillars on fruit trees and others.

9. Neem (Azadirachta indica)

Different parts of the neem tree can affect more than 200 insects species and some nematodes, fungi, bacteria and viruses. Neem contains several active chemicals, which work in different ways: as a result of this, pests are unlikely to become resistant to neem. The most well known chemical in neem is azadirachtin. Neem is easy to prepare and use, and is environmentally safe and not harmful to man and animals.

Neem protects itself from the multitude of pests with a multitude of pesticidal ingredients. Its main chemical broadside is a mixture of 3 or 4 related compounds, and it backs these up with 20 or so others that are minor but nonetheless active in one way or another. In the main, these compounds belong to a general class of natural products called triterpenses; more specifically, limonoids.

(a) Neem Extract

The 500 g Neem seed kernels are crushed and mixed with 10 litres of water. It is necessary to use a lot of water because the active ingredients do not dissolve easily. Stir the mixture well and leave to stand for at least 5 hours. Spray the neem water directly on vegetables using a sprayer or straw brush. The effect lasts for 3 to 6 days. If kept in the dark neem water will be effective for 3 to 6 days. It has been estimated that 20-30 kg neem seed (an average yield from 2 trees), can normally treat one hectare. If crops have to be watered, water should go directly on the soil because water running over the leaves of sprayed plants may wash off the extract. Its alcohol is available, 50 times more azadirachtin can be dissolved and extracted. This Neem spray is most effective against pests such as Cutworms. During the day the caterpillars stay on the ground and fed on plant roots. At night they eat young stems.

(b) Neem Oil

The kernels of the neem seed should be crushed and add a small amount of water until the mixture forms a firm paste that can be kneaded. Press firmly to extract the oil. The kneading and pressing should be continued in turn until the maximum amount of oil is removed. (The oil content of the seed kernel is about 45%).

In some areas there are traditional ways of removing oil from other seeds such as sesame or groundnut. It is a good idea to try these methods with neem. If the oil is heated in the process, its ability to control insects will not be affected.

(c) Neem Cake

The neem cake, which is left after the oil is extracted from the seed, is also useful for controlling, several pests, which live in the soil, particularly nematodes.

3.4. Integrated Pest Management (IPM)

Use of biopesticides is likely to be an integral part of disease management strategy for crops (include medicinal plants) in the coming years. Looking to the future, biopesticides offer a potential solution to the dilemma from both the environment and industrial viewpoint in integrated disease management system.

There is practically no risk of developing pest resistance to such products. They are less expensive having no adverse effect on plant growth, seed viability, active constituents (percentage value) and yield of medicinal plants. To meet the demand of herbal products or crude drugs for growing pharmaceutical industries, increase in herbal production is of utmost importance and hence pesticides seem to be indispensable. The use of pesticides should be managed in such a way that it will not pose any serious threat to environment and human life.

Integrated pest management (IPM) is their selection, integration and implementation of pest control based on predicted economic, and ecological and sociological consequences. The Government of India has adopted IPM as a major thrust area and the main plan of crop protection programme. IPM approach is eco-friendly. It aims at minimal use of hazardous chemical pesticides and instead employing alternate pest control methods and technique like cultural, mechanical and biological and biopesticides.

Soil conditioning, crop rotation and improved sanitation practices are some simple methods to reduce pest problems. The crops can also be protected by eliminating pests by methods of starvations, where a target crop is grown to lure insects away from the economic crop. Sometimes, to obtain desirable pest control, biological methods are supplemented with chemical pesticides. Employing the natural enemies of crop pest like parasites and predators and a variety of bio-control agents, such as beneficial bacteria and viruses, have been successful in the management of pests. Among the cases of IPM success are the package for staple crops *viz.* sugarcane, cotton, citrus, vegetable etc.

Chapter 4

ORGANIC FARMING: AN APPROACH FOR SUSTAINABLE HERBI-CULTURE

Organic Agriculture Systems are not a repudiation of the assets of modem agricultural technology, neither are they defined by the simple elimination of synthetic fertilizers or pesticides.

According to Lord Northbourne, it is biodynamic farming, had a vision of the farm as a sustainable, ecologically stable, self-contained unit, biologically complete and balanced-a dynamic living organic whole. The term thus did not refer solely to the use of living materials (organic manures, etc) in agriculture although obviously it included them, but with its emphasis on wholeness is encompassed best by the definition of pertaining to, or characterized by systematic connection or coordination of part of the one whole!

In reality, organic agriculture is a consistent systems approach based on the perception that tomorrow's ecology is more important than today's economy. Its aim is to stop degradation and reestablish natural balances. The economy must readjust to the primary production factors, and riot the other way around. Without ecology, there is no economy. If conventional agriculture had been made to pay for the degradation and environmental damage it is causing, the move towards organic farming system would have been made long ago. Besides this other possible aims of organic agriculture are:

- ☞ To work as much as possible within a closed system and draw upon local resources.
- ☞ To maintain the long term fertility of the soil
- ☞ To avoid all forms of pollution caused by agricultural techniques

↣ To provide a foodstuff of high nutritional quality in sufficient quantity

↣ To reduce the use of fossil energy in agricultural practices to the minimum tending of zero

↣ To give to all livestock the conditions of life that conform to their physiological needs

↣ To make it possible for agricultural families to earn a living through their work and develop their potentialities as human beings

↣ To maintain the rural environment and also preserve non-agricultural ecological habitats

4.1. Concept of Organic Farming

Organic refers to the agricultural systems used to produce food and fiber. Organic farming systems do not use toxic chemical pesticides or fertilizers. Instead, they are based on the development of biological diversity and the maintenance and replenishment of soil fertility. Organic foods are minimally processed to maintain the integrity of the food without artificial ingredients, preservatives, or irradiation. Organic farming describes two major aspects of alternative agriculture:

↣ The substitution of manures and other organic matter for inorganic fertilizers.

↣ The use of biological pest control instead of chemical pest control.

Organic agriculture is one of several to sustainable agriculture and many of the techniques used (*e.g.*, inter-cropping, rotation of crops, double-digging, mulching, integration of crops and livestock) are practiced under various agricultural systems. What makes organic agriculture unique, as regulated under various laws and certification programmes, is that: (1) almost all synthetic inputs are prohibited. and (2) soil building crop rotations are mandated.

The basic rules of organic production are that natural inputs are approved and synthetic inputs are prohibited, but there are exceptions in both cases. Certain natural inputs determined by the various certification programmes to be harmful to human health or the environment are prohibited (*e.g.* menic). As well, certain synthetic inputs determined to be essential and consistent with organic farming Philosophy. are allowed (*e.g.* insect pheromones). List of specific approved synthetic inputs and prohibited natural inputs are maintained by all certification programmes and such a list is under negotiation in Codex. Many certification programmes require additional environmental protection measures in adoption to these two requirements. While many farmers in the developing do not use synthetic inputs, this alone is not sufficient to classify their operations as organic.

4.2. Needs of Organic Farming

Organic agriculture is the key to a sound development and a sustainable environment. It minimise environmental pollution and the use of non-renewable natural resources. It conserves soil fertility and soil erosion through implementation of appropriate conservation principles.

Several reasons have been emphasized for the need of organic agriculture, like limited land holdings, poor socio-economic conditions of farmers, rise in input cost etc. The broadest view shows two major reasons *viz.*, population and environment, emphasised the ultimate need for ecofriendly technologies. FAO estimates that by the year 2005 the global population will be 25% higher than in the mid 1980s and that 900/b of this population increase -will be in the developing countries like India. It will be necessary to increase the supply of food and other agricultural products to meet their needs. At the moment, many of the methods being used to increase production are damaging to natural resources and the environment and farmers are supposed to invest heavily into inputs to improve yields and productivity. Further, the produce from organic farming has become inevitable. The food containing pesticides and other chemicals are increasingly made obvious by many research studies revealing the presence of pesticide residues in eatables. About one million people suffered from pesticides poisoning and 20,000 die every year due to the toxic effects of these chemicals worldwide but the crop loss due to pest is still 15.000 crores, apart from killing a portion of the human population by the same pesticide use. This perspective is also one of the needs for organic approach to farming. In order to avoid the deleterious effects of synthetic chemical fertilizers and pesticides, organic agriculture is needed as an alternative to provide ecologically safe methods of farming.

Organic agriculture is a production system, which avoids or largely excludes the use of inorganic fertilizers, pesticides, growth regulators and livestock feed-additives. To the maximum extent feasible, organic farming system rely on crop rotations, crop residues, animal manures, legumes, green manures, off farm organic wastes and aspects of biological pest control to maintain soil productivity and tilth, to supply plant nutrients and to control insects, weeds and other pests.

Sustainable agriculture, by definition, is that form of farming which produces sufficient food to meet the needs of present generation without eroding ecological assets and productivity of life supporting system of the future generation. The key characterization of organic farming in relation to sustainable soil fertility and organic farming include:

- ⮞ Protecting the long-term fertility of soil by maintaining organic matter levels, fostering soil biological activity and careful mechanical intervention.
- ⮞ Providing crop nutrients indirectly by using relatively insoluble nutrient sources which are made available to the plants by the action of soil microorganisms.
- ⮞ Nitrogen self-sufficiency through use of legumes and biological nitrogen fixation as well as effective recycling of organic materials, including crop residues and livestock wastes.
- ⮞ Weed, disease and pest control relying primarily on crop rotations, natural predators, diversity, organic maturing resistant varieties and limited thermal, biological and chemical intervention.
- ⮞ The extensive management of livestock, paying full regard to their evolutionary adaptations, behavioral needs and animal welfare issues with respect to nutrition, housing health, breeding and rearing.
- ⮞ Careful attention to the impact of farming system on the wider environment and the conservation of wild life and natural habitats.

Organic agriculture is viable alternative to conventional agriculture. It protects the soil from erosion, strengthens natural resources base and sustains biological production at levels commensurate with the carrying capacity of managed agro-ecosystem, because of reduced dependence of fertilizers and Plant Protection chemicals; problems of environmental pollution are greatly reduced if not totally avoided. The practice leads to regeneration of ecosystem. There is now a demand worldwide for organically grown foods, which command a premium in export markets. The demand for such safe foods is increasing annually and this opportunity needs to be exploited.

Organic agriculture is often associated with low yields. The demands for products of organic agriculture are growing in India. The European Union (EU) and the International Federation of Organic Agriculture Movement (IFOAM) have defined rules for organic agriculture producers. India too needs to establish its own certification scheme to facilitate exports of organic agriculture products and an accreditation agency to certify the produce of the farm. The essential elements of an organic quality assurance scheme are as follows:

- ☞ Development of standards
- ☞ Inspection and verification
- ☞ Certification
- ☞ Accreditation

As low yields are expected, the market's ability to pay a price premium is a key determinant of sustainable growth of organic agriculture. The European Union is currently world's leading market for organic products with various countries within the EU having deficits of fresh organic products. The legislation on organic agriculture varies across countries. At present certification is compulsory for the organic foods exported from India. To encourage mass production and consumption of organic products, there should be clear-cut policies and incentives to promote organic agriculture and necessary infrastructure and market.

National level efforts are needed to facilitate the successful adoption of organic agricultural by the farmers. The approach shall be farmer centered and programmes developed shall create conditions for the conservation and efficient use of locally available resources as inputs in agriculture. Research has also to play to the extent of developing appropriate agro-techniques in conformity with the basic principles of organic agriculture.

4.3. Economic and Market Status

Prices for organic foods reflect many of the same costs as conventional foods in terms of growing, harvesting, transportation and storage. Organically produced foods must meet strict regulations governing all these steps so the process is often more labor and management intensive and farming tends to be on a smaller scale. There is also mounting evidence that if all the indirect costs of conventional food production (cleanup of polluted water, replacement of eroded soils, costs of health care for farmers and their workers) were factored into the price of food, organic foods would cost the same, or, more likely be cheaper.

The organic agriculture is economically viable, that farmers can achieve more income as a result of premiums and that they need fewer inputs to maintain returns. Organic systems are based on the optima use of local resources and technologies and can give farmers greater independence and more control over their means of production. However, more comprehensive monitoring is needed to analyze their sustainability and impact.

Many farmers enter organic production because they want to farm in a more holistic way. Other major stimuli for developing organic agriculture include environmental and social concerns, economic necessity, a lack of chemical inputs, and market demands. Small farmers are also encouraged to take tip any stay in organic farming by the prospect of being able to produce more food at the subsistence level, having a larger surplus for local sale, or being able to cultivate a product of significant export value. Farmers are most responsive to organic agriculture when they have not been exposed to the 6 chemical message and their farming systems involve traditional or nil inputs. When production is relatively labour intensive and if farmers have the chance of developing the organic concept themselves they are also more inclined to convert to organic agriculture.

Farmers are less rely to take up organic farming in situations of high labour cost and labour scarcity and where there has been an over exposure to the chemical message. Farmers with relatively mechanized farms and a commitment to high input, high output strategies are also less likely to convert. Insecure land tenure means small farmers will be reluctant to plant permanent crops.

In developing countries, organic farming methods seems to provide similar outputs, with less external resources, supplying a similar income per labor day as high input conventional approaches. Large increase is observed where local farmers adopt organic farming systems, up to 400%, reaching levels "similar to those of high-input systems. Direct comparison of yields are difficult because of the differences in the farming systems adopted under high-input or organic management.

4.4. Organic Market

Consumer demand for "healthier" food has been considered one of the major factors influencing increased consumer demand for organic food. This perception is probably reflected in the lower consumption of livestock products among the consumers most likely to buy organic food. A survey of consumer choice in the United Kingdom showed that apart from health concerns, organic food was bought for environmental and animal welfare reasons. Approximately 70% of consumers buying organic food mentioned these two factors.

The trade in organic products is growing rapidly and becoming a reality throughout the world. Growth rates in the sector show organic products, that a few years ago supplied niche markets, have now entered mainstream marketing channels. Already 30,000 organic farmers are certified in Italy. In Scandinavian countries like Sweden and Finland and in Switzerland some 8% of farming is (certified) organic. Austria leads the world with 10% organic farming and in some Austrian provinces such as Salzburg and Tirol, the proportion is almost 50%. Although some in the organic movement might question whether we are straying from the path of 'healthy natural' growth, it cannot be denied that we are heading for a boom.

Approximately 1% of the U. S. food supply is grown using organic methods. In 1996, this represented over $3.5 billion in retail sales. Over the past six years sales of organic products have shown an annual increase of at least 20%. Organic foods can be found at natural foods stores, health food sections and produce departments of supermarkets and at farmers' markets, as well as through grower direct-marketing such as C. S. A. (Community Supported Agriculture). Many restaurant chefs across the country are using organic produce because they desire its superior quality and taste. Organic food is also gaining acceptance on a worldwide basis, with nations like Japan and Germany becoming important international organic food markets.

Though only a small percentage of farmers are expected to become organic producers, consumers demand for organically produced food and fibre products provides new market opportunities for farmers and businesses around the world. For many years, and with great success, the private sector alone has developed the concepts and markets for organic , products. However, the surge in consumer interest has created new interest from the public sector, and developing countries are particularly in need of good information.

The export opportunities offered by the so-called developed world have clearly been a major stimulus in getting organic farming established in many countries. It is unlikely that organic agriculture would have developed so successfully in the northern hemisphere if there had been no demand pressure. Supplying a market in which demand outstrips supply is a producer's dream. This has been the case with organic farming for a long time and figures show that this trend is set to continue. Yet, the sector has shown it is also vulnerable and that prices can collapse if there are sudden and exceptional increases in the supply of particular products. Therefore, Marketing with a clearly targeted strategy is a must for organic products. The organic market is a special one, partly because premium prices are involved. It requires a special effort. Attention should be given to meeting the requirements of a guarantee system that will ensure organic quality and allow consumers to develop their preferences for organic products with a feeling of trust.

4.5. Constraints and Opportunities

☞ In most cases farmers and post-harvest businesses seeking to sell their products in developed countries must hire an organic certification organization to annually inspect and confirm that these farms and businesses adhere to the organic standards established by various trading partners. The cost for this service can be expensive, although it varies in relation to farm size, volume of production, and the efficiency of the certification organization (*e.g.* IFOAM certification costs a maximum of 5 per cent of sales value). Few developing countries have certification organizations within their borders, and even when sufficient resources are available to pay for certification farmers often lack the information to find credible inspectors.

☞ While most developing countries have focused on export markets in the developed world, domestic market opportunities for organic food or eco-food may also be exploited. In China, for example, there is a growing market for "green food" which, according to government grading standards, is produced without certain pesticides and fertilizers and with biological methods. Chinese farmers also product organic food for export (*e.g.* tea to the Netherlands, soybeans to Japan).

☞ Whether the intent is to sell organic products domestically or abroad, reliable market information is difficult to obtain. There is virtually no systematic production or market survey data being collected with which to assess the role-and pattern of organic market growth. In particular, no projections for the market in the developing world have been made, nor have markets systematically been identified for developing country exports.

☞ There is a growing consumer demand in the developed countries for healthy food products. This change in the consumer preference reflects the increasing concern about the dangers associated with the consumption of food containing small quantities of food additives and pesticide residues. As the organically produced food is devoid of such harmful additives and residues, it is regarded as healthy food and, as such, enjoys special consumer preference. To the average consumer in the developed countries, organically produced food is more acceptable than the food produced under the conventional agricultural system because of the special qualities possessed by the former in regard to taste and nutrition.

☞ In the developing countries organically produced food has not yet gained consumer acceptance to any appreciable extent. This is despite the presence of harmful chemical residues at higher concentrations in vegetables, fruits and other farm produce. These pesticide residues lead to all kinds of chronic diseases of the heart, brain, kidney and liver including cancer.

☞ There is lack of awareness among the people of the developing countries about poisons in the food they are consuming daily. Even for those who are aware of the dangers of poison in fact, they do not have easy access to organic products. Wherever such products are available, they are costlier than the conventionally produced items mainly because organic agriculture does not enjoy the benefit of subsidy. However, there are still parts of India and other developing countries where people have access only to organic (chemical-free) foods because agro-chemicals have not yet penetrated into interiors and hinterlands.

> By introducing a wholly different segment of the consumer population to organic foods, supermarkets can help increase sales of organic products, improve the consistency of supply by creating greater demand and increase overall customer awareness of organic issues. These supermarkets can plough the vast masses of consumer, turning nonusers into light user, and light users into regular customers of organic specialty stores. A problem for specially stores and supermarkets is consumer resistance to the higher price of organic products. In spite of this, consumers who are aware are willing to pay higher price for such products not only because they taste better, but also because of the realization that, in the process, they are supporting organic farming and thus protecting the environment.

Very often organic farming systems begin with the production of 'cash crops' and gradually, with the knowledge gained in producing for export, the whole farming unit is converted to organic systems. Thus, planting herbs in plantations crops and using compost is often followed by introducing organic methods into the cultivation of medicinal herbs.

There has always been agricultural trade. Trade remains a necessity whether it is barter between neighbors or long-distance trade facilitated by money. Trade makes it possible for house-holds to supplement their own production and satisfy their need for food and money whilst exploiting comparative.

Farm inputs such as fertilizers and pesticides labels 'Organic' need to be verified. The farmer should have some way of knowing that the material purchased are acceptable brands to be used in sustainable farming practice. Guidelines need to be developed concerning acceptable and non-acceptable inputs.

4.6. Suggestions

> It is essential to set up and enlarge organic herbal raw material bases to satisfy people's growing demand for organic raw material that purpose, bases should be selected on the basis of the investigation and analysis of the environment quality background of the locations. Measures should be taken to ensure the production bases free environment pollution and in sound agricultural environment.

> It is important to develop organic crude drug processing industry by setting various types of organic processors producing various organic items, thus increasing the variety of organic products in the market as well as the economic benefit by processing.

> Actively opening up organic herbal market both at home and abroad stimulate development of the trade of organic herb. Market determination production. Only market can stimulate development of the production and processing of organic herb and production of quality of organic herb turn the development of the market.

> In developing the market for organic herb, the world market demand will usually stimulate development of a country organic herb. Therefore, demanded by the world market will help develop domestic market. But to build a sustainable organic herbal marketing system, it is essential develop and world market simultaneously.

> To ensure the quality of organic, it is essential to formulate organic herbal production and processing standards inline with the basic standards of IFOAM, to set up certifying organizations and to strictly enforce the procedures for inspection, and certification organic herbs. Inspection and certification or organic herb supply system as well as the guarantee of the quality control of organic produce. Most of the countries have not yet worked out their own country's standards nor set up any certifying organizations, which has significantly affected the development of organic herbs more attention should be paid to this work.

4.7. Indian Needs

To place organic farming on a sound footing in India, urgent action is necessary on the following fronts:

- ☞ Knowledge development and training of farmers in grading, curing and post-harvest practices according to international standards to ensure quality outputs. Enhancement of labour efficiency by developing intermediate technologies. Value addition process at farm-level or community level to improve the income of farmers.

- ☞ Arrangement for easy credit for working capital, supported by buyback arrangement, will encourage farmers to take up innovative cropping system.

- ☞ A separate Organic Development Wing under the Ministry of Agriculture could bring greater synergy to the effort to promote organic farming and the export of organic products.

AGRO-TECHNIQUES OF
MEDICINAL PLANTS

AONLA (*Phyllanthus emblica* Linn.)
Syn. *Emblica officinalis* Gaertn

Vernacular Names

San.–Amalaki, Dhatri; *Eng.*–Emblica myrobalan, Indian gooseberry; *Hin.*–Amlika, Amalak, Amvala; *Kan.*–Nellka; *Mal.*–Nellimaram, Nellikka; *Tam.*–Nelli; *Tel.*–Usirikaya, Amalakamu; *Guj.*–Amali; *Beng.*– Amlaki; *Mar.*–Awla; *Trade*–Aonla.

Introduction

Aonla is in cultivation in India since time immemorial. It belongs to family Euphorbiaceae. It finds mention in Vedas, Ramayan, Charak Samhita, Sushrut Samhita, literature of Kalidas, Kadambari and other ancient literature. Vitamin C and renders it a valuable anti-ascorbutic in the fresh as well as in the dried and processed condition.

Botanical Description

A tree of medium height and prefers grouped as evergreen in tropics but in subtropics it exhibits deciduous nature due to abscission of determinate shoots. Phyllanthoid branching habit with two types of shoots; determinate (short shoots) and arranged on determinate shoots that they look like pinnately compound leaf. However, the leaves are simple 1 to cm long. Flowers are borne in the axil of the leaves on determinate shoots as axillary cymules.

Geographical Distribution

It is native to Tropical South-Eastern Asia. It is naturally growing in the forest and drier regions of Central and. Southern India. It is more popular as a backyard fruit throughout the country. However, commercial orcharding can be seen in U.P. and Gujarat.

Medicinal Uses

Aonla fruit is very rich in vitamin C and pectin, therefore, regarded very important for medicinal value in Ayurvedics. Undoubtedly, it is a richest natural source of vitamin C known to mankind. A tannin, containing gallic acid, allagic acid and glucose in its molecule, which is naturally present in the fruit, prevents or retards the oxidation of vitamin C and renders it a valuable antiscorbutic in the fresh as well as in the dried and processed condition.

AONLA (*Phyllanthus emblica*)

This fruit is useful in haemorrhanges, diarrhoea, dysentery, anaemia, jaundice, dyspepsia and cough. It is an important ingredient of Triphala and Chavanprash in Ayurvedic medicine system. Fruits are commonly used for preserve (murabba), pickle, candy, jelly etc. It can be dried and powdered to be used subsequently. Used in the preparation of inks, hair dyes, hair oils. It is a great health and vitality restorer.

Chemical Constituents

It is major source of vitamin C. Seeds contain fixed oil, phosphatides and an essential oil. Fruit and leaves contain tannins, polyphenolic compounds. Leaves and stem yield lupeol and beta-sitosterol. Some other major amino acid present are; alanine, aspartic acid, glutamic acid, lysine, and proline.

Soil and Climate

It grows well in sandy loam to clay soils in India. It has great tolerance to salinity and sodicity and cultivated in pH range of 6.0 to 8.0 very successfully. However, production shall be highly benefited in deep and fertile soils. It prefers subtropical region with distinct winter and summer. However, in India it is being grown near sea coast upto 1800 m altitude.

Improved Varieties

'Banarasi' , 'Chakaiya', Kanchan, Krishana, N.A-6, N.A.-7, N.A-10, B.S.R.-1, Anand-1, Sanshar gold and 'Francis'.

AGRO-TECHNIQUE

Land Preparation

The field should be prepared to make good seed-beds, by light ploughing followed by two harrowing before the onset on the monsoon season.

Mode of Propagation

It can be propagated by seeds and shoot cuttings.

Nursery Raising

Seeds are collected from fully matured. The seed germination can be quickened by soaking dried seeds in 500 ppm GA solution for 24 hours. Seedlings can be raised in seedbeds or polythene bags. It will take them about 4 months to attain beddable size. Both inarching and budding methods can be used for raising plants but budding is more successful. Among three methods of budding attempted, Forkert and Patch budding done from June to September gave maximum success. However, T-budding gave 80% success in rejuvenation of old orchards. Even soft-wood grafting can also be used successfully for which seedlings are raised *in situ* and headed back severely in month of May for forcing new vigorous side shoots. On these new shoots, soft wood grafting method can be employed by way of wedge grafting. Use of new apex of root stock and activated scion is necessary.

Transplanting

Garft's or buddling of aonla are best planted in the beginning of monsoon in the months of June to July. Since the tree grows to a huge size a distance to 8 to 10 m both ways is recommended. In areas with Irrigation facilities, planting can also be done in spring (February to March). Before planting, it is necessary that planting place is appropriately marked, 1 m size pits are made well in summer, kept open for about a fortnight. In each pit 3 to 4 baskets of FYM are mixed with dug soil and filled. After first rain the plants are planted in the center of these pits and staked properly.

Pruning

Aonla tree does not require regular pruning. However, pruning in early year for giving proper shape and development of strong frame may be necessary for which tree should be trained. to single stem upto the height of about I m and then primary branches can be allowed at regular space all around the trunk.

Intercropping

Aonla tree is quite fast growing type. However, in initial 3 to 4 years sufficient space is available which could be advantageously used for raising intercrops. Since this crop does not require any Irrigation in summer due to dormancy of fruit, only scope of raising intercrops is in rainy season or in post monsoon period provided Irrigation facilities are available. For this legumes and vegetable crops *viz.* blackgram, tomato, gaur, sunflower, groundnut can be taken as intercrops.

Fertilizers and Manures

The aonla crop does not require heavy doses of manure and fertilizers. Organic manures @ 10-15 tonnes/ha should be applied along with the chemical fertilizers. Ammonium sulphate @ 60-80 kg/ha, when added to the soil shows positive results. The phosphorous fertilizer should be added to the field before sowing @ 40-50 kg/ha. Nearly 50 Kg of nitrogen should be added as a seed dose to the crop 1-2 year after sowing.

Irrigation

Aonla trees are hardy and stand very well against drought. Therefore, hardly any Irrigation is practiced. However, the crop shall be benefited by giving two/three Irrigations at the time of full bloom and set. During summer, when the fruit is dormant, there may not be any benefit to irrigate trees.

Pest and Diseases

Insect Pests

Gall caterpillar (*Betonsa stytophora*)

Young caterpillars bore into the apical portion of the shoot during rainy season and make tunnel. Due to this, damaged region bulges out abruptly into a gall which provides space for movement of the caterpillar. Due to this apical growth is checked, side shoots develop below the gall and subsequent growth in following season is greatly hampered. This can be controlled by cutting away the infected apices and prophylectic spray of systemic insecticide like Rogor 0.03%.

Leaf rolling caterpillar (*Garcillaria acidula*)

This caterpillar rolls the leaf and feed inside reducing the photosynthetic: capacity of leaves and causes subsequent leaf shedding. It can be controlled by spraying 0.08 per cent Malathion or 0.04% Monocrotophos.

Bark eating caterpillar (*Inderbela tetraonis*)

It damages stem and branches of grown up trees by eating bark. This insect is the problem of neglected orchards. Affected portion should be cleared of fraps and a few drops of kerosene should be applied in holes to keep this in control.

Mealy bug (*Nipaecoccus vastator*)

Both nymphs and adults are reported to feed on aonla tree from April to November. Organophospholitcs provide excellent control of this insect. Monocrotophos 9.04% or Malathion 0.08% or methyl para.thion 0.03% are effective as sprays.

Some other insects like aphids (*Cerclaphis emblica*), aonla butterfly (*Virachola isocrates*), and plant bug (*Scutellaria nobilis*) have also been reported to damage aonla trees.

Diseases

Ring rust (*Ravenelia emblicae*)

Ring rust appears as circular or semi-circular, reddish solitary or gregarious spots on leaves from the beginning of August. Generally, one or two pustules measuring 10 to 20 mm in diameter appear on infected fruit. Occurrence can be prevented by spraying 0.2% Dithane Z-78 at the interval of 7 to 28 days during the months of July to September.

Fruit rot (*Penicillium oxalicum; P. islandicum; P. oxalicum & Aspergillus niger*)

A number of organisms are associated with this rot in which a few affected fruits are found in the orchard. However, the major loss takes place during transit to the market. The earliest symptom of infection is seen as water soaked lesion on the fruit surface, which enlarges in size followed by development of small pin head size colonies of golden yellow colour. The older colonies turn olive green. It is recommended that fruits showing such symptoms are discarded for marketing. Bruising and injury at the time of harvesting should be avoided as a preventive measure.

Leaf rust (*Phakospora phyllanthi*)

This rust is commonly seen on leaves in the months of July and August and Bordeaux mixture 0.1% will control this rust.

Harvesting and Storage

A vegetatively propagated free starts fruiting commercial crop After 6 to 8 years of planting, while seedling begin trees may take 10 to 12 years, to begin bearing. Productive life of trees is estimated to be 50 to 60 years under good management. Generally aonla fruits are ready for harvest in November/December. Their maturity can be judged either by the change of seed colour from creamy, white to black or by the development of translucence exocarp. Maximum vitamin C content is observed in mature fruits. While immature fruits are acrid and low in vitamin C content and minerals. There are no post harvest standards available but grading of fruits on the basis of size, elimination of diseased and injured fruits and proper packing before marketing will always attract better market.

Production and Yield

Improved varieties of Aonla started fruiting in the age of 4-5 years old but yield is very low. The 6 or 7 years old plants produce nearly 60-70 kg fruits per tree. The Kanchan variety produces maximum 120 kg/tree fruits in compare to other varieties. The improved varieties of aonla yield nearly 5 tonnes/ha fruits after 4th years which will be continues increases upto 20 tonnes/ha (after 8th years).

Marketing and Trade

The current market of fresh fruit is Rs. 4 per kg while dried with seeds and without seeds is Rs. 10-14 and Rs. 20-25 per kg respectively.

Economics of Cultivation

The market of Aonla and their products are highly volatile. The economics (per ha) worked out here are subject to fluctuations, depending upon time and place. (Economic life of plant 50-60 years).

Expenditure (in Rs.)	—	15,000.00
Returns (in Rs.)	—	50,000.00
Net Profit* (in Rs.)	**—**	**40,000.00** (after 4th year)
Net Profit* (in Rs.)	**—**	**1,50,000.00** (after 8th year)

* Estimation of the net profit is analysed on the basis of grower/collectors price. The grower/collector price is always 25-40% less than wholesale market price due to various stakeholders *viz.* Agent, middle man, Brokers, Commission agents.

ASHWAGANDHA (*Withania somnifera* (L.) Dunal)
Syn. *Physalis flexuosa* L.

Vernacular Names

San.–Asavgandha; *Eng.–Winter Cherry*; *Hin.*–Asgandh; *Beng.*–Ashavgandha; *Guj.–Ason, Ashvagandha, Ghoda asor*; *Mar.*–Asgandha; *Tel.*–Aswagandha, Pannerugadda ; *Tam.*–Achuvagandi; *Kan.*–Ashvagandhi, Sogade-beru; *Mal.*–Amukkiram, Pivate; *Trade*–Ashwagandha.

Introduction

It is mentioned as an important drug in ancient Ayurvedic literature. Several types of alkaloids are found in this plant, out of which, withanine and somniferine are important. Ashwagandha is an important cultivated medicinal crop of India. Ashwagandha belongs to the family Solanaceae. Seven types of alkaloids are found in this plant, out of which withanine and somniferine are important. Withaferin-A a Steroidal Lactone, possesses growth inhibitory and radio-sensitizing effects on experimental mouse tumors. Ashawgandha could prove to be a good natural source of potent and relatively safe radio-sensitizer/chemotherapeutic agents. These are mainly used in Ayurvedic and Unani preparations. The similarity between the properties of ashwagnadha roots and the restorative properties of ginseng roots have led to it being called Indian Ginseng.

Botanical Description

An erect, herbaceous, evergreen, tomentose shrub, 13-150 cm high. All its parts are clothed with whitish, stellate hairs. Leaves ovate, entire, thin, base cuneate and densely hairy beneath. Flowers are greenish or lurid yellow, axillary, in clusters of about 25 forming umbellate cymes, sessile or subsessile. The fruit is a berry, 7 mm across, red, globose, smooth, enclosed in an inflated, membranous, somewhat 5-angled, pubescent, persistent calyx. The fruits turn orange-red in colour when they mature. The seeds are yellow in colour and reniform in shape. The flowering season is from July to September and the ripe fruits are available in December.

Geographical Distribution

It is native to Mediterranean region in North Africa. It is found wild in grazing grounds in Mandsaur and the forest lands in the Bastar district of Madhya Pradesh, all over the foothills of the

ASHWAGANDHA (*Withania somnifera*)

Punjab and Himachal Pradesh and Western Uttar Pradesh, in the Himalayas. Other than India, it is also found in the wild in the Mediterranean region in North Africa, Spin, Island, Morocco, Jordan, Baluchistan (Pakistan), Sri Lanka. The crop is cultivated in an area of about 4000 ha in India, mainly in the drier parts of Manasa, Neemuch and Jawad tehsils of the Mandsaur District of Madhya Pradesh, in Punjab, Sind, Rajasthan and South India. In Karnataka, its cultivation has been reported in the Mysore district.

Medicinal Uses

The drug is mainly used in Ayurvedic and Unani preparations. Withaferine-A has been receiving a good deal of attention because of its antibiotic and anti-tumor properties. It is used for curing carbuncles in the indigenous system of medicine. The paste prepared out of its leaves is used for curing inflammation of tubercular glands and that of its roots for curing skin diseases, bronchitis and ulcers. In some areas, the warm leaves are also used for providing comfort for eye diseases. However, the roots are mostly used for curing general and sexual debility.

Its fruits and seeds are diuretic in nature. The leaves are reported to possess anti-helmintic and febrifuge properties. An infusion of leaves is given for fevers. For the treatment of piles, a decoction of the leaves is used both internally and externally. The leaves are also used as a hypnotic in alcoholism. Externally, the leaves are used as fomentation for sore eyes, boils and swelling of hands and feet. As an insecticide, they are useful for killing body lice. An ointment prepared by boiling the leaves, is useful for bed-sores and wounds. The fresh leaf-juice is also applied on anthrax pustules.

An infusion of the bark is given for asthma. For the treatment of scrofula and constipation, the root is given in the form of a decoction as well. The decoction mixed with long pepper (*Piper longum*), butter and honey (25-50 g) and the powdered root with milk or clarified butter is used as an aphrodisiac and in seminal debility. For chest complaints, colds and chills, its decoction is recommended. The decoction along with milk and clarified butter is considered as a cure for female sterility, if taken for a few days after the menstrual period. An enema of the roots, with their bark removed, is given to feverish infants. The root is also said to have been used to treat snakebite. It is also used in chest complaints. A paste made of the green barriers with leaves and small twigs is useful for treating saddle-sores and girth-galls in horses.

Chemical Constituents

The total alkaloid content in the roots of the Indian types has been reported to vary between 0.13 and 0.31%, though much higher yields (upto to 4.3%) have been recorded elsewhere. In all, 13 components have been obtained chromatographically. The include choline, tropanol, pseudotropanol, cuscokygrene, 3-tigloyloxytropana, isopelletierine, anaferine, anahygrine, withasomnine and several other steroidal lactones. In addition to the alkaloids, the roots are reported to contain starch, reducing sugars, hentriacontane, glycosides, dulcitol, withanicil (0.08%), an acid, and a neutral compound.

In addition, the leaves are reported to contain five unidentified alkaloids (yield 0.09%), withanolides, glycosides, glucose and many free amino acids. The occurrence of chlonogenic acid, condensed tannin (also in the stems), and flavonoid are also reported.

Commercial Drug

The commercial drug consists of its dried roots which occur in small pieces, 10.0 to 17.5 cm long and 6.12 mm in diameter. The base of the stem is also used. The pieces are dark brown with a creamy interior. They are straight, unbranched and conical. The main roots bear fibre-like secondary roots. Their outer surface is buff to grayish-yellow with longitudinal wrinkles. The stem bases are thickened, cylindrical and green and have longitudinal wrinkles. the roots have a short and uneven structure, a strong odour and a mucilaginous, , bitter and acrid taste. The young tuberous and older roots have distinct macroscopic and microscopic characters.

Soil and Climate

It grows successfully in sandy loam or light-red soils (pH 7.5-8.0) with good organic matter and drainage. It prefers a subtropical dry climate. Annual rainfall of 660-750 mm is best suited its growth.

Improved Varieties

Jawahar Asgandh-20 (WS-20) and WS-22, WS 134 and Rakshita.

AGRO-TECHNIQUE

Land Preparation

Before cultivation, the land should be ploughed twice and the field should be cleaned thoroughly of weeds. If required, small canals may be prepared for drainage. About 25 t/ha of cowdung manure is also added.

Mode of Propagation

It can be propagated by seeds.

Direct Sowing

In this case, the seeds are sown directly in the main field by broad-casting. Since it is largely grown as a rainfed crop, the sowing is determined by the monsoon. After receiving one or two showers, the field is thoroughly prepared, divided into plots of convenient sizes and the seeds are sown during the second week of July. A seed rate of 10-12 kg/ha is required for this method of planting.

Nursery Raising

When the seedling are to be raised for Transplanting, they should to be sown in well-prepared, raised nursery beds. About 5-6 kg of seeds are required to provide enough seedlings for sowing one hectare. To avoid nursery diseases, the seeds are treated with common fungicide such as Dithane

M-45 @ 3 g/kg of seeds before sowing to protect the plant from seeds born fungal disease or seedling rot or wilt. The seeds in the nursery-beds are sown in lines spaced at 5 cm and covered with light soil in the month of May-June. The germination commences within 6-7 days of sowing and in about 10 days after sowing it is complete.

Transplanting

When the seedlings are 6 weeks old and sufficiently tall they are transplanted in 60 cm-spaced rows, 60 cm apart in well-prepared land. The optimum time of Transplanting by August-September.

Fertilizers and Manures

The crop does not require heavy doses of manure and fertilizers. In Madhya Pradesh, where it is grown on a commercial scale, no fertilizers are applied and the crop is cultivated on only residual fertility. Organic manures @ 10-15 tonnes/ha should be applied along with the chemical fertilizers. Ammonium sulphate @ 30 kg/ha, when added to the soil shows positive results. The phosphorous fertilizer should be added to the field before sowing @ 18-20 kg/ha. Nearly 10 Kg of nitrogen should be added as a seed dose to the crop 60-80 days after sowing.

Irrigation

The crop does not require any Irrigation. However, in areas where rainfall in restricted to a few months in a particular period, about 2-3 Irrigations will help the plants to give optimum yield.

Interculture

In broadcast plots only hand weeding is possible where as in line intercultural implements such a way that they do not over the crops. Weeds cut or plucked also can be used for mulching. The directly sown crop is thinned at 25-30 days. Hand-weeding at 30-day intervals helps to control the weeds effectively.

Pests and Diseases

Seed rotting, seedling blight and leaf blight are common diseases affecting Ashwagandha. They reduce the plant population drastically, ultimately reducing the yield. Their incidence can be minimized by treating the seeds before sowing with Captan at the rate of 3 g/kg of seeds, followed by spraying the crop with Dithane M-45 at the rate of 3 g/l of water when the crop is 30 days old. The spray should be repeated at an interval of 7-10 days if the disease is not controlled.

The other diseases reported on this crop are seedling mortality or die-back. When the conditions are humid and the temperature is high, seedling mortality becomes serious. The disease can be minimized by the use of healthy seeds and pretreatment of seeds with Thiram of Dettan (3-4 g/kg of seeds).

Harvesting and Storage

Harvesting starts from January and continues till March (150-170 days after sowing). The maturity of the crop is judged by the drying out of the leaves and the berries turning red. The entire plant is uprooted and the roots are separated from the aerial parts by curing the stem 1-2 cm above the crown. These are then transversely cut into smaller pieces of 7-10 cm for drying. Occasionally the roots are dried as a whole. The berries are plucked from the dried plants and are threshed to obtain the seeds.

Grading

The dried whole roots undergo cleaning, trimming and grading before dispatch. They are beaten with a club to remove the adhering soil and the thin lateral roots and rootlets. The main taproot may be cut into transverse pieces. The entire product is then carefully hand-sorted into four grades, based on the thickness and uniformity of the pieces.

(i) A-grade

Root pieces up to 7 cm in length, solid, with 1-1.5 cm diameter; they should be brittle and pure white on the inside.

(ii) B-grade

Root pieces up to 5 cm in length, solid, with a diameter of less than 1 cm, the roots should be brittle and white on the inside.

(iii) C-grade

Root pieces upto 3-4 cm in length, side branches solid, with a diameter of 1 cm or less.

(iv) Lower grade D

Small root pieces, semi-solid, very thin and yellowish on the inside.

Production and Yield

An average yield of 3 to 5 qts/ha of dried roots and 50 to 75 kg/ha seeds can be obtained.

Marketing and Trade

The only Neemuch (M.P.) market on an average receives about 5-6 tonnes of dry roots every year for auction and sale which costs about Rs. 20-30 crores. From here the roots are send to Kolkata, Chennai, Mumbai, Bangalore, Cochin, Amritsar, Siliguri and to other countries. Presently cultivated areas produce about 2000 tonnes of dry roots of this crop annually. Neemuch in mandsaur district is a whole sale market for this crop. The current market price of dried roots and seeds of Ashawgandha are Rs. 80-90 and Rs. 40-50 per Kg subsequently.

Economics of Cultivation

The market of Ashawgandha and their products are highly volatile. The economics (per ha) worked out here are subject to fluctuations, depending upon time and place.

Expenditure (in Rs.)	—	8,000.00
Returns (in Rs.)	—	35,000.00
Net Profit* (in Rs.)	**—**	**27,000.00**

*   Estimation of the net profit is analysed on the basis of grower/collectors price. The grower/collector price is always 25-40% less than wholesale market price due to various stakeholders *viz.* Agent, middle man, Brokers, Commission agents.

BABCHI (*Psoralea corylifolia* L.)

Vernacular Names

San.–Vakuchi, Bakuchi; *Eng.–*Babchi seeds; *Hin.–*babchi; *Beng.–*Latakasturi; *Guj.–*bavachi; *Mar.–*Bavanchi, Bavchi; *Tel.–*Bavanchalu, Kala gija; *Tam.–*karpokarishi; *Kan.–*Bavancigida; *Punj.–*babchi; *Urdu–*Bakuchi; *Trade–*Babchi seeds, Psoralea.

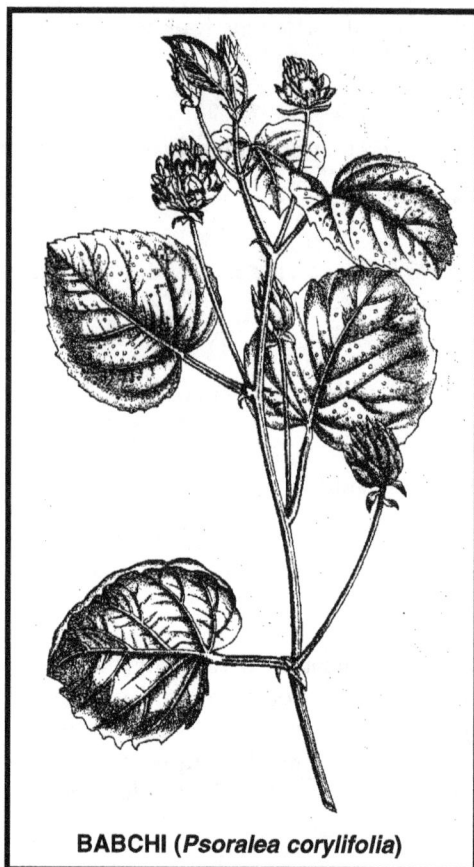

BABCHI (*Psoralea corylifolia*)

Introduction

The seeds of babchi are used in the indigenous system of medicine in the treatment of leucoderma, leprosy and psoriasis. Its belongs to the family Fabaceae. The seed is surrounded by a sticky, oily pericarp which contains coumarins, of which Psoralen and Isopsoralen are therapeutically important. Besides treating psoriasis, psoralin is being investigated as a cure for several diseases including AIDS. It is also used in the treatment of intestinal amoebiasis and the healing of wounds and ulcers. There are several reports in literature on the anti-microbial, anti-feedant and insecticidal activities of babchi, suggesting other possible uses.

Botanical Description

An erect, annual herb, that grows 30-60 cm tall under natural conditions and up to 160 cm under cultivation. Branches profusely; its stem and branches are covered with conspicuous glands and white hairs. Leaves, broadly elliptic, rounded and mucronate at the apex. The axillary, solitary inflorescence (raceme) comprises 10 to 30 flowers with a hairy pedicle. It bears a single-seeded pod which is indehiscent and the pericarp is usually found adhering to the seed. The flowers open in two flushes, once early morning and again in the evening. Anther dehisces about one hour before the stigma becomes receptive.

Geographical Distribution

It is widely distributed in the tropical and subtropical regions of the world. In India, it occurs in the South-eastern districts of Madhya Pradesh, Uttar Pradesh and wider distribution in parts of Rajasthan, Andhra Pradesh, Bihar and Gujarat.

Medicinal Uses

The seeds are laxative, stimulant and aphrodisiac. Seeds are useful in bilious affections and are also used to make perfumed oil and its powder is specially recommended by Vaidas in leprosy and leucoderma internally, and are also applied in the form of paste or ointment externally. The seeds contain an essential oil which is very effective against certain bacteria causing skin diseases. It is also useful in treating ulcers, scabies, leprosy and serves as a good hair tonic. Roots of this plants are reported to be useful in treating carries of teeth and the leaves in treating diarrhoea.

Chemical Constituents

The seeds contain an alkaloid bakuchiol, a brown fixed oil (10%), a dark brown resin (8.6%), psoralen/isopsoralen (1%), an essential oil (0.05%), a non-volatile terpenoid oil, a monoterpenoid phenol and raffinose.

Soil and Climate

In nature, it is found growing on a variety of soils ranging from sandy medium loam to black-cotton soils. However, sandy loam soil with good organic matter is the best for its growth and yield. It also tolerates a wide variation in soil pH, but too acidic and too alkaline soils should be avoided. The crop prefers dry tropical regions with a comparatively warmer type of climate for its successful growth and yield. It grows well in areas with low to medium rainfall during the summer months.

AGRO-TECHNIQUE

Land Preparation

The field should be prepared to make good seed-beds, by light ploughing followed by two harrowing before the onset on the monsoon season.

Mode of Propagation

It can be propagated by seeds.

Pretreatment of Seeds

The seeds have a problem of dormancy due to the hard seed-coat and germination is invariably poor (5-7%). The germination percentage can be improved by breaking the dormancy by the mechanical puncturing of seed coverings or by treating the seeds with sulphuric acid for 50 minutes. A pretreatment of seeds in 1% sulphuric acid has been found efficacious; the sulphuric acid should be washed off by repeated washings in water before the seed is sown.

Sowing

The seeds are dibbled in rows, preferably 45-60 cm apart, keeping plant-to-plant spacing around 30-45 cm, depending upon the fertility of the land. A seed rate of 7 kg will be enough to cover a one hectare area.

Fertilizers and Manures

The crop responds well to the use of organic and inorganic fertilizers. It has been noted that the basal application of Farmyard Manure (FYM) at 20 t/ha gives a good initial growth and increases the seed yield significantly. Besides, a fertilizer dose of 100 kg N, 60 kg P and 50 kg K/ha is recommended. Of this, half a dose of N and a full dose each of P and K is given as a basal dose, while the remaining 50 kg N/ha is applied in one dose as a top-dressing after 45 days of sowing.

Irrigation

Since it is mostly grown as a rainy season crop, the Irrigation requirement of babchi is moderate. However, after the rainy season is over (June-September), the crop may be irrigated fortnightly. About 6 to 8 Irrigations may be given until the crop is finally harvested.

Interculture

The field must be kept free from weeds by regular weeding as and when required. In all, about 2 to 3 hand-weedings are necessary during the early period of growth after which, the plants cover the ground densely providing practically no opportunity for the growth of weeds.

Pests and Diseases

The plant is sufficiently hardy and free from pests. However, under Bangalore conditions, the Lepidopteron insects, *viz.* the leaf folder (*Archips micaceana*) and two defoliators (*Helicoverpa armigera* and *Papilio polytis*) have been found to cause considerable damage to this crop. These insects can be effectively controlled by spraying monocrotophos @ ml/l of water.

Among the diseases, the powdery mildew is the only disease of a serious nature noticed on this crop. This disease can be controlled by spraying wettable sulphus @ 3 g/lit. of water at weekly intervals.

Harvesting and Storage

The crop takes about 7-8 months from the time of sowing to reach maturity. As the seeds continue to mature over the growing plants continuously, the seed-picking will commence from mid-December onwards to the end of February. In all, 4 to 5 pickings are usually taken. At maturity, the single-seeded fruit turns brownish-black and emits a mild odour.

Production and Yield

An average yield of about 2 t/ha of dry seeds may be obtained.

Marketing and Trade

The current market price of babchi seed is Rs. 10-12 per kg.

Economics of Cultivation

The market of Babchi and their products are highly volatile. The economics (per ha) worked out here are subject to fluctuations, depending upon time and place.

Expenditure (in Rs.)	—	6,000.00
Returns (in Rs.)	—	28,000.00
Net Profit* (in Rs.)	**—**	**22,000.00**

* Estimation of the net profit is analysed on the basis of grower/collectors price. The grower/collector price is always 25-40% less than wholesale market price due to various stakeholders *viz.* Agent, middle man, Brokers, Commission agents.

BASELLA (*Basella alba* (L.) Stewart
Syn. *B. rubra* Linn)

Vernacular Names

San.–Upodika; *Eng.*–Indian spinach, *Hin.*–Poi, Lalbaclu, *Kan.*–Bansali; *Mal.*–Vasalaccira, Basalaccira; *Tam.*–Vasalakirai, Sivappu Vasalakkirai; *Tel.*–Baccali; *Trade*–Basella.

Introduction

Basella popularly known as *malabar night shade* or Indian spinach is grown in all parts of India.

Botanical Description

A fleshy, annual or biennial, succulent twining much branched herb with alternate, broadly entire leaves. Leaves are broadly ovate and pointed at the apex. Flowers are small, white or pink, sessile in clusters on elongated thickened peduncles in an open branched inflorescence. Fruit is enclosed in fleshy perianth.

Geographical Distribution

It is native to Asia. It is found in West Bengal, Assam, South India, U.P. and Punjab.

Medicinal Uses

The stem and leaves are sweet, cooling, emollient, aphrodisiac, laxative, haemostatic, appetizer, sedative, diuretic, demulcent, maturate and tonic. They are useful in vitiated conditions of *pitta*, burning sensation, constipation, flatulence, anorexia, sleeplessness, leprosy, ulcers, dysentery, gonorhoea, banalities, fatigue and general debility. They are especially useful as a laxative in children and pregnant women.

Chemical Constituents

The leaves contain Iodine, Fluorine, carotenoids, organic acids and vitamin K.

BASELLA (*Basella alba*)

Soil and Climate

In well manured sandy loam soils with good drainage and aeration, the crop grows luxuriantly. Adequate moisture and partial shade results in better growth of the plant. Cultivation of this crop should be avoided in regions affected by frost. The crop is usually grown during warm and moist seasons.

AGRO-TECHNIQUE

Land Preparation

The field should be prepared to make good seed-beds, by light ploughing followed by two harrowing before the onset on the monsoon season.

Mode of Propagation

It can be propagated by seeds, stem or root cutting.

Nursery Raising

In the northern and eastern plains of India, seeds are sown from March to May, while in the southern parts, it is grown twice, once sown in June and again in October to November. Late spring or early summer season is the best time for sowing in the hills. Nearly seeds @ 12-15 kg will be required to sow a hectare land. Seeds can be treated with 0.3M potassium nitrate for 24 hours to enhance germination and vigour of seedling growth.

Transplanting

Seedling are transplanted when they can easily be handed in beds with 45 cm spacing. The crop is also raised on bamboo stakes or trained in trellis where seeds are shown 20 to 25 cm apart in rows at the base of bamboo stake or trellis.

Manure and Fertilizers

For getting good crop, rich soil are preferred and only nitrogenous fertilizer application has been found more beneficial. The mustard cake @ 50 kg nitrogen/ha produced the highest yield. Recommended doses of fertilizers for basella for northern parts of Karnataka is 350 kg nitrogen, 50 kg phosphorus and 50 kg potash/ha.

Irrigation

The crop in general requires 5 to 6 Irrigation when grown in summer and frequency of Irrigation depends on the soil type. It requires uniform moisture supply for continuous growth. It may be advisable to irrigate the crop once in 4 to 5 days during hot weather and 7 to 8 days during cool season.

Interculture

Basella is a small herbaceous shrub and, therefore, the competition with weeds during its initial stages of growth should be avoided by keeping the land free from weeds. The first weeding is done after about 20-30 days of planting. After this, one or two weedings are done after 60 days of sowing to keep the weeds under control during the initial stages of crop growth. Later, weeding follows every harvest.

Pests and Diseases

Several diseases are reported to attack the crop but, the most important diseases affecting this crop are damping off (*Pythium amphanidermatum*), leaf spot (*Acrothecium basella, Fusarium maniliforme* and *Cercospora* spp.). The removal of infested leaves and plants with rust diseases will help in controlling rust disease.

Harvesting and Storage

The leaves and stems are harvested after 8 to 10 weeks of sowing. After harvesting, the plants are dried under shade for 3-4 days before storage. The harvested crops is chopped into pieces, particularly the young herbaceous shoots, and dried by spreading under the shale or sun or on wire-racks, taking care to see that the dried leaves remain green. The harvested leaves are spread in thin layers and turned frequently while drying. The dried leaves are packed in gunny bags and stored in moisture free places.

Production and Yield

A good crop gives yield nearly 130-150 quintals herbage/hectare.

Marketing and Trade

The current market price of herb is Rs. 12-18 per kg.

Economics of Cultivation

The market of Basella and their products are highly volatile. The economics (per ha) worked out here are subject to fluctuations, depending upon time and place.

Expenditure (in Rs.)	—	3,500.00
Returns (in Rs.)	—	15,000.00
Net Profit* (in Rs.)	—	**11,500.00**

*   Estimation of the net profit is analysed on the basis of grower/collectors price. The grower/collector price is always 25-40% less than wholesale market price due to various stakeholders *viz.* Agent, middle man, Brokers, Commission agents.

BELLADONNA (*Atropa acuminata* Royle ex Lind.)

Vernacular Names

San.–Suchi; *Eng.*–Deadly Nightshade; *Hin.*–Angurshafa, Sagangur; *Beng.*– Yebrui; *Guj.*–Dhatooro; *Kash.*–Sagangur, Mait-brand; *Mar.*- Girbuti; *Kan.*–Nati belladona; *Garh.*-Dhol; *Punj.*–Bantamaku;*Trade*–Indian belladona.

Introduction

The genus name Atropa has been derived from Atropos the Greek goddess of death, whereas, the word belladonna is derived from the Latin word Bella meaning beautiful and Donna meaning woman. Indian belladonna, belonging to the family Solanaceae, comprises four species of medicinal plants. Out of these four species, the commercial drug is obtained from the leaves, flowering tops and roots of *A. acuminata*, commonly called Deadly Night Shade. Belladonna leaves were included in the British pharmacopoeia in 1809. The roots, however, were accepted as a drug only in 1880.

Botanical Description

An erect, herbaceous plant, 70 to 150 cm high, with a long, cylindrical, dichotomously branched stem. The leaves are green or olive-green in colour, alternate, but on the upper branches, each leaf is usually accompanied by a small, stipule-like leaf. The blade is ovate, pointed, entire, soft, with conspicuous veins and frequently with a purplish mid-rib. The flowers, brownish-yellow with greenish veins and are borne singly or in groups of two to four. The fruits are globular and purple-black in colour.

Geographical Distribution

It is indigenous to Western Himalayas but grows more commonly under conifer forest of extending from Kashmir to Shimla and the adjoining areas of Himachal Pradesh at elevation of 2500-3500 m a.s.l. It is also cultivated in some places in the Himalayas. It is cultivated for its drug in Central Europe, England, the USSR, the United States and Northern India.

Medicinal Uses

Belladonna leaves are employed for extracting total alkaloids for use as proprietary pharmaceutical preparations to treat various diseases like gastro-intestinal hypermotility, hyper-secretion, peptic ulcer, spastic constipation, spastic dysmenorrhoea, nocturnal enuresis, bronchial asthma and whooping cough. Its leaves are widely used for the manufacture of tinctures, extracts and plasters. The drug serves as an anodyne, sedative stimulant, antigalactogogue, antidiuretic, mydriatic or anti-asthmatic, antispasmodic and anti-inflammatory.

BELLADONNA (*Atropa acuminata*)

It is also used for relief from whooping cough, in the treatment of renal and bilary colic, in stomach disorders, to stop sweating, in ophthalmology to cure gastric ulcers and as an antidote for depressing poisons such as opium, muscarine and chloral hydrate. The roots are primarily used in the external treatment of gout, rheumatism and other afflictions. It is also used in the treatment of epilepsy, Parkinson's disease and bradycardia.

Chemical Constituents

The leaves and roots of belladonna contain tropane alkaloids whose concentrations varies from 0.13 to 0.7% (the average is 0.45%). Leaves contain 0.45% hyoscyamine and non-alkaloidal volatile bases not greater than in *A. belladonna*. The roots of *A. acuminata* contain 0.20-0.8% hyoscyamine (0.47% is the average) and non-volatile alkaloids (0.4%).

Soil and Climate

It is a temperate plant and grows well in slightly acidic, deep, fertile soils of medium texture, which re rich in humus. Heavy clay soils which are water-logged should be avoided. It behaves as a perennial in temperate climates and gives maximum herbage and alkaloid yield. In subtropical areas it can be grown as a winter crop. However, the plant behaves as an annual as it dies during the summer months and hence the yield is poor.

Improved Varieties

There are no named varieties reported under this crop but the workers at the Regional Research Laboratory, Srinagar (J.K) have selected plants containing 0.60% alkaloids in *A. acuminata*.

AGRO-TECHNIQUE

Land Preparation

The land is prepared well and brought to a fine tilth with the help of tractors and organic manure at the rate of 25 t/ha, to begin with, is mixed with the soil.

Mode of Propagation

It can be propagated by seeds or cutting of shoots & rootstocks.

Nursery Rising

The crop propagated through seeds extracted from the berries, usually collected from September to November. Belladonna has been cultivated by direct sowing, but raising them in the nursery gives the best results. The nursery may be raised from the second week of May to the end of autumn (September to October) and in summer under sufficient shade. Before making the nursery beds, the land should be ploughed well so as to give a fine tilth. Raised beds of 3 m × 1 m size may be made, surrounded by drainage and Irrigation channels. Well decomposed FYM or sheep manure should be mixed in the top 10 cm of the soil. About 4 kg of seeds will be required to raise seedlings for one hectare of land while about 20 kg seeds are required for sowing one hectare of land by broadcasting.

Seeds are pre-treated with sulphuric acid, ethyl alcohol or petroleum ether five better germination than the untreated ones, but the plants from treated seeds are stunted and their alkaloids content is only 40% that of the plants from untreated seeds. Treating the seeds with gibberellin acid (40 ppm for 24 hours) or thiourea (0.5-2.0%) improves germination without affecting the alkaloid content. Surface sterilization of the seeds by fungicides like Captan, Agrosan GN, Dithane-45 or Agallol before sowing reduces seedling mortality due to damping off.

The beds must be watered immediately after sowing with a watering can and daily thereafter. The seeds normally germinate within 3 weeks with a good germination percentage, and the seedlings will be ready for planting in the main field when they attain a height of 15-20 cm after about 8-12 weeks.

Transplanting

The ideal time for planting in the field is early spring (March-April) or autumn (October-November). Generally, seedlings raised in autumn are planted in spring and those raised in summer are planted in autumn. Seedlings are planted at a spacing of 45-60 cm in rows about 45-60 cm apart. It is always safer to plant the seedlings on raised beds with 1 m-wide strips or ridges as it avoids water-logging and facilitates Irrigation. The field may be irrigated immediately after Transplanting.

Fertilizers and Manures

It is always better to give a balanced fertilizer of nitrogen, phosphorus and potash in order to get a good crop. Normally, a basal dose of 25-30 kg nitrogen, 50-60 kg phosphorus and 40-50 kg potash per ha should be applied before planting. An additional dose of 60-80 kg N is applied in 3-4 split doses as a top-dressing at monthly intervals after every harvest. The application of heavy doses of nitrogen gives a good crop of leaves and also increases their alkaloid content.

From one year after planting, a basal dose of nitrogen and phosphorus should be repeated every year in the beginning of the spring season (March-April), which is followed by a top-dressing of nitrogen. Micronutrients have been found to affect the growth. It is better to irrigate the field after each application of fertilizers, if there is no rain.

Irrigation

Belladonna has a high water requirement and it should be irrigated frequently, once in 10-15 days during the dry period. Normally, 6-7 Irrigations are required during the dry months. Care should be taken to avoid water-logging.

Interculture

The crop should be kept free from weeds by repeated weeding and hoeing. One hand-weeding is necessary when the plants are young. Normally, 3-4 weeding and hoeing are required during the growing season. A tractor-drawn cultivator or hoe may be employed for later weeding. However, hand weeding is practicable in hilly or small terraces.

Pests and Diseases

Among the insects, cut-worms (*Agrostis flammetra* & *A. ypsilon*) and white grubs cause considerable damage to young plants. The cut worms are most virulent during the dry months of June and July. Early Transplanting during April-May help to reduce the incidence of the pest. Application of Aldrin @ 5% of nursery bed before sowing, protects the crop. Drenching of seedling or Transplanting with a solution of Chlorodan reduces the pests in the field. The common potato beetle and flea beetle is prevalent on the crop during early summer months (April to May), when the seedlings and young plants are cut at the ground-level. Sprays or dusts containing Rotenone @ 0.5-0.75% are effective in reducing the damage.

Harvesting and Storage

Belladonna yields two times of leaves during the first year of growth. In the second and subsequent years, it gives three or four times leaves. The roots are, however, dug out after the third or fourth year of planting. Harvesting begins with the first sign of flowing when the alkaloid content in the plant is at its peak. The leaves are cut with the help of a cutter (pruning scissors) about 30 cm above ground-level as against 7.5 cm above ground-level, done earlier. The harvested crops is chopped into pieces, particularly the young herbaceous shoots, and dried by spreading under the shale or sun or on wire-racks, taking care to see that the dried leaves remain green. The leaves may also be dried with artificial heat, with or without fans for air circulation. The harvested leaves are spread in thin layers and turned frequently while drying. The woody stems are discarded before drying. Under humid conditions, the material absorbs moisture resulting in the lowering of alkaloid content. Lumps of calcium chloride kept in the store rooms help to avoid such as moist conditions.

Production and Yield

The present production of belladonna leaves from cultivated source is around 25-40 tonnes while the quantity of leaves collected from forest areas hardly accounted for 3-5 tonnes per annum. The yield of leaves from a uniform belladonna crop varies between 5 to 6 quintals/ha in the first year reaches up to 15 qts/ha in the subsequent years. The yield of dried roots is also varies vary from 1.5 to 3.0 q/ha.

Marketing and Trade

The current market price of dried leaves and dried roots is Rs. 20-30 and Rs. 50 per kg respectively.

Economics of Cultivation

The market of belladonna and their products are highly volatile. The economics (per ha) worked out here are subject to fluctuations, depending upon time and place. (Economic life of plant is 3 years).

Expenditure (in Rs.)	-	11,00.00 (in Ist year)	5,000.00 (in IInd year)	5,000.00 (in IIIrdyear)
Returns (in Rs.)	-	18,000.00 (in Ist year)	45,000.00 (in IInd year)	45,000.00 (in IIIrdyear)
Net Profit* (in Rs.)	**-**	**7,000.00 (in Ist year)**	**40,000.00 (in IInd year)**	**40,000.00 (in IIIrd year)**

*   Estimation of the net profit is analysed on the basis of grower/collectors price. The grower/collector price is always 25-40% less than wholesale market price due to various stakeholders *viz.* Agent, middle man, Brokers, Commission agents.

BISHOP'S WEED (*Ammi majus* L.)

Vernacular Names

Eng.–Bishop's weed, greater ammi.

Introduction

It belongs to the family, Apiaceae, is an important medicinal plant indigenous to Egypt. Its fruit is the principal source of coumarins 2% (consisting mainly of xanthotoxins, and imperation). The Xanthotoxin is mainly used in treatment of vitilago or leucoderma and also in the formulation suntan lotion.

Botanical Description

An erect, annual herb attaining a height of 80 to 120 cm. The leaves are decompound, light green, alternate, variously pinnately divided, having lanceolate to oval segments. It bears auxiliary and terminal compound umbels of white flowers. The fruits are ribbed, ellipsoid, green to greenish-brown when immature, turning reddish-brown at maturity. The odour of the fruit is slightly characteristic of terebinthinate and becomes strong on crushing.

Geographical Distribution

The plant have been first introduced in India along with *A. visnaga* by the Forest Research Institute, Dehradun, in 1955, by the courtesy of the UNESCO. It is indigenous to Egypt and is widely distributed in Europe, the Mediterranean region, Abyssinia and West Africa. It is being successfully grown in the temperate and subtropical regions of Himachal Pradesh, Uttar Pradesh, Gujarat, Tamil Nadu and Karnataka.

Medicinal Uses

Xanthotoxin is commonly used for various ailments of human suffer in especially for the treatment of vitiligo or leucoderma, psoriasis and in formulations of suntan lotions. It is also used as an expectorant, diuretic, lithontriptic, deobnstruent, and used in vitilago, jaundice, asthma and angina and as an antidote for various chemical and insect poisons.

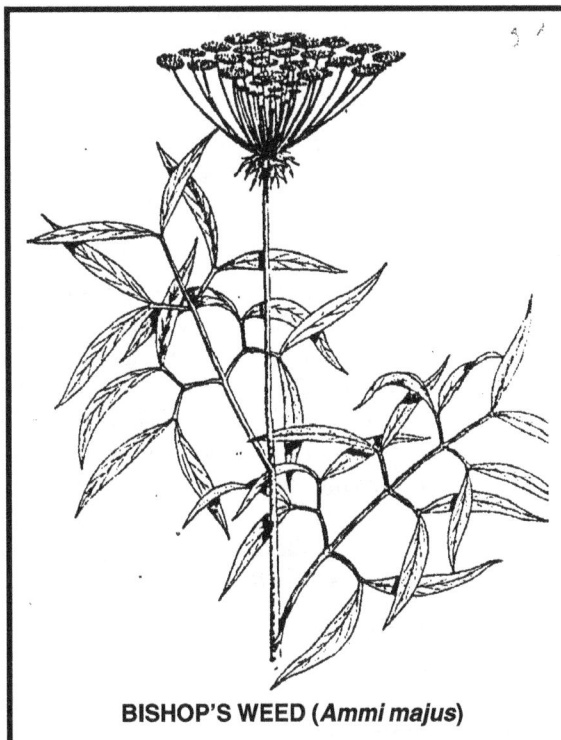

BISHOP'S WEED (*Ammi majus*)

Chemical Constituents

The fruit contains about 1% of an amorphous glucosidal principal (Xanthotoxin), 0.45% tannin, 4.76% oleoresinous product, 3.2% acrid oily liquid, 12.5% fixed oil, 0.2% glucose, 14% protein etc. The other furocoumarins are bergapten isopimpinellin, isoimperatorin, oxypeucedanin, heraclenin, oxypeucedanin hydrate, sexalin, pabulenol, etc.

Soil and Climate

It can be grown on a variety of soils, but thrives best in well-drained sandy loam to clay loam soils containing a moderate amount of organic matter. Any soil which is saline, alkaline or water-logged is unsuitable for its cultivation. It requires a mild, cool climate in the early stages of growth, and warm and dry weather at maturity.

Improved Varieties

Sutton and Monoica.

AGRO-TECHNIQUE

Land Preparation

Before cultivation, the land should be ploughed twice and the field should be cleaned thoroughly of weeds. The field is prepared well, divided into convenient sized plots and thoroughly mixed with organic manure. About 25 t/ha of cowdung manure is also added.

Mode of Propagation

The crop can be propagated by seeds.

Direct Sowing

The seeds can be sown in furrows at 45 cm apart, from August to the middle of September. A seed rate of 12 kg/ha is recommended when the crop is to be raised by direct seeding. Since the seed is very small, it is mixed with fine soil or sand in a ratio of 1:10. If there are no rains within a week of sowing, the field should be irrigated liberally. The seed starts germinating after a fortnight and it takes about a month to complete the germination for the whole field. The ideal time for Direct Sowing in the field is September to middle of October. Late sowing *viz.* last week of November does not allow the seed to germinate.

Nursery Raising

The seedlings are raised in well-prepared nursery beds of 1x3 m in size. Before sowing, a mixture of equal quantities of calcium ammonium nitrate and single super phosphate at the rate of 150 g per bed is added and mixed well with the surface soil. About 1.5-2.0 kg of seeds will be enough to raise seedlings for 1 ha of land. The best time for Nursery Raising is September. After sowing, the seeds are covered with a mixture of FYM and soil, mulched and watered. The beds are kept moist by light Irrigations every day. The seeds start germinating after 7-8 days of sowing. Application of 100 g calcium ammonium nitrate to each bed is recommended after 10 days of the emergence of the seedling. Within a month almost all the seeds have germinated. At this stage, the mulch is removed and the beds are kept free of weeds and watered continuously. Nearly 50-60 days old seedling with true leaves and 8-10 cm in height are ready for Transplanting in the field.

Transplanting

For Transplanting, 8-10 cm in height seedlings are carefully removed to avoid any damage to the roots. They are transplanted n well-prepared land at a spacing of 45x30 cm or 45x60 cm in the main field by end of October to the end of November.

Fertilizers and Manures

The crop has been found to respond well to the application of manures and fertilizers. A fertilizer dose of 90-100 kg nitrogen, 115-120 kg Phosphorus and 25-30 kg potash/ha has been reported as optimum for the crop. This is in addition to 20-25 tonnes of FYM. The manure and half of the nitrogen, the full dose of P and K are applied in furrows before Transplanting. The rest of the nitrogen is given as a top-dressing after thinning, when the plants have started branching. Dose of nitrogen at the earlier stage suppressed growth of the plant but increased seed yield. Use of 5-10 kg Super phosphate/ha increased the seed yield by 23-29%.

Irrigation

It is an herbaceous crop and the availability of moisture has a direct effect on its growth and productivity. Hence, it should be irrigated frequently during the dry period. The first Irrigation is given immediately after Transplanting and afterwards at an interval of 7-10 days during dry season. Since the crop is quite sensitive to excess moisture, care should be taken to avoid water-logging.

Interculture

As the crop is very sensitive, it is necessary to keep the plots weed free at the initial stages. Once they grow, they cover the whole plot and thus there is very little scope for the other weeds to complete. About 2-3 hand-weedings will be enough to keep the weeds down. In large-scale plantations, two hoeings should be given to keep the soil mellow and free from weeds.

Crop Rotation

The cultivation of *Heracleum candicans* was also tried in Palampur (H.P). The seed was sown in the nursery in March and seedling were transplanted towards the end of April/beginning of May in those very fields where *A. majus* was grown. The growth was vigorous during rainy season and the plant started flowering in July. The plant was harvested after 21-22 months (in December following year) and the fresh weight of root on an average was 1.2 kg per plant with a total coumarin (active principle of *Heracleum candicans*) content of 8.5%.

Pests and Diseases

Several fungi are known to cause damage to the fruits during storage. *Aspergillus ochraceus* was observed to cause heavy damage to xanthotoxin during the first month of storage, while *Fusarium oxysporum* causes damage to *A. niger* and *A. flavus* and during the first three months of storage and drastically reduces the xanthotoxin content.

Harvesting and Storage

It is a 8-9 months duration crop. When sowing is done during the months of September-October, the crop starts flowering during the months of March and fruiting by the end of April. The flowering and maturity of seeds is spread out over about a month. The best stage for harvesting the fruit when Xanthtoxin is maximum. Mature brown fruit contain 0.4%, mature but green fruit 0.72% and immature green fruit 1% of Xanthotoxin. If harvesting is delayed, the loss is yield due to shattering of the fruits. The immature green fruit (1% Xanthotoxin) stage of harvesting by the end of April, is recommended. While harvesting, the umbels are hand-picked individually by cutting the plant at its middle. The cut umbels are kept in loose bundles and staked till the fruit are dry. The fruits are later separated by threshing and winnowing.

Production and Yield

On an average, the crop may yield about 9-12 quintals/ha dry seeds.

Marketing and Trade

Although the recent statistics on the area, production, import and export details of this crop are not available, but according to an estimation, European countries mainly France, Switzerland and Germany require about 1000-1200 tonnes of *Ammi majus* fruit annually, part of the requirement being met from export from Indian. On the other hand, India has been importing Xanthotoxin worth about Rs. 15-20 lakh. The current market rate of its seed is Rs. 10-15 per kg.

Economics of Cultivation

The market of Bishop's weed and their products are highly volatile. The economics (per ha) worked out here are subject to fluctuations, depending upon time and place.

Expenditure (in Rs.)	—	5,500.00
Returns (in Rs.)	—	45,000.00
Net Profit* (in Rs.)	—	**39,500.00**

* Estimation of the net profit is analysed on the basis of grower/collectors price. The grower/collector price is always 25-40% less than wholesale market price due to various stakeholders *viz.* Agent, middle man, Brokers, Commission agents.

BRAHMI (*Bacopa monnieri* (L.) Penn.)
Syn. *Lysimachia monnieri* Linn.

Vernacular Names

San.–Bramhi; *Eng.*–Thyme leaves gratiola, Water hyssop; *Hin.*–Jalnim, brahmi; *Beng.*–Brahmi-sak, jalanimba; *Tel.*–Sambrani, chettu; *Tam.*–Piram, vivitam; *Kan.*–Neeru brahmi, kiru brahmi; *Mal.*–Brahmi, Nirbrahmi; *Trade*–Brahmi.

Introduction

Brahmi has been used in indigenous system(s) of medicine since ancient times in India. It is mentioned in Atharvaveda, Charak Samhita and Ayurvedic Materia Medica for its healing properties. It is valuable nervine tonic for curing insanity, epilepsy, other mental disorders and increasing memory. It belongs to the family Scrophulariaceae.

Botanical Description

A prostrate, succulent herb with branches spreading or ascending and rooting at the nodes. Leaves obovate-oblong or spathulate, upto 18 mm long. Flowers campanulate, bluish-purple or white with bluish veins, solitary, axillary, short-or-long-pedicilate and blooming in the month of August to continue upto October.

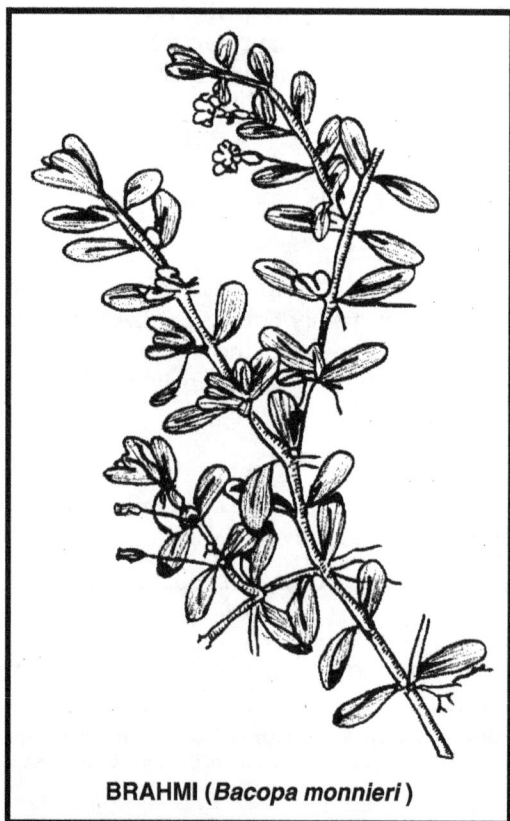

BRAHMI (*Bacopa monnieri*)

Geographical Distribution

Globally, it is distributed in humid and warmer parts of the world. In India, it occurs in damp and marshy tracts in Sub-tropical region, upto 1200 m elevation. It also found in West Bengal.

Medicinal Uses

The whole herb is used in Ayurvedic, Unani and Siddha systems of medicine as a nerve tonic, diuretic, astringent, bitter, cooling, intellect promoting, anodyne, carminative, anti-inflammatory, anti-convulsant, emmenagogue, sudorific, depurative, blood purifier and used for the treatment of asthma, hoarseness, insanity, epilepsy, ulcer, tumours, dyspepsia, flatulence, constipations, elephantiasis, sterility, fever, general debility and various skin diseases.

Chemical Constituents

The herb contains the saponins, monnierin, hersaponin, bacoside-A. The others are D-mannitol, betulic acid, b-sitosterol, stigmasterol and its esters.

Soil and Climate

It requires a well-drained, moist sandy loam soil, rich in organic matter. The crop prefers a moist, warm climate. It performs best in a temperature ranging between 30°C-40°C and humidity 65-80%. For the luxuriant growth of the crop, well-distributed rainfall and good sunshine during winter are ideal.

Improved Varieties

Subodhak and Pragyashakti.

AGRO-TECHNIQUE

Land Preparation

The selected field should be plough 4 to 5 times using a disk and harrow, remove the weeds and bring the soil to a fine tilth. Incorporate the 5-6 tonnes/ha FYM during the final ploughing for thorough mixing with the soil.

Mode of Propagation

It can be propagated by runners.

Planting

It is cultivated summer-rainy season crop. The plant cuttings about 4-5 cm long, each containing a few leaves, nodes and roots are ideal planting materials. These can be obtained by cutting mother plants into small pieces with roots. The cuttings are transplanted in wet soil at spacings of 40 cm × 40 cm. Ideally, the plants should be transplanted in March-June.

Fertilizers and Manures

For good yield, 100-120 kg nitrogen, 60-80 kg phosphorus and 60-80 kg potash should be given at the time of planting. The nitrogen dose should be applied in 3 split doses.

Irrigation

Irrigation immediately after Transplanting is essential. Subsequently, the field is irrigated by flooding as per requirement usually every 7-8 days. There is no need for Irrigation during monsoon.

Interculture

Initially hand weeding is required every 15-20 days. Later as the plant proliferates and forms a dense mat of vegetation, weeding may be required sporadically.

Pest and Diseases

Defoliating insect such as grasshopper and tobacco cutworm, *Spodoptera litura* are a menace during the summer months. These cause mass defoliation and need to be controlled in time. Neem-based pesticides should be applied for controlled the larvae. Alternatively, spray of 0.1% Dimacran or Nuvan is very effective in controlling pest infestation.

Harvesting and Storage

The crop can also be maintained in a perennial state with 2 harvests in a year, the first one in June and the other one after monsoon, in October. The plant can be ideally harvested by rationing so that the upper portion of the stem 4-5 cms from the base are removed and the rest left for subsequent regeneration.

The herbage is dried by spreading on the ground under shade at room temperature. The material is to be cleaned free of any external matter. The dry matter should be stored in a cool dry room packed in bags/boxes having concrete flooring, away from walls.

Production and Yield

The fresh and dry herb yields of brahmi go upto 200 q/ha and 50 q/ha, respectively, when harvested after September while bacoside-A yield can be as much as 85 kg/ha. After the first harvest, 30 q dry herb yield in a year will be obtainable.

Marketing and Trade

It is estimated that about 10,000-12,000 tonnes of fresh herb is collected every year from wild habitats mainly from Tamil Nadu and West Bengal for the manufacture of various herbal formulations available in the market. The current market price of the dry herb is Rs. 40-60 per kg.

Economics of Cultivation

The market of brahmi and their products are highly volatile. The economics (per ha) worked out here are subject to fluctuations, depending upon time and place.

Expenditure (in Rs.)	—	17,000.00
Returns (in Rs.)	—	1,20,000.00
Net Profit* (in Rs.)	—	**1,03,000.00**

*   Estimation of the net profit is analysed on the basis of grower/collectors price. The grower/collector price is always 25-40% less than wholesale market price due to various stakeholders *viz.* Agent, middle man, Brokers, Commission agents.

BUCKWHEAT (*Fagopyrum esculentum* Moench)

Vernacular Names

Eng.–Buckwheat; *Hin.*–Kotu, Phaphra; *Him.*- Ogla, Bhares; Gargh.-Ogal, Phaphra; *Punj.*-Ogal, phaphar; *Ass.*-Doron; Unani-Anjubar; *Trade*–Buckwheat.

Introduction

Buckwheat is an important medicinal plant, commercially grown for Rutin. It belongs to family Polygoneaceae. Buckwheat came into the limelight in 1946, as the most promising and economical source of rutin. It is considered as an excellent crop for soil improvement, particularly for virgin lands or for poor soils which do not support any other cereal crop. By its spreading habit and vigorous growth, it smothers weeds and forms an excellent cover. It acts as a soil-binder preventing soil erosion during heavy rainfall. It also produces a large quantity of green biomass which can be incorporated into the soil as green manure. In USA and Russia, the grains are mostly used in the manufacture of livestock and poultry feed.

Botanical Description

It is an annual herb growing up to 80-160 cm in height. The leaves are alternate, acute and 2.5-7.5 cm long; flowers are produced in auxiliary or terminal cymes and are pinkish-white in colour, fragrant, diamorphic and highly self-sterile. The fruit is of 3 colours, ranging from silvery-grey to brown or black.

Geographical Distribution

Backwheat is said to be a native of Central Asia and is cultivated in various countries as a food or fodder crop. In India, it is grown as a grain crop in the higher altitudes (600-3650 m) of the Himalayas (from Kashmir to Sikkim) and in the Nilgiris, but more often as vegetable crop in the lower mountain ranges.

Medicinal Uses

With the growing importance of rutin as a medicine to treat the harmful effects of X-rays, it may be of use to persons exposed to dangerous atomic radiations. In addition, rutin is used in the treatment of increased capillary fragility which is associated with hypertension - a condition which sometimes results in the bursting of blood vessels in the brain, leading to apoplexy or retinal haemorrhage and causing partial or complete blindness. It is useful in a variety of

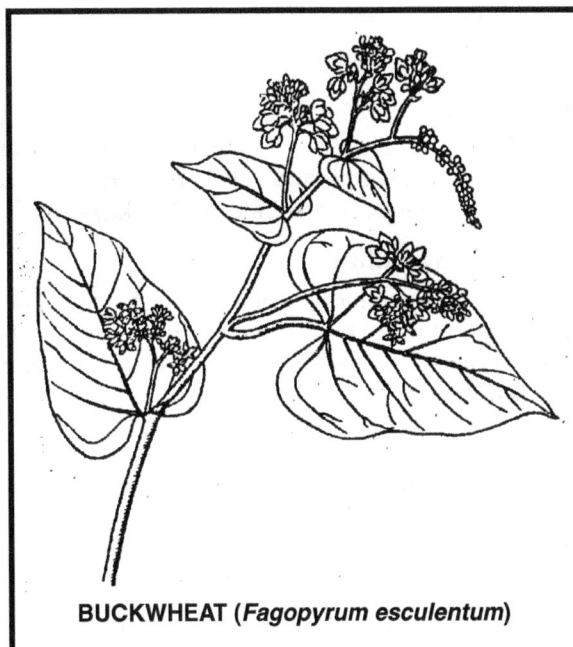

BUCKWHEAT (*Fagopyrum esculentum*)

haemorrhagic conditions which include certain types of purpura, *i.e.*, bleeding from the kidney, hereditary haemorrhagic telangiectasia and haemophilia. It has been used with success as a prophylactic treatment against gangrene due to frost bile. Rutin injection with certain miotics is useful in glaucoma, and in combination with dicummarol in the treatment of retinitis. It also prevents the weakening of capillaries due to drugs such as salicylates, arsenicals, thiocyanates, sulphadiazine and gold salts. The decoction of roots is used in rhematic pains, lung diseases and typhoid; while the juice is useful in urinary disorders. Rutin is valued as a cure for capillary fragility of the heart.

Chemical Constituents

It contains bioflavonoids, especially rutin ($C_{27}H_{30}O_{16}$) is a strong antioxidant, a greenish-yellow, crystalline powder obtained from the leaves (3.5-4.7%) and fruit (1.3%). Also contain fagopyrin, flavanol, fagomine, alanine, rhamnodiastase, enzymes hydrolyse and various flavanone glycosides.

Soil and Climate

The crop can be grown on a variety of soils, but it thrives best on well-drained sandy or loamy soils. It does not grow well on poorly-drained, heavy or acidic soils. A moist, cool climate is ideal for its optimum growth. Normally, 18-30°C temperature seems to be satisfactory.

AGRO-TECHNIQUE

Land Preparation

Before cultivation, the land should be ploughed twice and the field should be cleaned thoroughly of weeds. The land must be thoroughly ploughed 3 to 4 times and brought to a fine tilth by harrowing. If required, small canals may be prepared for drainage. About 25 t/ha of cowdung manure is also added.

Mode of Propagation

It can be propagated by seeds.

Sowing

The seeds are line-sown with a spacing of 20 cm between rows and 10 cm within each row. The most favourable temperature for germination is found to be 35°C. In hills, the crop is shown during May-June. It requires a seed rate of 60-80 kg/ha.

Manures and Fertilizers

The field is moderately manured with well-rooted FYM. An application of 50 kg nitrogen and 40 kg/ha each of phosphorous and potash is found to be adequate for a good crop. Half the quantity of nitrogen along with the full dose of phosphorus and potash are applied as a basal dose. The remaining half-dose of nitrogen is applied 4 to 5 weeks later.

Irrigation

It is generally grown as a rainfed crop with one or two protective Irrigations, if required.

Interculture

The crop is kept weed-free by frequent hoeing and weeding in the early stages of growth.

Pests and Diseases

No major insects affect this crop. A few bacterial and viral diseases are noticed on this crop and are not of a serious nature. However, leaf blight which appears during flowering or even at earlier stages of growth, causes some damage to the crop. Initial symptoms start with minute spots on leaves which quickly develop and coalesce to from bigger patches, which turn into blight and finally result in leaf-fall and the drying up of the plant. The leaf blight disease can be controlled effectively by spraying 0.2% copper oxycholoride, whenever the disease is noticed.

Harvesting and Storage

The best stage for harvesting is 40 to 50 days after sowing and when the plants begin to bloom. The best harvesting time is September-October. Harvesting should, however, be done well before grain-formation, if the crop is intended for rutin production. Delayed harvesting causes a sharp decline in rutin content from 6.0 to 3.8%. The period between blooming and seed-setting being short, close observation is required to enable harvesting the crop at the proper stage. The crop at this state which would have generally attained a height of 60 to 70 cm is cut and harvested above ground-level. As the stems are hollow and the leaves contain 88 to 90% moisture, sun-drying is slow and may also lead to fermentation and enzymatic action leading to a loss of rutin. Dehydration technique by artificial rapid drying of the leaves is the best suited method. A process for rapid dehydration of leaves using an alfalfa drier has been developed for obtaining leaf-meal rich in rutin and stable under ordinary storage conditions for at least a year.

Production and Yield

An average yield of 40-50 quintal fresh herb/ha is considered satisfactory. The dry herb yield is around 5 quintal/ha.

Marketing and Trade

The Current market price of seeds is Rs. 50-60 per Kg.

Economics of Cultivation

The market of Buckwheat and their products are highly volatile. The economics (per ha) worked out here are subject to fluctuations, depending upon time and place.

Expenditure (in Rs.)	—	6,500.00
Returns (in Rs.)	—	1,20,000.00
Net Profit* (in Rs.)	—	**1,13,500.00**

*   Estimation of the net profit is analysed on the basis of grower/collectors price. The grower/collector price is always 25-40% less than wholesale market price due to various stakeholders *viz.* Agent, middle man, Brokers, Commission agents.

CHEBULIC MYROBALAN (*Terminalia chebula* Retz.)

Vernacular Names

San.–Abhaya, Amrita, Hemavati, Haritaki, Jeevanti, Sudha, Vijaya, Harra; *Hin.*–Har, Hararh, Harh, Harra harrar; *Ass.*- Halikha, Hilikha, Silikha; *Beng.*–Haritaki; *Guj.*– Haradi, Hirde, Hardo; *Kan.*–Alalai, Allale, Aralaikai, Arili, Halle, Herrda; *Mal.*–Kadukukka, Pulincakku; *Mar.*–Habra, Harla, Hirada, Hirda, *Ori.*– Harida, Horada, Horitoki, Karedha; *Punj.*– Har, Harar; *Tam.*–Kadukhai, Illagucam, Kadakai, Kolaippakku, *Tel.*–Karakai, Kraka; *Eng.*–Chebulic Myrobalan.

Introduction

Chebulic myrobalans have a great significant in the Indian national economy due to their immense value in different industries *viz.* indigenous medicines, dye and tanning of leather, etc. Its fruit extract is one of the major export items and its demand is increasing by leaps and bounds due to the superior quantity of leather by it and that vegetable tans are less hazardous than the inorganic tanning materials and thus preferred all over the world. It belongs to family Combretaceae.

Botanical Description

A medium sized to large deciduous tree, attaining a height of 15 to 24 m. with usually a cylindrical bole of 4-9 m in length, rounded crown and spreading branches. Bark dark brown, often with shallow longitudinal cracks, exfoliating in irregular woody scales. Leaves sub-opposite or sometimes alternate, ovate or elliptical, glabrous. Flowers with an offensive

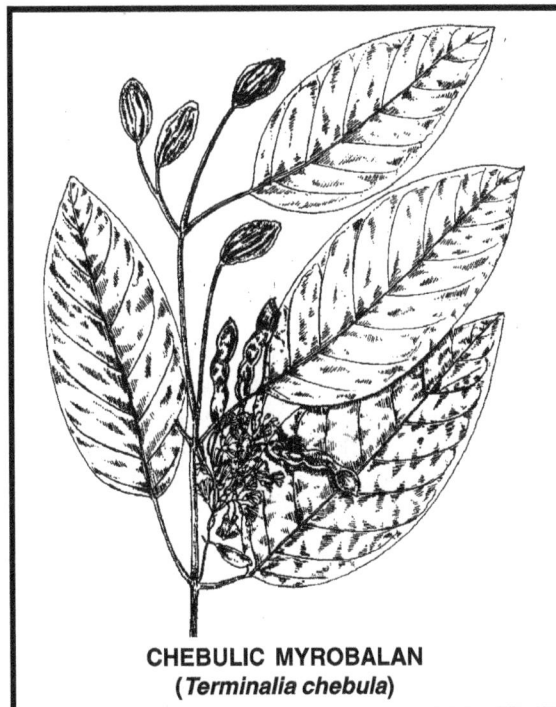

CHEBULIC MYROBALAN
(*Terminalia chebula*)

smell, pale yellowish white in colour, in axillary and terminal blooming in month of April-June. Fruit drupe, ellipsoidal, usually obovoid, or ovoid, yellow to orange brown sometimes tinged with red or black, 5 angled when dry, glabrous, shining, hard when ripe.

Geographical Distribution

It is found throughout the greater part of India including sub-Himalayan tract from the Ravi, Punjab eastwards to West Bengal, Bihar and Assam and south ward to Madhya Pradesh, Maharashtra, Orissa, Andhra Pradesh, Tamil Nadu and Karnataka. It is ascending upto 1500 m in the outer Himalayas and upto 900m on dry slopes in the Western Ghats. It is also found in Uttar Pradesh, Uttaranchal, Punjab, Rajasthan & Himachal Pradesh (Kangra). In Assam, it is found in Goalpara, Garo hills and Kamrup. It also occurs in Sri Lanka and Myanmar.

Medicinal Uses

The fruits provide a rich source of tannin that constitutes one of the most important vegetable taning material. The dried flesh surrounding the seed contains 30-32% tannin. They are also used in medicines as laxative, astringent, stomachic and tonic; applied on chronic ulcers, as gargle in stomatitis, as dentifrice for toothache, bleeding gums; bark diuretic, cardiotonic. Fruits & bark are used in about 20 Ayurvedic preparations. In combination with the fruits of *Terminalia bellirica* and *Emblica officinalis* they form a purgative constituent the important Ayurvedic medicine, triphala, used for several ailments. Fruit extract is largely used for internal treatment of locomotive feed-waters, and as an additive in oil-drilling compositions.

The utilization of extract in various processes, such as in petroleum purification, in cement manufacture, for colouring state-stone; as a flocculent for washing waters in the preparation of coal by the wet method. Myrobalans are also employed in the preparation of ink and in dyeing as a mordant for the basic aniline dyes. Crushed myrobalans from which the stones are removed is a regular article of trade. It is also marketed within the country, or exported, in the form of whole nuts, crushed nuts and solid and spray dried extracts, which are used in tanning.

Chemical Constituents

Myrobalans contain various types of pyrog-hydrolysable tannins. Chebulagic acid ($C_{41}H_{30}O_{27}$) and corilagin, are the major tannin constituents. The usual commercial samples of chebulic myrobalans contain around 30 to 40% tannin but some fruits contain as low as 20-25% and the best samples contain 40-50%. The usual average assumed by tanners is 33%.

Soil and Climate

It is capable of growing on different soils ranging from poor rocky ground to sandy, clayey, deep or shallow loam, lateritic loam, gravelly fertile alluvium of the Indo-gangetic plains and the bhabar and tarai region of the Sub-Himalayan tract. The plant thrives best in areas with an annual rainfall varying between 100-150 cm. It requires maximum shade temperatures are in range 35 - 47.5°C and minimum temperature ranges from 0 to 17.5°C.

AGRO-TECHNIQUE

Land Preparation

The field should be ploughed and harrowed several times, leveled properly and drainage channels should be made. Since yams have a high requirement of organic matter for good tuber formation, a sufficient quantity (20-25 t/ha) of FYM is incorporated at the time of Land Preparation.

Mode of Propagation

It can be propagated by seeds or vegetatively by shoot cutting.

Nursery Raising

For raising seedlings in the nursery. Pre-treatment of seeds is common. The depulped seeds should be treated by fermentation process as discussed above and then sown in the nursery beds. The nursery should be shaded against the sun. In Punjab, the fruit stones after removing the outer pulpy portion are dried and sown either in wooden boxes or in nursery beds that are covered with soil and regularly watered. Ordinary clayey loam or sandy loam will suffice and no manuring is required. The young plants may require watering during the first hot season. Plants suitable for Transplanting are obtained in the second rains. Shelter is desirable in the early stages of growth i.e. in the nursery and also after Transplanting. Optimum spacing in nursery is around 15 cm × 15 cm.

Natural Regeneration

Natural regeneration is found scarce due to poor germinative capacity as also due to perishing of seeds due to insects, rats & squirrels. The tree propagates by natural regeneration in some localities but it is affected adversely to a great extent when rats, squirrels and rodents destroy the seeds. The seeds germinate better if it is covered up with the earth or debris than, if it is lying in the open. Germination takes place in the rainy season.

Vegetative Propagation

Vegetative propagation has been found advantageous over seed propagation as the former technique reduces the juvenile period and subsequently facilitates early maturing. Bud - grafting and cleft grafting would be better option. Grafting of young shoots on seedlings of the same species resulted success in root induction (30%).

Tending

Regular weeding is carried out at least for the first 3 years or until the plants are successfully established. When the crowns of trees begin to touch with each other, the crop should be thinned out to a wider espacement, removing about 50 percent of the trees. In the final crop, each tree should be given about 7.5 to 9.0 m of room in all directions.

Pests and Diseases

Pathgenic fungi recorded on the species are Uredo terminaliae attacking leaves; *Phyllactinia terminaliae* causing powdery mildew and *Cercospora catappae* causing leaf spots.

Animals

Fruits are much eaten by squirrels, rats, porcupines, hares and peacocks. Rodents do a lot of damage in nurseries. Land crabs eat off the cotyledons and cause a lot of damage. Young plants are liable to damage by cattle-browsing.

Insects

A few pests, the larvae of *Traala vishnou* (Castor Hairy Caterpillar) feed on the leaves. *Selepa celtis* also sometimes causes widespread defoliation, the midrib and stouter side veins being left untouched while feeding on large leaves. *Acathopsyche moorie* also causes considerable damage to the leaves. Other minor defoliators are: *Asura dharma, Brassa alopha, Ascotis infixaria, Hyblaea puera, Teleclita strigata* and *Polyprychus trilineatus.* The beetle and larvae of Attagenus alfierii and of A. gloriosae.

Harvesting and Storage

The fruits fall on the ground soon after ripening. The crop yield varies from year to year. The fruits should be collected before maturity otherwise there is apt to be a variation in their tanning strength. The astringent principle is found in the outer pulp of the fruit, the stone-like kernel containing hardly any. The season of collection has a bearing on the tannin content and value for tannin. January is considered to be the best time for the collection of Chebulic myrobalan in many areas, the later collections are slightly inferior, but earlier collections are inferior. The collection, however, starts in December and continues up-to the end of March in India and the assembling markets also start functioning simultaneously. However, there is no conclusive data still to indicate the actual/proper stage of ripeness for the best collection. The fruits are generally collected by shaking the trees and picking up from the ground. The January to March is the best period for fruit collection. Fruits should be collected in the first half of January from the ground as soon as they have fallen. The harvested seeds are dried in thin layers, preferably in shade and graded for marketing.

Production and Yield

No data seem to have been collected regarding the yield of myrobalans as nowhere collection is done in an organized manner and the distributed is scattered and so the yield per hectare could also not be ascertained. The annual yield of fruits ranged from 2 kg to 10 kg/tree/year depending on girth classes from 0.28 m to 1.78 m. The average yield nearly 15-17 tonnes of fruits per year.

Marketing and Trade

India holds the monopoly in export of chebulic myrobalans in the form of whole fruits or in crushed form or as extracts to the world market. UK,USA, Australia, Belgium, Pakistan and Malaya Federation are the main importers of crushed myrobalans whereas Australia, Bangladesh, France, Pakistan, U.K. and USA are import whole myrobalan. Pakistan, New Zealand, Australia and Japan are the chief importers of myrobalan extract. The current market of fresh fruit is Rs. 6-10 per kg while dried and without seeds is Rs. 50-60 per kg.

CINCHONA (*Cinchona officinalis* Linn.)

Vernacular Names

San.–Sinkona, Kunayanah; *Hin.*–Quinine; *Kan.*–Sinkona, Barkino; *Mal.*–Sinkona, Koyina; *Tam.*–Cinkona, Konia; *Tel.*–Cinkona, Jaddapatta; *Eng.*–Quinine bark, Peruvian bark.

Introduction

Cinchona is used commercially for its bark which is the source of quinine and other anti-malarial drugs. Some times the barks of *C. lancifolia, C. ovata, Remijia pedunculata* and *R. purdieana* are often used as substitutes for cinchona bark. The alkaloids are extracted from the powdered bark. Quinine is isolated from the total alkaloids of the bark as quinine sulphate. It belongs to family Rubiaceae.

Botanical Description

An evergreen shrubs or trees 6-9 m in height with brown bark marked with black and whitish lines. Leaves simple, opposite, lanceolate or ovate-lanceolate, shining, petiole reddish, entire; the stipules are interpetiolar and deciduous. The flowers are small, pink or purple coloured in short corymbiform terminal and axillary compound cymes and fragrant. The fruit is a capsule, dehiscing from the base upwards with 40-50 small, flat, winged seeds.

Geographical Distribution

Cinchona is indigenous to South America, and occurs in the Andes, Colombia, Eucador, Costa Rica, Peru and Bolivia. Some species are also cultivated in Java, Sri Lanka, India, Burma, Tanganyika, Bohemia, Panama, Indonesia, Guatemala, Uganda, Philippines, Tanzania, Kenya, Zaire and Costa Rica. In India, it is confined to some parts of West Bengal, Tamil Nadu (the Annamalais, Naduvattam, Shevaroy Hills, Palani and parts of Tirunelveli) and Karnataka (South Kanara and Coorg) in an area of about 6 000 to 8 000 ha.

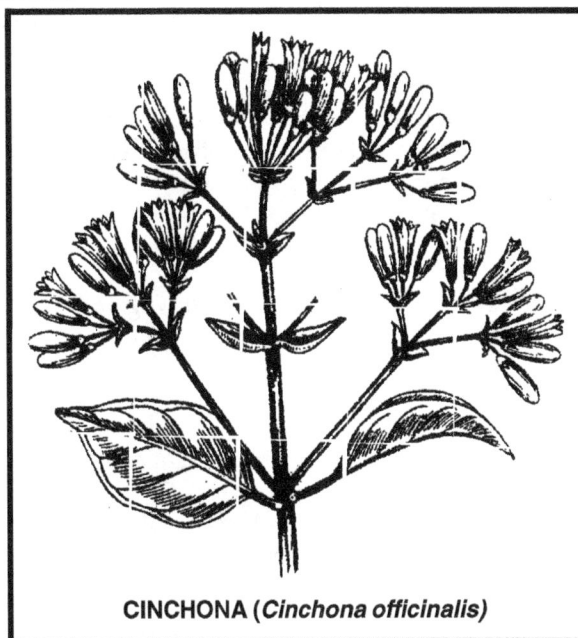

CINCHONA (*Cinchona officinalis*)

Medicinal Uses

Quinine, as a cure for malaria, has been known of for a long time and, even today, expert medical opinion regards quinine as a safe and sure remedy for malaria. Besides, quinine was in use as an anaesthetic as a substitute for coasine. Its anaesthetic action is prolonged. It has been used as a sclerosing agent in the treatment of internal haemorrhoids and varicose veins. Quinine protects the skin against sunburn. It is a bitter tonic, stomachic and appetizer. In small doses, in the form of a solution, it is a mild irritant and a stimulant of the gastric mucosa and other mucous membranes. It is a weak non-medical uses also. Salts of quinine are employed in beverages, as an addition to hair-oils, as a vulcanisation accelerator in the rubber industry, for making polarized lenses and has various other uses in photography and optics.

Chemical Constituents

Its bark contains a potential alkaloid, quinine (1.5 to 2.0%). In addition, more than twenty other alkaloids have been isolated from cinchona, of which chinchonidine, quinidine and cinchonine are the most important. The alkaloids exist chiefly as salts of quinic and cinchotannic acids, and their relative concentrations vary in different species. The leaves contain 1% total alkaloids.

Soil and Climate

Cinchona prefers a light, well drained, virgin forest soil, rich in organic matter with no possibility of subsoil water-logging and with a high moisture holding capacity. It prefers acidic soils (pH 4.6 to 6.5). The calcium requirement of this crop is high. It grows best in tropical climates at altitudes of 1800 m (6000 ft.) with an average temperature of 13.5°C-21°C. It grows well in places where the annual rainfall is a little less than 200 mm. It is susceptible to frost and, hence, is not grown on very high hill ranges.

Improved Varieties

Clone No. 701 is released by the Tamil Nadu Forest Department, Chennai.

AGRO-TECHNIQUE

Land Preparation

For raising cinchona plantations, virgin forest soils are best suited. The forest is cleared and the ground is levelled and dug to a depth of 1.5 ft to improve the soil structure and to remove stones, if any. About a fortnight prior to planting, pits of 60x60x60 cm are dug and filled up with top soil and well decomposed organic matter.

Mode of Propagation

It can be propagated by seeds or vegetatively *viz*. cutting, stooling, layering, grafting & patch-budding.

Nursery Raising

The seeds are generally sown during April in sloping beds, 12x4 ft, and covered with a thatched roof. Cinchona seeds are small and light and lose viability on storage. The nursery area is prepared in such a way that the top layer, up to a depth of 2-3 inches, is composed of a mixture of leaf-mould and sand in equal proportions, and is carefully pressed by hand so that it is uniformly firm all over. Cattle manure is usually not applied to the nursery beds.

While sowing, the fresh seeds are scattered thickly on the surface and covered with a thin layer of fine sand. The beds are then lightly watered using a spray can. The seeds germinate in about 20-40 days after sowing. About 50% germination is noticed, out of which only about 10% of the seedlings will be suitable for Transplanting after eliminating all the weak, lean and lanky seedlings. The seedlings grown in the nursery can be transplanted in baskets or polythene bags when they are about four months old, with two pairs of leaves. The seedlings will be ready for planting in the main field during mid-May of the succeeding year when they are about 14-18 months old and 30-60 cm in height.

Vegetative Propagation

Among the vegetative methods of propagation, patch-budding, soft terminal cuttings and layering have recorded the best results. A high percentage (85%) of success is obtained in patch-budding in the period from March to the middle of June. Patch-budding is usually done in the nurseries or plantations, in situ on plants which are one to two years old. Even the budding on coppices has given a better growth of budded plants. The cuttings are made to root by cincturing and etiolation. During May-June, the shoots are treated, from which cuttings are taken after 50-65 days and planted in the nursery. In layering, the East Malling method with some modifications has given good results, each shoot giving annually about 100-200 shoots in 2 or 3 coppicing.

Transplanting

Planting is done before the onset of heavy rains. The soil should be sufficiently moist at the time of planting. The planting is done in open pits at a spacing of 110 × 110 cm or 150 × 150 cm or dense planting of about 8000 plants per hectare is done. Young cinchona plants need shade which is provided by planting shade plants like *Alnus nepalensis, Erythrina indica, Albizzia stipulata* and *Grivella robusta*, 15-20 ft. apart.

Manures and Fertilizers

Application of a fertilizer mixture containing nitrogen, phosphorus and potash gives marked results. Liming (dolomite or limestone) is done for the soil if the pH is 5 or lower. Nutrients are supplied @ 100 kg nitrogen, 95 kg phosphorus and 125 kg potash/ha in the form of triple superphosphate, muriate of potash and ammonium sulphate. The quinine content in cinchona is known to increase with the age of the trees, under favourable nutritional conditions.

Interculture

Weeds have to be removed at regular intervals, particularly in young plantations.

Pests and Diseases

The grubs of cockchafer beetles (*Holotrichia repetita*, *Rhizotrogus refus*, *Serica nilgiriensis*, *Popilla chlorion*, etc.) cause serious damage to the seedlings in South Indian plantations. The adults of these beetles come to the surface during summer and should be flushed by irrigating the seed-beds with water mixed with crude oil emulsion. The tea-fly, *Helopeltis antonii* and *H. cinchonii*, infested leaves in nurseries and also in the main field. These insects cause leaf-curl by sucking the juice from the tender leaves. If heavy damage is noticed, sodium arsenate mixed @ 28 g with 113 g of molasses in 9 litres of water may be sprayed on the plants for controlling these insects. The leaf bug, *Disphinctus humeralis*, is occasionally found on tender foliage.

Stem blight caused by *Sporotrichium* and *Verticillum* spp., seedling light due to *Phytophthora parasitica*, root-rot due to *Phytophthora cinnamomi* and *Sclerotium rolfsil* and damping-off due to *Pythium vexans* and *Rhizoctonia solani* are some transplant bed-diseaes affecting cinchona seedlings. The spread of these diseases can be checked by scooping out the soil in the diseased patches all around, to a depth of 10 cm and throwing it away. The cavity, so formed, may be filled with sterilized, dry soil. The nursery-beds should also be drenched with mercurial fungicide to check the spread of the diseases. Die-back or pink disease is caused by *Pellicularia salmonicolor*. This organism attacks the tender tips of the stems and branches and gradually spreads. The branches should be pruned to prevent the spread of the disease. Besides, the fungus *Armillaria mellea* causes root rot and *Rosellinia* spp. Cause leaf spot diseases.

Harvesting and Storage

The trees are coppiced when they are 6 to 8 years old. Coppicing involves pruning the trees at a height of 5 cm (2 inches) from the ground-level. The second coppicing is done 8-10 years after the first coppicing, where only about 2 to 3 shoots are left to grow further. The plants are uprooted in the 30th year when their vigour declines. The major harvests are obtained at the time of the first two coppicing and only little yield of bark is obtained from the dead and drying trees and pruning. The first set of yields are obtained in the third year after planting. The bark is separated from the coppices by beating it with a mallet and is then peeled by hand or a knife. The peeled bark should be dried immediately to prevent the loss of alkaloids, preferably in the shade. In rainy weather, drying is done in special sheds or by means of artificial heat. In well-established plantations, drying is done in well-regulated ovens. For this purposes, hot air ovens, regulated by 70°C, are employed. The long, strips of bark are cut into small pieces and fed into the upper end of a long, slightly inclined, rotating, cylindrical oven. The dried product contains 10% moisture. The dried bark is then packed in gunny bags. The dried bark is called Druggists bark (quinine content 1.8-2%) in trade.

Production and Yield

During the first two coppicing, a yield of 40 quintals dry stem bark/ha is obtained. At the final stage of uprooting the tree, the yield of the bark may be about 60 quintals /ha.

Marketing and Trade

The salt of Cinchona alkaloid like quinine sulphate, quinine hydrochloride, quinidine sulphate etc. are important drugs and in demand within the country and aboard. The export demand for these products started in 1964-65 and both the Government (West Bengal and Tamil Nadu) put together would have exported Rs. 15 crores worth of quinine products so far. The export revenue from quinine sulphate is nearly Rs. 34 lakhs in 1996-97. The export during 1978-79 have gone down due to the partial ban imposed by Government of India on export of Quinine products as well as due to competition from synthetics.

It is anticipated that there will be sustained demand for quinine due to resurgence of malaria and for the manufacture of quinidine (by conservation of quinine to quinidine) and for other uses in Soft Drinks, Champagne, etc. the Cinchona Industry should be nurtured as a national welfare industry to combat malaria and other fevers as a valuable foreign exchange earner for the country.

COLEUS (*Coleus forskohlii* (Willd.) Briq)
Syn. *Plectranthus barbatus* Andr.

Vernacular Names

San.-Pashan Bhedi; *Hin.*–Patharchur; *Beng.*–Patharchur; *Guj.*– Garmar; *Kan.*–Mangani beru; *Mar.*–Maimnul, *Panoiva*; *Eng.*–Coleus.

Introduction

Coleus is the one of the most significant potential medicinal crops of the future as its pharmacopeial properties have been discovered only recently. It is belonging to the family Lamiaceae. It has been used as a condiment in India for a long time. The tuberous roots of this plant, resembling a carrot in shape and brown in colour, are the commercial parts. The tuberous roots are identified as a rich drug for glaucoma, congestive, cardiomyopathy, asthma and certain cancers.

COLEUS (*Coleus forskohlii)*

Botanical Description

An aromatic perennial plant upto 0.5 m tall with square stems, branched, the nodes are often hairy. The leaves are usually pubescent, narrowed into petioles. The roots tuberous, fasiculate, conical fusiform, straight organise and strongly aromatic. Flowers very showy bluish to pale lavender colour, arranged in whorls on a long spike like racemes.

Geographical Distribution

The crop has been distributed all over the tropical and sub-tropical regions of India, Pakistan, Sri Lanka, tropical Easy Africa, Brazil. Egypt, Arabia and Ethiopia. In India, it grows on the dry slopes of the Indian plains and shady open hilly slopes of foothills or subtropical Himalayan regions from Kumaon to Nepal, Bihar and the Deccan Plateau of Southern India. It is cultivated in parts of Rajasthan, Maharashtra, Karnataka and Tamil Nadu in an area of about 2500 ha.

Medicinal Uses

In Egypt and Africa, the leaves are used as an expectorant, emmenagogue and diuretic while its foliage is employed in treating intestinal disorders, The tuberous roots are found to be a rich source of forskolin (syn. Coleonol) which is being developed as a drug for hypertension, glaucoma, asthma, congestive heart failures and certain types of cancers. It is also useful in lowers blood pressure, dilates the bronchioles, dilates the blood vessels and as heart tonic.

Chemical Constituents

It contains forskolin, babar tusin, gallic acid, tannic acid etc. It also contain volatile oil and diterpenes.

Soil and Climate

It thrives best in well-drained soils with a pH (5.5-7.0). It does not require very fertile soils. The red, sandy loam soils of Karnataka are ideal for the cultivation of this crop. The climate is humid (83 to 95%) and temperature (10-25°C). The annual rainfall is 100-160 cm.

Improved Varieties

'K-8', Karnataka Clone (0.5% forskolin).

AGRO-TECHNIQUE

Land Preparation

The field is ploughed deep soon after the pre-monsoon showers and brought to a fine tilth. Further, the land is prepared into ridges and furrows at a spacing of 60 cm.

Mode of Propagation

It can be propagated by seeds and shoot cutting.

Nursery Raising

The viability of the seeds being very poor (8-10%), a sufficient quantity of fresh seeds has to be sown in well-prepared nursery-beds (May-June) to obtain good germination. Regular care about watering, weeding and plant protection of the nursery should be taken. In about 15 to 20 days, the germination is completed. When the seedlings are 45 days old and have attained about 8-10 cm height, they are ready for Transplanting.

Vegetative Propagation

Vegetatively, the crop is propagated through terminal cuttings in month of November-December. Normally, 10-12 cms long cuttings, comprising of 3-4 pairs of leaves are sown in already prepared nursery-beds in the month of October-November and regular care about shading and watering is taken. The cuttings establish well in the nurseries and there is no problem in their rooting. After about a month's time, when the cuttings have produced sufficient roots, they are transplanted to the main field.

Transplanting

In most areas the crop is planted June-July at the onset of the Southwest monsoon (in case of seedling developed by seeds) but when the plants are developed by vegetatively, transplanted in month of February-March. In prepared field, seedling are transplanted into ridges and furrows at a spacing of 60 cm and the rooted cuttings or seedlings are planted 20 cm apart within the row.

Manures and Fertilizers

A combination of 35-40 kg nitrogen, 50-60 kg phosphorus and 40-50 kg potash/ha was found to be optimum for obtaining the maximum crop. Half the dose of nitrogen, the whole phosphorus and whole potash may be applied as the basal dose followed by the remaining half nitrogen, 25-30 days after planting as top-dressing.

Irrigation

The first Irrigation is given immediately after transplanting, if there are no rains. During the first two weeks after planting, the crop is irrigated once in three days and, thereafter, weekly Irrigation is enough to obtain good growth and yield.

Interculture

Due to the frequent Irrigations during the initial stages, there is a lot of competition from weeds. In order to obtain economic yields, frequent weeding during the early growth period is desirable.

Pests and Diseases

The leaf-eating caterpillars, mealy bugs and root-knot nematodes are the important pests that attack this crop. These insects can be controlled by spraying the plants and drenching their roots with 0.1% methyl parathion, while nematodes can be controlled by the application of carbofuran granules @ 20 kg/ha. Among the diseases, bacterial wilt is the major one. The spread of wilt can be controlled by spraying and drenching the soil adjoining the affected plants with 0.2% Captan solution immediately after the appearance of the disease and later after a week's interval

Harvesting and Storage

The crop is ready for harvest 5-6 months after planting. Flowers, if any, should be nipped off during the growing period to obtain more biomass of roots. The plants are loosened, uprooted, the tubers separated, cleaned and sun-dried for storage.

Production and Yield

On an average, a yield of 15-20 quintals/ha of dry tubers may be obtained.

Marketing and Trade

The current market price of coleus root is Rs. 60-100 per kg. Sometimes these price may be goes upto Rs. 140 on the basis of quality.

Economics of Cultivation

The market of coleus and their products are highly volatile. The economics (per ha) worked out here are subject to fluctuations, depending upon time and place.

Expenditure (in Rs.)	—	7,500.00
Returns (in Rs.)	—	60,000.00
Net Profit* (in Rs.)	**—**	**42,500.00**

*   Estimation of the net profit is analysed on the basis of grower/collectors price. The grower/collector price is always 25-40% less than wholesale market price due to various stakeholders *viz.* Agent, middle man, Brokers, Commission agents.

COPTIS (*Coptis teeta* Wall.)

Vernacular Names

Hin.–Mamiran; *Guj.*– Haladievachang; Ass.- Mishmi tita; *Punj.*– Mamira; *Sind.*– Mahmira; *Eng.*–Gold thread Mishmi tita; *Chin.*–Hawnglien; *Trade*–Coptis

Introduction

Coptis is an important medicinal plant and is highly valued in Indian and foreign markets. It belongs family Ranunculaceae. The rhizome is yellowish brown. It is odourless and non-astringent. It is bitter in taste and the specific name of the plants Tita - is in fact derived from its bitter taste. Various records mentions its use and occurrence in China.

In the past, there was no regular cultivation of Coptis, though efforts have been made from time to time to grow it. However, over the last few years, with the fast dwindling of natural growing stock, cultivation of it has been taken up by the local people in some area like Anini, Hunli, Disali, Changlagam and Melinja of Arunachal Pradesh. Cultivation of it has also been taken by Forest department of Aruncachal Pradesh at an elevation of 2000m from the MSL in Dibang Valley Forest Division on the Roing-Hunli Road, Coptis nursery was raised in 1976 and plantations were taken up from 1977 and onwards.

Since the Coptis occurs scattered and collection has been going on since time Immemorial one has to spend quite some period in search and collection of Coptis. The collection is generally done in September December after the rains. The roots of Coptis are sometimes adulterated with those of *Picrorhiza kurroa* and *Thalictrum filiolosum* and *Actoea soicata*.

Botanical Description

It is a small evergreen stemless perennial herb, with an obliquely to almost horizontally growing rhizome covered thickly with brownish yellow fibrous roots and, in some cases, with remains of dried up persistent stem-clasping leaf-petioles and the inflorescence. The small compressed stem in a grown up plant is crowned with purp1ishly coloured persistent involucre. The rhizome is crow-quill to pencil or a little over that in thickness. It is yellowish brown externally and yellow and golden yellow internally, exhibiting, when broken, radiating structure. The rhizomes specially the Mature ones, have jointed appearance. Some of them are branched at the crown into two or three heads.

Geographical Distribution

It occurs naturally in the interior, temperate, inountancous regions of Arunachal Pradesh and is cultivated in small pockets of Lohit and Dibarig Valley

COPTIS (*Coptis teeta*)

Districts. Coplis is indigenous to India. It occurs in localised patches scattered over extensive area in Lohit, Dibang Valley, East Siang, West Siang and Upper Subansiri Districts of Arunachal Pradesh at an elevation of 2000 to 3000 m altitude. The Kutki or Katki (*Picrorhiza kurroa*) has been confused with Coptis. Outside Arunachal Pradesh, is reported to be cultivated Oil scale, in certain area of adjoining District of Nagaland.

Medicinal Uses

Coptis is bitter, cooling and very potent bacteriostatic herb. The drug is a febrifuge, and all round tonic and stomachic. It has been used for treatment of various types of fevers, baciliary dysentry, nausea, jaundice, flatulence and viscerel obstruction, thirst, hemorrhages (either from dysentery or from hemoptysis) and conjunctivitis. It is also an effective tonic in debility during convalescence after fever and dysentery. Patients treated with this, after suffering from acute diseases, are reported to have restored their appetite, increased digestive powers and improved in strength. When used as a collyrim it clears the sight and as a snuff, the brain, it is also reported to relive tooth aches. In China, it is said to have been used as an anti-diabetic. In Indochina, it has been used to treat leucorrhea, amenorrhea, aphthae, oral ulcers and conjunctivitis. With fresh ginger, it has been used to care very painful pimples in the mouth. After being pounded it has been applied as a poultice to boils. It is also reported to be an effective drug cure of bacillary dysentery, inflammation of alimentary canal and a diabetes.

Chemical Constituents

The rhizome contains moisture (7.78%), berberine (7.1-8.6%), resin (2.7%), and ash (3.1-3.3%). The other alkaloids are coptisine, plamatine and jatorrhizine.

Soil and Climate

It does not grow in all types of climate and soil. It does well in temperate areas which gets covered with snow during winter. It prefers well drained peaty sandy loam soil with lot of loose well-rotten leaf litter.

AGRO-TECHNIQUE

Land Preparation

The field should be ploughed and harrowed several times, leveled properly and drainage channels should be made. Since Coptis have a high requirement of organic matter for good tuber formation, a sufficient quantity (20-25 t/ha) of FYM is incorporated at the time of Land Preparation.

Mode of Propagation

It can be propagated by seeds.

Nursery Raising

Raised mother beds are prepared with two part of soil rich in leafy mould and one part of sandy soil. The nursery can be raised by two ways. One by collected seed and sowing in the beds and another by planning old plants in the bed which will shed seeds *in situ* and will serve as mother beds. Fruits ripen and seeds are dispersed in May to September. Seed is very small @ 1150 seeds/gm. Seed is broadcasted over the bed which germinates within 30-35 days. Viability of seed is short, initial growth of seedling is slow, 5-5 during first year. Plant is slow growing new leaves and flowers start appearing (flower appear first) from December to February and continue upto March/April depending upon altitude and intensity of snow fall.

Transplanting

The seedlings are pricked out after one year at a spacing of 25x25 cm and will be ready for planting in the field after one year of pricking out when they will attain a height of 8 to 9 cm. Seedlings or wilding are planted out forest area in April-May. Ground is prepared before planting by removing brush wood, soil is worked to make it loose and over head shade is manipulated so that filtered light reach the ground. If the soil is hard then rhizome will not develop well. No spacement trial has been conducted. Spacing varies from 30x30 cm to 60x60 cm. The spacing adopted by forest department is 60x60 cm. close spacing gives higher yield but will require more seedlings.

Manure and Fertilizers

Since soil is rich in humus manuring is not required. Chemical fertilizer no tried but will certainly help development of rhizome.

Irrigation

No watering is required as the area remains moist naturally.

Interculture

Two weedings in a year one during May/June and the other during August/September.

Pest and Disease

No Pests and Diseases have been noticed.

Harvesting and Storage

Collection of rhizomes is done from September onwards before snow fall. This period is suitable for drying rhizomes, is over for natural regeneration and the plant has put on growth of the season. The farmers dug out the plant with rhizome, remove the rhizome keeping half cm rhizome with the old plants and then the plant is planted there itself for seedling purposes for raising new plantations and at the same time new rhizome will also develop from this plant. This is done in individual cultivation. Whereas from natural areas, the leaves are removed and the rhizomes along with portion persistent leaf are washed and dried. With the removal of entire plant, natural areas are shrinking rapidly. The rhizomes are dried in the sun for a week and then sold in the nearest market. Well-dried materials remains usable upto twenty years.

Production and Yield

The local people do not keep any record of yield harvested from their plantation and the plantations raised by Forest Department have not been harvested yet. However rhizomes samples 3, 4 and 6 years old plantations of local people were taken. The 6 to 10 years old plantations yield depending on the planting materials will be around 124 kg/ha and 55 kg/ha rhizomes for 30x30 cm and 45x45 cm placement respectively.

Marketing and Trade

Coptis is in demand and trade from a very long time. The plant continues to be collected and sold till today. There was a very good market till today. It Tezu and Roing the Coptis used to fetch around Rs. 600 to Rs. 750 per kg. While in Kolkata market the price was around Rs. 1200 to Rs. 1500 per kg. But due to ban imposed by the Government of India on its export, the prices fell down to Rs. 200 to Rs. 300 per kg at Roing or Tezu. The materials is reported to have been exported to Japan and South Eastern Countries.

COSTUS (*Costus speciosus* (Koe.) Smith.)

Vernacular Names

San.–Kemuka; *Him.*-Keu, kemuk, kushtha; *Beng.*–Keu; *Kan.*–Chengalvakoshtu; *Tam.*–Kuiravam; *Tel.*–Chengalvakoshtu; *Eng.*–Elegant costus, eylon calumba root, turmeric tree.

Introduction

Diosgenin, widely used as a starting material in the commercial production of steroidal hormones, is chiefly obtained from its rhizomes, belongs to family Zingiberaceae. Diosgenin is further synthesized into sex hormones and steroidal drugs which are widely used for their anti-fertility, anti-anabolic and anabolic, properties in family planning and health programmes all over the world.

Botanical Description

An erect, stout herb attaining 2 m in height. Rhizome stout, creeping horizontally. The leaves are broad, lanceolate, alternate, lower part surrounding the stem, tomentose, dorsally silky, borne spirally on stems. The flowers are white, with a large in-curved lip. These are clustered in terminal globose heads, characterized by large and shiny brown or red bracts. The flowering commences during July and continues till the end of September and the fruits ripen during the middle of November, after which the leaves are shed and the majority of the canes start drying up.

Geographical Distribution

It is indigenous to the Indo-Malayan region. It grows wild in wet places in Assam plains, Tripura, Foot hills of Kangra (H.P) and Garhwal (U.A), Khasi and Jaintia hills and coastal regions of Goa and Kerala. It is encountered in the evergreen forests and bamboo brakes in North Bengal, Arunachal Pradesh, Manipur and Meghalaya.

Medicinal Uses

The rhizome possesses anti-fertility, anti-anabolic and anabolic properties. It is indicated in the treatment of cough, catarrhal fever, dyspepsia, skin disease, snake bite, neuralgia, anemia, rheumatism and inflammation. The juice of the fresh tips of young branches is instilled in the ear of otitis. The plant can be used as raw material for diosgenin extraction.

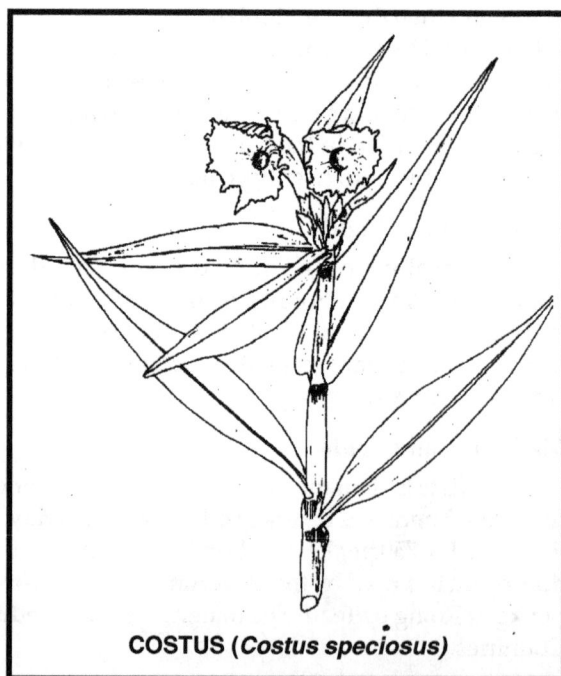

COSTUS (*Costus speciosus*)

Chemical Constituents

Rhizomes contain diosgenin (2.12%) out of total sapogenins (3.86%).

Soil and Climate

The plant can be grown on a variety of soils ranging from coastal alluvium to the heavy, brown, forest type. It grows more luxuriantly on alluvial soils which have a sandy to clay loam texture (pH 5.7-7.5). It can be grown from sea-level to about 1 500 m elevation. It requires subtropical climate with an annual rainfall ranging between 1 000 and 1 500 mm, produce good quality material. Coastal regions and regions with high annual rainfall and humidity throughout the year bear poor quality raw material, in terms of diosgenin content.

Improved Varieties

Two varieties of *C. spacious, var. nepalensis* found only in Nepal and Arunachal Pradesh and *var. argyrophyllus* having a wide distribution in India.

AGRO-TECHNIQUE

Land Preparation

The land is ploughed 2-3 times and the soil is brought to a fine tilth. FYM @ 30 t/ha is applied and mixed well with the soil and furrows are made 50 cm apart.

Mode of Propagation

It can be propagated by seeds or stem and rhizomes cuttings.

Planting

The cuttings of rhizome pieces (nipple-shaped buds) are suitable for propagation. The rhizome pieces are placed at depth of 8-10 cm taking care to place the eye-buds facing upwards, horizontally, in rows 50 cm apart and covered with soil. The weight of the seed piece with two eye buds varies from 50-80 g. The crop irrigated immediately after planting. The best period of planting is from April to May. The active vegetative growth period of the crop is from July to mid-September and the maximum tuberization takes place between September to November. The underground portion remains dormant from December to March. After 70-75 days about 90-95% sprouting should be obtained. About 2.0 -2.4 tonnes of fresh rhizomes are required for planting a hectare of land.

Manures and Fertilizers

It is a rhizomatous crop and, to compensate the biomass production, heavy manuring is required. Normally, for the maximum yield of diosgenin the optimum doses are required 45-50 kg nitrogen, 25-30 kg phosphorus and 25-30 kg potash along with 15 t/ha of FYM. The FYM and a half-dose of phosphorus and potash are applied in 2 split doses at 20 and 60 days from the time of planting, the remaining half-dose of phosphorus and potash is given along with the second dose of nitrogen after the 60 day of planting.

Irrigation

The crop requires a liberal supply of water for its successful growth. The crop planted during April and May requires Irrigation at least two to three times a month till the outbreak of the monsoon. As September to November is the period of maximum tuberization, at least two Irrigations should be given in this period. During the dormancy period (December to March), if winter rains are scanty, there is a need for Irrigation.

Interculture

One weeding during the sprouting period for the crop followed by two more, keeps the crop fairly free of weeds. During the period of active vegetative growth (July to September), most of the weeds are suppressed. Weeding once or twice during the dormant season helps in better sprouting of the crop during the next season.

Pests and Diseases

There are no major pests which affect this crop. However, crop is affected by *Phytophthora solani* and *Pythium spinosum*. In this disease, the infection in the plant starts from the injured portion of the rhizomes and in later stages the rhizomes turn dirty brown in colour, the leaves become yellowish-brown and finally dry off. These pest are controlled by the spray of fungicidal solutions like Benlate, Bavistin or Dithane-Z-78. Sometimes the leaves are also affected by leaf blight diseases, caused by *Curvularia paradisii*. It is very severe from July to September. This pathogen can be effectively controlled by spraying 0.3% Dithane M-45 at fortnightly intervals.

Harvesting and Storage

The crop is harvested 16-20 months after planting. The harvested rhizomes are spread out for a few days and the mud adhering to the rhizomes is removed or, alternatively, the rhizomes are washed under a jet of water from the tube well discharge pipe. The rhizomes have to be chopped and dried before the diosgenin can be exacted. Very finely chopped material dries quickly (2-3 days).

Production and Yield

Nearly, 28-30 t/ha of fresh rhizomes may be obtained in the 1st seasons crop (8-9 months) harvested in the dormant season, but the diesgenin yield will be poor on account of low recovery. The crop should, therefore, be harvested after 17-19 months of planting for a better yield of fresh rhizomes and diosgenin content. From the second season's crop (17-19 months), there is a yield of 50 tonnes of fresh rhizomes from which 75-125 kg of crude diosgenin may be obtained.

COWHAGE (*Mucuna pruriens*) (L.) DC
(Syn. *M. prurita* Hook)

Vernacular Names

San.–Atmagupta, Kapikacchu, *Vanari*; *Hin.*–Gonca, Kaunc, Kivacc; *Kan.*–Nasuganni, Nayisonanguballi; *Mal.*–Naykkorara, Niakkurana, Carivalli; *Beng.*–Alkushi; *Guj.*–Kivanch; *Tam.*–Punaikkali, Punaippidukkan; *Tel.*–Pilliadugu; *Garh.*–Kaiwach; *Eng.*–Cowhage, *Cowitch*.

Introduction

It is an annual, herbaceous climber belonging to the family Leguminoceae. It enjoys an important place among aphrodisiac herbs in India since ancient times. The literal meaning of the word "kapikacchu" is, one who scratches the body like a monkey. The surface of the pods of this climber is covered with bristles, which are allergic to the skin. On touching, it gives intensive itching sensation. The seeds of this plant (black-seeded variety) contain L-DOPA. The L-DOPA is an anti-parkinsonism drug and used in various therapy.

Botanical Description

An annual, half-woody, twining herb with long cylindrical branches. Leaves are alternate, 3-foliate; petioles are 6.3-11.3 cm long; are ovate or rhomboid, membranous and glabrescent above, 15-20 cm long and are on short, thick, hairy stalks with stipulate at their base. Its flowers are 2.5-3.7 cm long, purple coloured and are borne in racemes and blooming in the month of July continue upto September.

The fruit is a pod, 10-12 cm long, covered with brown to grey stinging hairs. The seeds are 5-6 mm in size, small, white, black or mottled and number 4-6 in each pod.

Geographical Distribution

In India, it is found in the foothills of the Himalayas and the plains of West Bengal, Madhya Pradesh, Karnataka, Kerala, Andhra Pradesh, Uttaranchal, Uttar Pradesh, the Andaman and Nicobar Islands and Sri Lanka.

Medicinal Uses

It is seldom used externally. The seeds, roots and the bristles of its pods have great medicinal value. It is the most commonly used ingredient in many tonics for impotency and for enhancing sexual vitality. It also works well as a restorative for conditions of debility and weakness. It is commercial importance for the production of L-dopa, an anti-Parkinsonian and hypertensive drug. The plant is said to have anti-diabetic properties also. The seeds are astringent, nutritive and have shown hypoglycaemic activity in albino rates. They are also used as an aphrodisiac, as a nervine tonic, in scorpion stings, for leucorrhoea, spermatorrhoea and menstrual disorders. The roots are bitter, sweet, thermogenic, emollient, febrifuge,

COWHAGE (*Mucuna pruriens*)

pungent, stimulant, diuretic, purgative and tonic. They are recommended for diseases of the nervous system such as facial paralysis, etc. an ointment made from the root is used for the treatment of elephantiasis. Besides, the roots are prescribed in fever powdered and made into a paste and applied to the body in dropsy a strong infusion mixed with honey is given for cholera also.

The bristles of the pods, filled in gelatin capsules, are used with benefit in intestinal worm infestations, especially in round worm. The seed power, combination with honey, is commonly used as general tonic. In bronchial asthma, the seed powder, honey and *ghee* is an effective combination.

Chemical Constituents

The seeds contain a red viscous oil, and the alkaloids mucunine and mucunadine. The seeds also contain L-DOPA (L-3, 4 digydroxy phenylalanine) which is a non-protein amino acid. The seeds also contain glutathione, mulhingallic acid, a number of alkaloids including nicotine, purienine, prurienidine and five bases designated as P, Q, R, S and Y. The seed-kernel oil contains sitosterol and lecithin.

Soil and Climate

The crop can be grown on a wide range of soils, but well-drained sandy to clayey loam soils are best suited. This plant can withstand drought and can be grown in tropical and subtropical climates. Under favourable conditions, the plant grows vigorously. It is sensitive to frost and should not be grown in frost prone areas.

Intercropping System

Maize or Jawar are grown as intercrop.

AGRO-TECHNIQUE

Land Preparation

The field is ploughed well 2-3 times and brought to a fine tilth. About 10 t/ha of FYM is applied to the soil at the time of the last ploughing. Finally, ridges are made along the contours with a spacing of 60 cm between the ridges.

Mode of Propagation

The plant can be propagated by seeds.

Sowing

The crop can be grown both under rainfed and irrigated conditions. The rainfed crop is sown in June with the onset of monsoon. As an irrigated crop, it can be grown throughout the year. But the best season for sowing the crop is August. The seeds are sown on the inner sides of the ridges (opened at every 60 cm) 60 cm apart, provided that support is given to the plants. If support is not given to individual plants, the seeds can be sown 45 cm apart in 60 cm rows. A basal Irrigation may be given before sowing. The seeds germinate in about 7-10 days after sowing. About 50 kg of seeds are required for sowing one hectare of land. Due to twining nature, artificial supports are provided to the vines to get the best yields.

Manure and Fertilizers

The crop should be supplied with 100 kg N, 80 kg P and 40 kg K/ha. Phosphorus is applied as a basal dose along with FYM. The quantities of N and K are applied in two equal, split doses when the crop is 30 and 60 days old.

Irrigation

For an irrigated crop, before sowing, a basal Irrigation should be given. A second Irrigation should be given after the germination of the seeds. Further, the crop is to be irrigated after each top-dressing of fertilizers. The crop needs about 4-6 Irrigations from November to December at intervals of 20-25 days.

Pests and Diseases

No serious incidence of pests of diseases has been reported in this crop, except towards the end of the season. At that time, occasionally, leaf eating insects and *Cercospora* leaf spot are noticed. The crop can be protected from these insects and diseases by two sprays of Malathion at 2 ml alongwith Dithane M-45 at 1.5 g/l at 30 days of growth and another at 90 days of growth.

Harvesting and Storage

The plants start flowering after about 45 days of growth. Flowering in this crop is continuous and the pods mature periodically. Best time for harvesting the crop is January- February. Hence, the dry pods are harvested 3 to 4 times in the season. The pods are sun-dried after harvest. Due to the sun-drying, the pods will break open. Then, the seeds are separated from the husk by winnowing.

Production and Yield

Under rainfed conditions, if the crop is left on the ground without staking the plants, a seed yield of 15-17 quintals/ha can be expected. Whereas, under the same conditions, if staking is provided to the crop a seed yield of 30-38 quintals/ha can be obtained.

Marketing and Trade

According to an estimation, the demand of cowhage by the herbal industries and ayurvedic drug producers only in Maharashtra is nearly 120 tonnes annually. The annual production of cowhage *vis-a-vis* their percentage consumption by Indian Ayurvedic Pharmacies in Maharashtra only is 5000 mts and 10% respectively (*Source:* ADMA, Mumbai). The annual consumption of Cowhage seeds by CHEMEXCIL (Leading pharmaceutical company), Mumbai, is nearly 200 t in 1999-2000 and received their supply from Gujarat and Maharashtra. The seeds current market price is nearly Rs. 12-15 per kg.

Economics of Cultivation

The market of Cowhage and their products are highly volatile. The economics (per ha) worked out here are subject to fluctuations, depending upon time and place.

Expenditure (in Rs.)	—	4,500.00 (without staking)	6,500.00 (with staking)
Returns (in Rs.)	—	12,000.00 (without staking)	24,000.00 (with staking)
Net Profit* (in Rs.)	—	**7,500.00 (without staking)**	**17,500.00 (with staking)**

*   Estimation of the net profit is analysed on the basis of grower/collectors price. The grower/collector price is always 25-40% less than wholesale market price due to various stakeholders *viz.* Agent, middle man, Brokers, Commission agents.

DANTI (*Baliospermum montanum* (Willd.) Muell.-Arg.
Syn. *B. axillare* Bl.

Vernacular Names

San.–Danti, Nikumbha, Udumbaraparmi; *Hin.*–Dante, Hakum; *Kan.*–Kaduharalu, Naaga dnati; *Mal.*–Nagandanti, Niratimuttu; *Tam.*–Nakatanti, Peyamanakku, Appaiccevakam; *Tel.*–Adavi Amudamu, Kondaamudamu; *Guj.*–Dantimul, Jamalgota; *Mar.*–Danti; *Ori.*–Denti; *Beng.*–Danti, Hakum; *Eng.*–Wild castor, Red physic nut.

Introduction

It is also known as *jangli jamalgota*. It belongs to Euphorbiaceae family.

Botanical Description

An erect, leafy stout, undershrub, upto 2 m tall. Leaves are sinuate toothed, the upper small, lanceolate, penninerved and the lower large. Flowers are greenish, numerous in axillary racemes or contracted panicles (all male or with a few females below). Fruit 3-lobed capsule with ellipsoid, quite smooth mottled seeds and oily endosperm.

Geographical Distribution

It is found throughout India near water streams and riverside in its natural habitat. It occurs abundance in North Bengal, Chhatisgarh, Bihar, Jharkhand, Nagpur, outer Himalayas from Kashmir to Bhutan.

Medicinal Uses

The seeds are purgative and used externally as stimulant, rubificant and in snake bite. Oil from seeds is used in the external application in rheumatism. The roots are acrid, thermogenic, purgative, anti-inflammatory, anodyne, digestive, anthelmintic, diuretic, diaphoretics, rubefacient, febrifuge and tonic. It is also useful in inflammations, constipation, anaemia, leucoderma, jaundice, piles, wounds and enlarged spleen. It is digestive and appetizer and is best for ascites, hernia, oedema and other diseases due to vitiation of *Vata* and *Kapha*. The leaves are good for asthma and bronchitis.

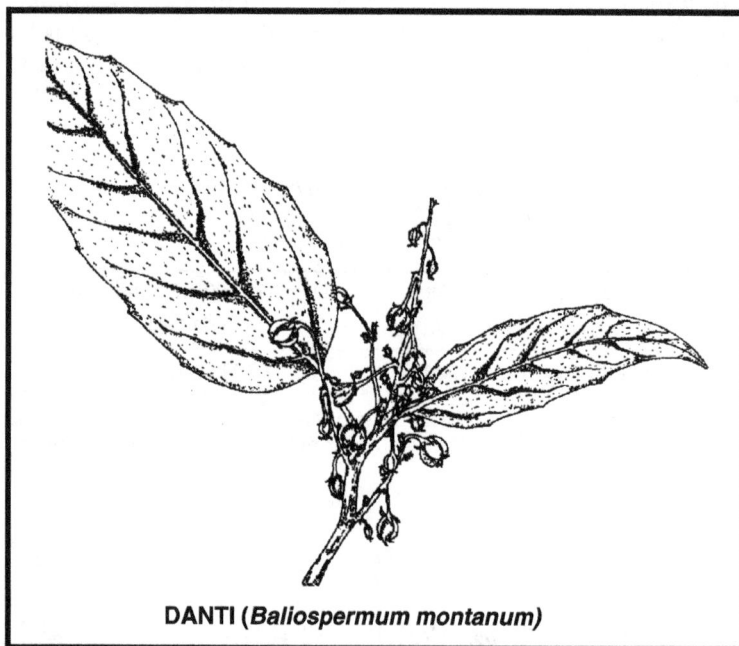

DANTI (*Baliospermum montanum*)

Chemical Constituents

Root contains an alkaloids like mountannin and baliospermin etc. Leaves contain beta-stisterol, B-D-glucoside and hexacosanol.

Soil and Climate

It thrives better in porous and well-drainage soils (pH 5.5-7.0), however also tolerant to alkaline conditions. Sandy loam and black cotton soil are well suited for its successful growth.

Improved Varieties

No varieties.

AGRO-TECHNIQUE

Land Preparation

The field is ploughed 2 to 3 times and brought to a fine tilth. Farm Yard Manure (FYM) @ 25 t/ha is applied and mixed well. Ridges and furrows are opened at 1.5 m apart and planted 1.2 m within the row.

Mode of Propagation

It can be propagated by seeds and terminal cuttings.

Nursery Raising

The seeds are sown in well prepared, raised nursery beds of 1x6 m in June-July. They are then covered with a light soil and leaf-mould mixture and watered to keep the bed moist. About 5-6 kg of seeds will be enough to raise seedlings to cover a one hectare area. About 75% germination is obtained.

Transplanting

The nursery raised seedling are ready for Transplanting after 15-20 days. Seedlings are planted before the onset of monsoon or seeds direct sown in the main field. If the upper portion of the shoot is taken, the inflorescence should be removed and only terminal bud is to be retained for a satisfactory results.

Manures and Fertilizers

The farm yard manure (FYM) @ 25 t/ha is recommended however the recommended fertilizer schedule is not yet been standardized so far.

Irrigation

The crop required once in three days during summer and weekly interval during winter season.

Interculture

As danti requires more water for its growth and development frequent weeding is required initially and after the shrubs have put enough vegetative cover, it suppress weed growth. However, weeding is required after a fourth-nightly interval.

Pests and Diseases

No major diseases and pest are observed under field conditions, expert ants which could be easily controlled by dusting Malathion powder.

Harvesting and Storage

It is ready for harvesting 6-7 months after plantings. The crop is harvesting manually by digging and uprooting the individual plants. The roots are cleaned and drided.

Production and Yield

On an average a yield of 15 tonnes/ha of dried roots can be obtained yielding 60-65% of dry matter under proper management conditions.

Marketing and Trade

An estimated demand of Danti roots by herbal industries and ayurvedic crude drug producers only in Maharashtra is 100 t annually. The supply of Danti roots is required nearly 18-25 tonnes annually but their supply got reduced since 1998. The current market price of dried root is Rs. 6-10 per kg.

Economics of Cultivation

The market of Danti and their products are highly volatile. The economics (per ha) worked out here are subject to fluctuations, depending upon time and place.

Expenditure (in Rs.)	—	7,500.00
Returns (in Rs.)	—	42,000.00
Net Profit* (in Rs.)	**—**	**34,500.00**

*   Estimation of the net profit is analysed on the basis of grower/collectors price. The grower/collector price is always 25-40% less than wholesale market price due to various stakeholders *viz.* Agent, middle man, Brokers, Commission agents.

DATURA (*Datura stramonium* Linn)

Vernacular Names

San.–Dhustura; *Eng.*–Datura, Thorn-apple; *Hin.*–Dhatura; *Beng.*– Dhatura; *Kan.*–Ummatti, Dattura, *Mal.*–Nella ummam, Neela matulam; *Mar.*–Datura; *Tam.*–Umattai, vellum-mattai; Tel. Ummetta, Tellavummetta; *Guj.*–Datura; Garhwali-Dhouturo; *Trade*–Datura.

Introduction

Datura is an important medicinal plant in the local system of medicines as well as under allopathic system. Its importance is due to the presence of alkaloid scopoloamine and hyoscine in higher concentration. It belongs to the family Solanaceae. The alkaloid scopolamineis used as pre-anaesthetic in surgery and child-birth, in ophthalmology and in prevention of motion sickness. It is narcotic, poisonous when used in large doses, and medicinal in small doses.

Botanical Description

It is erect, glabrous or farinose annual, usually 90 cm high. The leaves are pale-green, ovate or triangular-ovate, irregularly toothed. The flowers are large, white or violate, trumpet-shaped; the capsule is erect, ovoid, thickly covered with sharp spines and dehiscing into 4 valves.

Geographical Distribution

It is believed to be a native of South America. It is found growing on the hills throughout India up to an altitude of 2400 m. The plant occurs as weed around habitations in west Himalayas, Nepal, and Nilgiri and Pulney hills in south India. This crop is commercially cultivated in the USA and Europe with a view of obtains a drug of uniform potency. In India, it is cultivated in the hills of Uttar Pradesh, Himachal Pradesh, the Kashmir Valley and some parts of Karnataka.

Medicinal Uses

It has a characteristic disagreeable odour and a bitter, unpleasant taste. Stramonium is a narcotic, an antispasmodic and anodyne and is used chiefly to relieve the spasms of bronchitis or asthma. The leaves may be smoked into cigarettes or smoked in a pipe, with or without tobacco, to relieve asthma. They are also used in the treatment of Parkinson's disease. It relaxes the muscles of the gastro-intestinal, bronchial and urinary tracts, and reduces digestive and mucous secretions. Useful in boils, breast inflammation caused by excessive formation of milk; asthma, salivation, travel sickness (sea or air), diarrhoea, enlarged testicles, insanity and itch. The juice of the fruits is applied to the scalp for curing dandruff and falling hair.

Chemical Constituents

It contains 0.3 to 0.5% of tropane alkaloids, chiefly hyoscyamine (0.3-0.8%) and small quantities of atropine and scopolamine (hyoscine), flavonoids, withanolides, coumarins and tannins.

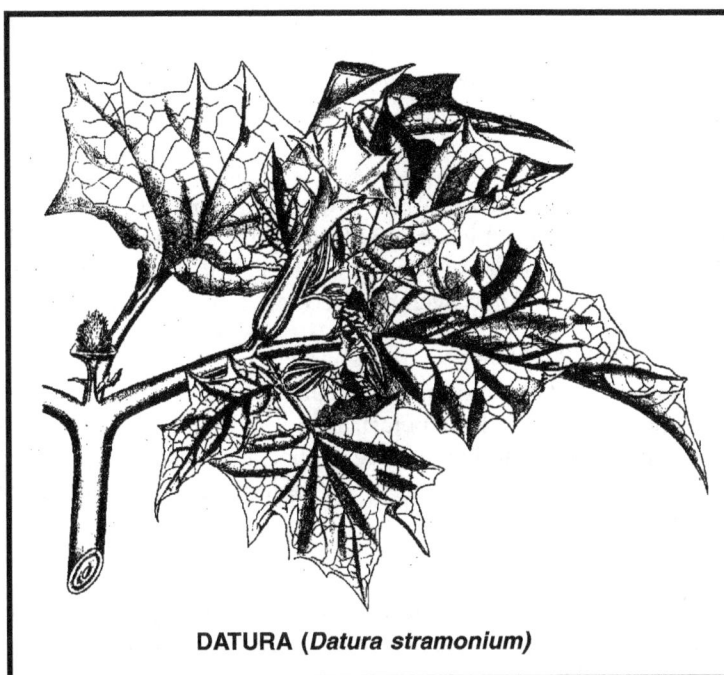

DATURA (*Datura stramonium*)

Soil and Climate

It can be grown on a variety of soils but prefer rich alkaline or neutral, clay, loamy soil rich in organic matter, but water-logged soils do not suit this crop. It can be grown up to an altitude of 2400 m. The plant is sensitive to frost and sheltered situations. Locations with an annual rainfall of 100 cm and a temperature range between 10-15^{0}C in winter and 27-28^{0}C in May-June are ideal for its cultivation.

Improved Varieties

RRL Neel Lohit and RRL Green are developed by Regional Research Laboratory (RRL) Jammu.

AGRO-TECHNIQUE

Land Preparation

The land is ploughed 2-3 times followed by planking. Weeds and stubble are removed and FYM with a basal dose of fertilizers is incorporated into the soil during the Land Preparation.

Mode of Propagation

It can be propagated by seeds.

Direct Sowing

The seeds are sown in rows 45-60 cm apart and covered with soil. The seeds start germinating within a fortnight and in a month's time the germination is complete. About 7-8 kg of seeds are sufficient for sowing one hectare. Weeding and thinning is done when plants are 10-12 cm high, keeping a plant-to-plant distance of 30-45 cm within the rows.

Nursery Raising

The nursery-beds are prepared by mixing well-rooted FYM into the beds. The seeds are sown in March-April and in November in the North Indian plains. The seed germination can be enhanced by soaking the seeds overnight in water and washing them repeatedly 2 to 3 times with fresh water before sowing. The seeds are sown by the broadcast method on the beds and covered with a layer of sand and FYM. The beds are kept moist by spraying water. About 2 kg of seeds are required to raise seedlings for planting one hectare.

Transplanting

The seedlings are transplanted when they are 8-12 cm tall and possess four leaves, at a distance of 30-45 cm in rows which are 45-60 cm apart. A spacing of 75-75 cm is recommended for obtaining higher yield and alkaloid content of seeds, for the crop transplanted to the field in May-June in temperate areas and in November-December in subtropical areas.

Manures and Fertilizers

It was found that the application of 10 tonnes of FYM along with 60-90 kg of nitrogen, 30-40 kg of phosphorus and 40-50 kg of potash are optimum for good growth. A half-dose of nitrogen and the full dose of phosphorus and potash are applied at Transplanting, and the remaining half-dose of nitrogen can be given after two months of Transplanting. The alkaloid content is reported to be affected by micronutrients.

Irrigation

The first Irrigation is arranged immediately after the sowing or Transplanting, if there are no rains. The subsequent Irrigations may be given at intervals of 8-10 days depending upon the weather conditions, till the final harvest.

Interculture

The first weeding and hoeing are done when the plants attain 10-12 cm height, *i.e.,* about one month after Transplanting, and the second hoeing and weeding are done after two months. Dimid (diphenamid) @ 5 kg/ha applied a few days before the crop emergence, provides effective weed control and improves crop growth and yield.

Pests and Diseases

No major pests have been reported on this crop in India. However, thrips act as vectors in transmitting the mosaic virus, which can be controlled by spraying a suitable insecticide like Metasystox (1.5 ml/lit.). In India, the plant is very hardy and is not affected any serious disease. However, some of the important diseases observed elsewhere are:

Wilt

This disease is caused by *Sclerotium rolfsii.* Infested plants show initial symptoms of drooping leaves and finally wilting of the entire plant. The characteristic browning and rotting has been observed in the roots and the collar region of the infested plants. No control measure is suggested for this disease. However, field sanitation and crop rotation are recommended.

Root rot

The causal organism for this disease is *Corticium solani.* Seedlings show symptoms of damping-off and the plants are killed within few days in the nursery. In manure plants, the leaves become yellow and there is drooping of apical parts. The infested portion at the soil surface becomes dirty brown in colour and the tissues get separated from the bark. Young seedlings can be saved by drenching them with a solution of copper fungicide (0.3%) in the nursery.

Viral diseases

Several viral diseases like Distortion mosic, Enation mosaic, Rugose leaf curl, Little leaf and Mosaic have been reported under this group of crops. Some of the control measures recommended to control them are: uprooting the affected plants from the nursery as well as from the main field, the use of seeds from disease-free fields and control of vectors like thrips (mosaic virus) and plant hopper (rugose leaf curl) by spraying any suitable insecticide.

Harvesting and Storage

The crop is ready for harvest 6-7 months after sowing. Leaves, flowering tops and seeds are collected from June to August. While harvesting, the entire plant is cut when the fruits are mature but green, and partially dried in the sun or in the shade. The leaves are stripped and separately dried. The seeds are shaken off from the capsules when the fruits begin to burst.

Production and Yield

The annual availability of the raw material from forest areas may range around 40-60 tonnes only. An output of 12-17 q/ha of dry seeds may be expected. Nearly 1 kg of dry herb yield about 0.4-0.5 g of total alkaloid annually.

Marketing and Trade

In its pure form it is expensive drug and costs about $ 20 per once in USA. An estimated demand of Datura by the herbal industries and crude drug producers in Maharashtra is nearly 3000 t annually. In Kerala, the total requirement of this crude drug (Punchang) was nearly 2.5 Mts in 1995 (*Source*: AFC study on Medicinal plants Farm Project, 1995). The current market price of panchang and seeds of Datura are Rs. 10 and Rs. 100-150 per kg respectively.

Economics of Cultivation

The market of Datura seeds and their products are highly volatile. The economics (per ha) worked out here are subject to fluctuations, depending upon time and place.

Expenditure (in Rs.)	—	6,500.00
Returns (in Rs.)	—	50,000.00
Net Profit* (in Rs.)	**—**	**43,500.00**

* Estimation of the net profit is analysed on the basis of grower/collectors price. The grower/collector price is always 25-40% less than wholesale market price due to various stakeholders *viz*. Agent, middle man, Brokers, Commission agents.

DIGITALIS (*Digitalis purpurea* L.)

Vernacular Names

Eng.–Common Foxglove, Purple Foxglove; *Trade*–Digoxin.

Introduction

Digitalis, commonly known as Foxglove, belongs to the family Scrophulariaceae. It is the most important for their glycoside content, used in life-saving medicines. The leaves of the recommended varieties are a rich source of digoxin, an active cardiac glycoside. The word 'Digitalis' owes its origin to latin world 'digitu' meaning finger and attributed to the plant because corolla of its flowers is finger-like. The purple colour of corolla has earned the species name 'purpurea'.It was recognized as a pharmacopocial drugs in UK in 1650, in France in 1732 and in Germany in 1771. Sometimes few species *viz. Verbascum, Inula, Synophytum* are used as adulterants of digitalis.

Presently, the major portion of the raw material required by leading pharmaceutical industries in India, is being met by imports. The price of digitoxin amounts to crores of rupees. In view of the importance of the crop for its glycoside content, improved cultivation practices have been developed through a series of experiments conducted at Solan and Kodaikanal. The northern states like Jammu and Kashmir, Himachal Pradesh and the hilly and elevated areas in South India have vast potential to produce a sufficient quantity of leaves and seeds to meet the demands of the pharmaceutical industries in the country.

DIGITALIS (*Digitalis purpurea*)

Botanical Description

An erect, branched biennial herb attaining a height of 60-90 cm. Leaves are simple, exstipulate, alternate, opposite or whorled, lance-shaped. Flower are bell-shaped, purple-pink or white arranged in long spikes.

Geographical Distribution

Digitalis is a native of Western Europe, the British Islands and the erstwhile USSR countries. The crop is being cultivated in France, Germany, UK, Hungary, USA and other Asian countries over centuries for medicinal purposes. It has been introduced as an ornamental plant throughout North America, Canada, Mexico, Central America and Asia. In India, it is cultivated in Kashmir at Tangmarg and Kishtwar, Darjeeling district (W.B), the Nilgiris (T.N), Solon (H.P) and Kodaikanal (Tamil Nadu).

Chemical Constituents

Leaves contain 0.2-0.45% of a mixture of cardioactive glycosides. The principal glycoside is digitoxin belonging to the lanatoside-A series. Other important glycosides are digitoxigenin, glucodigitoxigenin-bis-digitotoxoside, glcogitaloxigein-bisdigitoxoside, glucovatoromonoside, glucogitoroside, glucolanadoxin, varadoxin, stropeside and anthraquinone.

Medicinal Uses

The glycosides of Digitalis act mainly on the cardiovascular system, increasing the excitability of cardiac muscles thus stimulating the vagus nerve which leads to more forceful concentration of the heart without increasing the rate in congestive heart failure, it can increase cardiac output and thus relieve venous congestion, without increasing oxygen consumption. In the case of burns; it is more selective in preserving the cells severely injured by heat.

Digitalis has a cumulative effect; the margin between therapeutic and toxic dosage is small. For urgent digitalization, oral administration of prepared (powdered) digitalis may be given. Digoxin is also used in animals as a myocardial stimulant and in treatment of atrial fibrillation.

Soil and Climate

It requires a well-drained, sandy loam soil, rich in organic matter. Clay and sandy soils are also harmful, While acidic soil are essential for proper growth. The crop prefers a warm climate. It performs best in a temperature ranging between 20^0C and 30^0C. For the luxuriant growth of the crop, well-distributed rainfall and good sunshine during winter are ideal. It is found growing successfully up to an altitude of 200 m above mean sea-level (MSL).

Improved Varieties

Various strain E.C. 115996, D-76, D-21, DYF and DPF.

AGRO-TECHNIQUE

Land Preparation

As the crop is a biennial, the field preparation should be thorough. Plough the land 4 to 5 times using a disk and harrow, remove the weeds and bring the soil to a fine tilth. Incorporate the FYM during the final ploughing for thorough mixing with the soil. Divide the field into plots of convenient size (2 to 3 sq m) for distributing the water evenly.

Mode of Propagation

It can be propagated by seeds.

Direct Sowing

The field, thus prepared, can be used for direct seeding by line-sowing at a depth of not more than 2 cm to avoid failure of seed germination. About 1500 seeds weigh one gram. A spacing of 45 cm between rows and 30 cm between plants in a row are maintained in both methods of planting. About 5-8 kg of seed is required for directly seeding one hectare.

The time of planting is very critical for this crop. The best planting period is mid-April in cooler regions like Solan (Himachal Pradesh) and the second fortnight of July and Kodaikanal (Tamil Nadu). Planting in August in the other hilly regions with a higher elevation in South India, registered a higher yield of leaves and glycoside content.

Nursery Raising

Fresh seeds should be used for sowing, since the seeds lose viability after two years of harvesting. For raising nursery, nearly 3 m × 1 m sized beds enriched by organic matter are prepared. The seeds are sown in lines 4-6 cm apart (April-May); the surface of the bed is covered with a 1 cm layer of sand. The beds are irrigated lightly every evening with rose-fitted cans. The seeds germinate in 15-20 days.

Transplanting

The seedlings are Transplanting about 35 to 45 days after sowing, when they have attained a height of 8-10 cm. The field is kept ready by preparing ridges and furrows at a 45-60 cm distance. When the seedlings attain a height of 8-10 cm at a spacing of 30 cm apart, in rows.

Manures and Fertilizers

FYM is applied at the time of Land Preparation while fertilizers are applied in the furrows and incorporated in the soil thoroughly immediately after Transplanting. The application of 40-60 tonnes of well-decomposed FYM along with 100-150 kg nitrogen, 50 kg phosphorus and 25-30 kg potash is the optimum. One-fourth of the nitrogen, supplemented in the form of calcium ammonium nitrate, along with a full dose of P and K should be applied at the time of planting. The remaining nitrogen is applied in 3 split doses, *i.e.*, first top-dressing at 60 days after Transplanting, the second and third dose is applied at 1.5 to 2 month intervals thereafter. The application of N, P and K is repeated following the above schedule during the second year also. The glycoside content has been found to increase by application of trace elements like molybdenum and magnesium.

Irrigation

During summer (April, June and July), the crop is watered at an interval of 7 to 10 days, then 2 to 3 Irrigations will be sufficient each month until the dormant period which sets in by November. The first harvest is done just before that. Then one or two Irrigations a month will be required till the new leaves are formed. Weekly Irrigations are recommended till the leaves mature. However, during the monsoon, watering can be suspended as there will be sufficient precipitation to take care of the crop.

Interculture

Weeds, by competing for light, moisture and nutrients and also acting as host for many diseases and pests cause a drastic reduction in leaf yield. Hence, the crop must be kept completely free of weeds by either chemical or manual weeding during the rainy season, hand weeding is preferred. During summer as well as after each harvest hoeing should be attended to. Atleast two hoeings before the first harvest and another afterwards are important.

Pests and Diseases

The occurrence of insect pests is not serious. Leaf blight caused by *Alternaria spp.* is a common disease of foxglove. Brown spots appear on the leaves, which enlarge, covering the entire leaf surface and killing the foliage in a few days. Spraying any copper fungicide at the rate of 0.1 to 0.2% effectively controls this disease.

Occasionally, there is an occurrence of curly top, a mosaic and a broad-bean wilt virus. In addition to the chemical control, destroy the alternate hosts and control the vector in the vicinity. The affected digitalis plants should be pulled out and destroyed.

Harvesting and Storage

The glycoside content in the leaves is highest around mid-day and the lowest at mid-night, since the glucoside quickly disintegrates in the dark. The leaves are well developed when they attain 8-10 cm length and are ready for the first harvest about 5 months after planting. The subsequent harvests will be at a 1.5 to 2 month's interval then, afterwards, during the first year of the crop. The plants undergo dormancy after the last harvest. In the second year, the crop can give two harvests, besides seed production. The first harvest of leaves in the second year will be in March-April, approximately about 3 to 4 months after the first harvest of the previous year. The final harvest will be about 2 months later. Although the crop is biennial and the leaves are potential glucoside yielder during both the years, it is suggested that the crop is grown as an annual for leaf production, unless meant for seed production in which case it can be grown as a biennial.

In the leaves dried under normal conditions, increase in lanatosides was reported in first 24 h. Drying in shades or artificial drying at 30-40°C proves to be most appropriate treatments. Drying at

110°C give the highest value of lanatoside C. Drying in lowers the quantity of active constituents. It is recommended that large scale drying of digitalis leaves in an air-drier. The maximum contents of lanatoside C in the leaves was obtained while drying at 40°C.

Production and Yield

It gives a higher foliage yield of 15 t/ha on a fresh-weight basis and about 3 t/ha on a dry-weight basis. Generally, it also gives glycoside content of 0.6 -1.2%.

Marketing and Trade

Exact figures on the production of Digitilis are not available. The volume of world trade in digitalis is estimated at some 1,000 tonnes per annum, with the digitalis being cultivated in the country of extraction as well as being imported in large quantities from Eastern Europe. It is interesting to note that most of the digitalis glycosides originate from European countries, mainly the federal Republic of Germany and Switzerland, which export throughout the world. According to trade sources, an average of 22.2 million prescriptions used Digitilis for heart failure per year with digoxin being used in 20.7 million prescriptions in 1979 in United States. During 1978-79, 33 kg of digoxin worth Rs. 1,28,500 was imported into India from U.K. However, digoxin worth Rs. 4,000 only was imported from France during 1979-80.

Prospects for the cultivation and trade of digitalis in developing country is limited as sufficient supplies are available. However, it can be grown in developing countries for internal demand as it is an essential drug in the treatment of heart failure. No real competitor to natural Digitalis glycosides is yet available in world market. A variety of plant constituents and semi-synthetic cardic glycosides and aglycones have been reported. The current market price of digoxin in International market is $ 4.0- 4.5 per gram. The current market price of dried leaves is Rs. 12-18 per kg.

Economics of Cultivation

The market of Digitalis and their products are highly volatile. The economics (per ha) worked out here are subject to fluctuations, depending upon time and place.

Expenditure (in Rs.)	—	5,500.00
Returns (in Rs.)	—	30,000.00
Net Profit* (in Rs.)	—	**24,500.00**

*   Estimation of the net profit is analysed on the basis of grower/collectors price. The grower/collector price is always 25-40% less than wholesale market price due to various stakeholders *viz.* Agent, middle man, Brokers, Commission agents.

DIOSCOREA (*Dioscorea deltoidea* Wall.)

Vernacular Names

San.–Varahi, *Kash.*–Kins, kildri; *Punj.*-Kniss, kirta.; *Him.*–Shingly-mingly; *Garh.*–Kiris.

Introduction

Steroidal drugs, alone constitute about 6% of the total production of pharmaceuticals. They are one of the costliest and most important medicines used throughout the world. Because of their wide use as anti-arthritic, anti-inflammatory, anabolic and antiferlity agents, they have gained tremendous importance in the health and family planning programmes of the developing countries. Diosgenin, a steroidal sapogenin obtained from the rhizomes of Dioscorea (Family: Diascoreaceae), is the major base chemical for several steroid hormones including sex hormones, cortisones, other corticosteroids and the active ingredient in the oral contraceptive pill. It is estimated that the world production of diosgenin is 1000 tonnes and of other precursors is 1200 tonnes. Mexico is the largest producer of diosgenin, producing about 750 tonnes annually. The annual diosgenin requirement of India in 1983-84 was estimated to be around 242 tonnes against the 25-30 tonnes produced indigenously.

Botanical Description

An extensive climber with stems twining to the left. The leaves are stalked, cordate, acute, often triangular. Fruit-triangular, trigonous membranous. Rhizomes are horizontal, borne close or often deep in the soil, grayish brown in colour, with rigid scattered roots on them.

Geographical Distribution

Occur throughout the world, mostly in tropical and subtropical regions and, to a limited extent, in temperate regions also. It is an indigenous species found growing wild in the North-western Himalayas. It is obtained from the forest of J.K., H.P. and Uttaranchal is approximately 15-20 tonnes annually.

DIOSCOREA (*Dioscorea deltoidea*)

Medicinal Uses

The corticosteroids are the most important group of steroidal drugs synthesized from the diosgenin. Adrenal cortex of the man secretes two types of hormones (adrenocortecoads) *viz.* the glucocorticoids and the mineralcorticods. The first group consists of hydrocortisone (cortisol) and corticosterone, which regulate carbohydrate and protein metabolism. The second group consists of aldosterone, which controls balance of potassium, sodium and water in the human body. The glucocorticoids in the form of cortisone and hydrocortisone, are used orally, for the treatment of rheumatoid arthritis, rheumatic fever, other collagen diseases, ulcerative colitis, certain cases of asthma and a number of allergic diseases affecting skin, eyes and the ear. These are also used for the treatment of gout and a variety of inflammation of skin, eyes and the ear, as replacement in addison's diseases.

Both male and female sex hormones are also synthesized from diosgenin, the main male hormone (androgen), which is produced from diosgenin, is testosterone. Testosterone are used to correct male hormone deficiency, specially in certain cases of importency, testicular inefficiency, pituitary, dwarfism, eunuchism, hypogonadism, benign prostrate hypertrophy and cases of undescended testes, mainly as a replacement therapy. Testosterone in combination with estrogen, is also used for the treatment of brest cancer, menopausal symptoms, various forms of uterine breeding and hypobleeding in women. Androgens also have anabolic action associated with nitrogen retention and protein deposition. Estrogen therapy is used to suppress symptoms of natural or surgical menopause, local atrophic changes in the adult vagina and vulva, resulting from estrogen deficiency, gonorrheal vaginitis in young children and suppression of lactation in painful mammary glands. It is also used to control menstrual disorders and post-menopause (vaginitis). In males, it is used for treatment of cancer of the prostate gland and abnormal sexual urge in certain individuals.

Chemical Constituents

The rhizome contains Diosgenin, a steroidal sapogenin is the major base chemical for several steroid hormones including sex hormones, cortisones. The other important sapogenins found are yamogein, botogenin and kryptogenin. Minor sapogenins like pannogenin and tigogenin are also found in certain cases.

Soil and Climate

It can be grown in light sandy loam soils to heavy clay soils with rich in organic matter. It is a temperate crop. The plant tolerate a wide range of pH but too acidic and highly alkaline soil should be avoided.

AGRO-TECHNIQUE

Land Preparation

The field should be ploughed and harrowed several times, leveled properly and drainage channels should be made. Since yams have a high requirement of organic matter for good tuber formation, a sufficient quantity (20-25 t/ha) of FYM is incorporated at the time of Land Preparation.

Mode of Propagation

It can be propagated either by seeds, rhizome pieces or stem-cuttings.

Nursery Raising

The seed has a wide membranous wing that can be removed without affecting germination. Seed progeny is variable and takes a longer time to start yielding tubers, compared to plants raised from tubers. The seeds can be sown either in raised beds in the shade (with a mixture of loamy soil and FYM) or in polythene bags (filled with sand, soil and FYM @ 1:1:1). The planting depth should not be more than 1.25 cm and frequent watering of the beds is essential. The seeds germinate within 21 days. Ideal time of planting in North India is May to August. In North India these can be planted at any time except during winters when the temperature is low for germination.

Transplanting

The seedling are ready Transplanting in 3-4 months. The seedlings should be supported immediately. It gives higher yields at spacing of 60x30 cm. The best season for Transplanting the seedlings to the field is just before the start of rains, *i.e.*, in June in South India, but in North India, this can be done at any time except during the winter. The vines need support for their optimum growth, as this exposes the maximum number of leaves to the sunlight.

Propagation from Tuber

In India, commercial plantations are raised from tuber-cuttings. The choice of propagating material will depend on the cost of planting and the prevailing climatic conditions of the region. This crop grows best from tuber pieces. Tubers or rhizomes are divided into approximately 50-60 g pieces for planting. The growth of plants is slow and the yield lower if smaller pieces are sued for planting. Crowns produce new shoots within 30 days of planting, while the others take nearly 100 days to sprout. The tuber pieces can be planted wither in February-March of June-July. In Karnataka, February-March planting is better. The new sprouts will grow vigorously during the rainy season which commences from June.

In order to avoid the rotting of tubers (before sprouting), only healthy tubers should be selected. The healthy tubers must then be dipped in benlate fungicide @ 0.3% for 5 minutes followed by dusting the cut ends with 0.3% benlate powder before planting or storage.

Propagation by Stem-cuttings

It can be propagated by stem cuttings with 80% success. The vines should be raised from 20-100 g tuber pieces in the green house. One or two month-old vines, are taken and cut into single node cuttings, each with one leaf. They are planted in sand-beds keeping the leaf blade above the sand. Before planting, the cuttings should be treated with 1 ppm 2, 4-D and 0.1% benlate for 4 hours. The beds should be watered regularly. After rooting, the cuttings are transplanted to polythene bags and produce about ten leaves in a period of two months.

In vitro Propagation

In Dioscorea, single nodes and apices have been successfully used as explants for propagation. Shoot apices explants are reported to produce full-fledged plants after 99-100 days. In vitro-raised single node cuttings were ready for planting in the field after 50-60 days.

Manure and Fertilizers

In Dioscorea, for a 4-year-old crop, the recommended fertilizers under Kashmir conditions are 30-40 kg nitrogen, 65-70 kg phosphorus and 60-70 kg potash in split doses at an interval of one month from the time of planting in the first year and the same quantity in the second year. In the third year, only nitrogen at 30 kg/ha has to be applied in two split doses. For increasing the tuber yield and diosgenin content, the application of sulphur, Calcium and Magnesium has also been recommended.

Irrigation

The crop needs Irrigation frequently during summer months. An interval of 4 to 5 days in summer and 7 to 10 days in winter is desirable for the proper growth and development of this crop and for economic yields.

Interculture

In the initial stages, it require more frequent hand-weeding than in the later stages. Chemical herbicides have also been tried, but are not recommended commercially. The practice not only minimizes the growth of weeds but also give additional returns. Apart from this, intercropping also helps in reducing the requirement of nitrogen as the crops recommended above are leguminous.

Intercropping

Cowpea, cluster beans and French beans.

Pest and Diseases

No major disease has been reported to affect this crop. The only disease which affects the crop is the rotting of crop tuber pieces during storage in sand-beds. This can be controlled by treating the tubers with 0.3% benlate solution. Besides, Cercospora, Remularia and late blight caused by *Phytophthora infestans* are reported from Tamil Nadu and can be controlled by Dithene Z-78 (0.3%) or any other copper fungicide. Collar rot of seedlings, which is a nursery disease, can be effectively controlled by spraying Brassicol (0.5%).

The two important pests affecting this crop are aphids and red spider mites which can easily be controlled by spraying 0.5-1.0% of Kelthane (1 ml/l) or Rogor (2 ml/l). Cut-worms are also reported to damage this crop.

Harvesting and Storage

The crop should be harvested only after three years to get the optimum yield from the crop with the maximum diosgenin content. The tubers are harvested when the plants are in a dormant condition to obtain the maximum yield of diosgenin. Generally, the tubers are harvested during November-March. Harvesting can be done by manual labour with pickaxes.

Production and Yield

The diosgenin content of the tubers tends to increase (within certain limits) with age. An average yield of 15 to 20 t/ha of fresh tubers can be obtained during the first year, and up to 40 to 50 t/ha during the second year. The diosgenin content of the tubers tends to increase, on an average, from 2.5-3% in the first year to 3.0-3.5% in the second year.

Marketing and Trade

The total turn over of bulk steroids in the world is estimated to be about 500 million US dollars and estimated world usage to be somewhere between 550-600 tonnes of diosgenin. Mexico, Guatemala, Costa Rico, Indian and China are the major diosgenin producing countries. The Dioscorea tuber today finds a national and international market. A number of drug companies, using diosgenin as raw material, are interested in buying this crop grown on a contract basis.

Economics of Cultivation

The market of Dioscorea and their products are highly volatile. The economics (per ha) worked out here are subject to fluctuations, depending upon time and place. On the basis of the prevailing market conditions with dry tubers, a net profit of about Rs. 40000/ha in a one-year-old crop and Rs. 60000/ha in a two-year-old crop can be obtained.

DUBOISIA (*Duboisia myoporoides* R . Br.)

Vernacular Names

Eng.–Australian mgmeo.

Introduction

Duboisia, belonging to the family Solanaceae, is a subtropical shrub endemic to the Australian region. It is commonly called the 'corkwood tree'. The leaves are rich in hyoscine and hyoscyamine.

The alkaloids are used extensively in medicines throughout the world because of their mydriatic, antichloinergic and antipasmodic action. Hyoscyamine and atropine are used in respiratory diseases like asthma, intestinal disorders, colic pain and peptic ulcers. Hyoscine is used for the treatment of motion sickness and asthma. The genus name 'Duboisia' in honour of the French botanist 'Dubois'. Presently, the commercial cultivation of Duboisia is done only in Australia. Mainly in Queenland area from where it is exported to European countries for further processing.

First attempt of its Introduction in India was made by the Forest Research Institute (FRI), Dehradun (UA). But, commercial cultivation has not yet been established. CIMAP, Lucknow is trying to establish it as a commercial crop in Northern Plains and Karnataka.

Botanical Description

A shrub or small tree, occasionally attaining a height of 40 ft. with numerous suckers. The bark is grey, corky, fissured. The leaves are alternate, petiolate, ovate-oblong to broad lanceolate, blunt-tipped and tapering at the base. The inflorescence axillary with terminal cymes. The flowers are small, bell-shaped, white or pale-lavender or lavender-streaked, pedicillate and actinomorphic.

Geographical Distribution

It grows wild on the borders and open woodlands along the east coast of Queensland and Northern and central new South Wales. It is indigenous also to New Caledonia. It is commercially cultivated in Australia, mainly in the Queensland area, from where it is exported to European countries for further processing. It is also grown in Japan and India. Duboisia was first introduced in India at the Forest Research Institute, Dehra Dun and is being grown as a commercial crop in Karnataka, Andhra Pradesh, Madhya Pradesh and Maharastra.

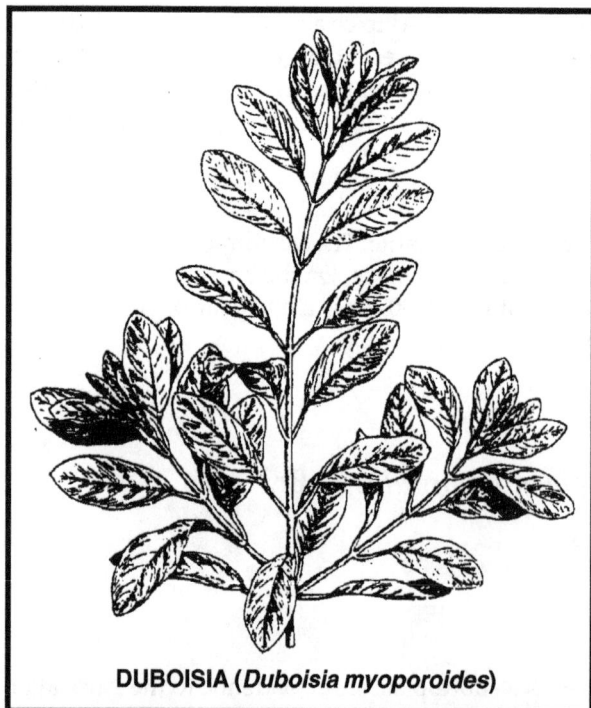

DUBOISIA (*Duboisia myoporoides*)

Medicinal Uses

It is never used as such in medicine, but these are the main raw materials for the production of tropane alkaloids in the world today. Their alkaloid (hyoscyamine, atropine and hyoscine) are used in modern medicine mainly because of their anti-cholinergic, anti-spasmodic and mydriatic activities. It is also used as a pre-medication before anaesthesia and employed in the treatment of sympatholytic paralysis, Parkinson's disease and as antidote in certain poisons like opium alkaloids and muscarine. It is also used in treatment of duodenal and gastric ulcers in combination with other drugs. Because of anti-spasmodic action, it is used in treatment of asthma, whooping cough, colic and urinary bladder pain. In ophthalmology, it is applied as mydriatic to dilate pupil of the eye. As local analgesic, it is used in ointments and eye drops, etc.

Chemical Constituents

The leaves contain a complex of tropane alkaloids including hyoscine (scopolamine), hyoscyamine, non-hyoscyamine, valeroidine, tigloidine, poroidine, isoporoidine, isopelletierine and anabasine. The constitutens vary, not only with the stage of growth, but also with the geographical location. The total alkaloid content of the leaves has been found to vary from 2.75 to 4.90%, of which hyoscyamine were 56.7%, hyoscine 23.3% and other alkaloids 20.0%. On an average, the plant gives a crop containing 3% of total alkaloids.

Soil and Climate

It can be grown on a wide range of soils, but it grows best in well drained, medium to light loamy soils, with the neutral pH. It is a subtropical plant. Hence, it requires a moderate climate, where the winter and summer temperatures are not severe. Areas with an average annual rainfall of 100 cm, well distributed throughout the year, are ideal. Heavy rainfall is harmful to the crop.

Improved Varieties

There are no varieties available in this crop. However, a hybrid of *D. myoporoides* has been developed. Higher alkaloid yields are observed in crosses, where the female parent is *D. myoporoides*.

AGRO-TECHNIQUE

Land Preparation

Land is ploughed to a depth of 20-25 cm about 2-3 months before planting. The soil should be worked to a good tilth to eradicate weeds.

Mode of Propagation

It can be propagated by seeds and vegetatively *viz.* root-cuttings, terminal shoot cuttings.

Nursery Raising

The seeds are usually sown from February to April. The seeds are washed to remove all the pulp and dried in the sun before being used for propagation. They are sown in sterilized soil in nursery beds under 60% shade, or seedlings can also be raised in polythene bags containing a mixture of fine sand and FYM in equal proportions, @ 2 seeds/bag. Pre-treatment of seeds with gibberellic acid reportedly gives better germination. The seeds are soaked in 250 ppm gibberellic acid at about 40°C for about 24 hours. They are then thoroughly washed in fresh water to remove the chemical and dried in the sun. Treated seeds are used for sowing after about six weeks. The germination starts from about the 14th day and complete within 40-60 days. Therefore, it is very essential to keep watering the nursery to ensure good germination.

Transplanting

The seedlings are ready for Transplanting when they attain a height of about 25 cm and are 3 months old. After the land is prepared, pits of 30x30x30 cm are made at 2mx2m spacing. The healthy seedlings are planted in each pit, taking care to see that the seedlings are planted at the same depth as in the nursery bed or polythene bag, to avoid crown rot. Planting is usually done in the dry period. The field must be irrigated immediately after Transplanting.

Manures and Fertilizers

At the time of Land Preparation, FYM @ 20t/ha may be incorporated into the soil. Prior to Transplanting, 20-25 kg nitrogen, 30-40 kg phosphorus and 45-50 kg potash/ha should be mixed well with the soil. Subsequently nitrogen @ 20 kg/ha is given to the crop at bimonthly intervals. Manure and fertilizers are supplied to the crop in the same dosage as above during the second and subsequent years of growth.

Irrigation

Plants should be watered frequently in the initial stages. The frequency is gradually reduced and when the crop is well established, Irrigation once in about 15 days is sufficient during the dry period.

Interculture

Clean cultivation is a prerequisite; weeds are usually controlled mechanically. Grass-cover crops are sown after the leaf harvest.

Pests and Diseases

The major pests are the flea beetle, leaf-eating ladybird beetle, brown olive scale, trunk borer and green caterpillar. The crop can be sprayed with 0.2% Malathion whenever the insects are observed. No serious diseases have been reported in the crop. However, a viral infection causing leaf yellowing has been noticed in the plains of North India.

Harvesting and Storage

The plants grow very fast and will be ready for harvested after 6 months of Transplanting. The main as well as the side branches are cut at a height of about 1 m, retaining about 20% of the leaves on the plant to encourage profuse branching. Two harvests are possible during the first year, and three harvests a year can be had from the second year onwards. The plants are likely to yield well for at least five years, after which, replanting should be done. The leaves are separated and spread out for drying in the shade, which takes 2-4 weeks in warm weather, and are packed in gunny bags after drying.

Production and Yield

The plants are likely to yield well for at least five years. The average yield of fresh leaves per harvest is about 2 kg/plant which, on shade drying, reduces to about 0.5 kg. The fresh herbage yield is 10 t/ha in the first year and 15 t/ha from the second year onwards. This, on shade-drying, gives about 2 tonnes of air-dried leaves in the first year and 3 tonnes from the second year onwards.

Marketing and Trade

The current market price of dried leaves is Rs. 10 to 15 per kg.

Economics of Cultivation

The market of Duboisia and their products are highly volatile. The economics (per ha) worked out here are subject to fluctuations, depending upon time and place. On the basis of the prevailing market conditions with dry leaves, a net profit of about Rs. 30000/ha in a one-year-old crop and Rs. 50000/ha in a two-year-old crop can be obtained.

ERGOT OF RYE (*Claviceps purpurea* (Fr.) Tul.

Introduction

Ergot consists of dried scleotia of the fungus *Claviceps purpurea* (Fries) Tulasne, parasitic on Rye (*Secale cereale* L.). Ergot as a disease of rye and other cereals and grasses has been known to Romans,

Greeks and Chinese for thousands of years, because of the toxic nature of the sclerotia when mixed with grain or hay. People who consumed infected grain suffered from the deadly disease 'Ergotism' which resulted in nausea, falling off of extremities, onset of gangrene and finally collapse and death. Pregnant mares when fed on the grass infected with this sclerotia used to abort. This observation led the midwives to use ergot for facilitating childbirth. It was first mentioned by Chou Kung in his book 'Thya' written in 1100 BC to be used for facilitating childbirth. During the middle Ages, the epidemic of ergot fungus used to be so serious that it was mentioned by Romans as 'Holy Fire' or St. Anthony's Fire' and people believed that it was because of the curse of the gods that the disease appears every year. The first serious epidemic of ergotism was probably recorded in Greece in 436 BC followed by one in Germany in 857 AD and Paris in 945 AD. According to ancient records about 40,000 people died in France in epidemic in 994 AD. This was followed by a more serious epidemic in various parts of Europe.

Although, the ergot powder was used to hasten childbirth by midwives for thousands of years, its use in modern medicine was first reported by German physician Lonicer in 1582. The first scientific report of ergot as an oxitocic agent was made by the American physician Stearns in 1808. However, as the drug was too toxic as an oxitocic agent, Hesek in 1824 recommended it to be used for stopping postpartum haemorrhage. The first alkaloids ergotixine was isolated by Berger and Carr in 1906 and ergotamine by Stoll in 1918. The alkaloid ergometrine was isolated by Dudley and Moore in 1935.

Botanical Description

A parasitic fungus belonging to the class ascomycetes. The ascospores of the fungus infect the rye flowers through the stigma. The fungus mycelium grows into the ovary producing microconidia (asexual spores) in a mixture of honey dew. These spores are carried by wind, rain and insects to healthy florets causing secondary infection. At the end of the season, the fungus mycelium produced is converted into a long horn-like structure called sclerotium or ergot in place of the grain. These sclerotia fall on the ground at the end of the season and the temperate areas like Europe, these overwinter and germinate during next spring to produce stromata containing perithecia with ascend ascospores. The ascospores shot up from the asci and blown by wind, fall on flowers of rye and grow on the stigma producing a new cycle of infection.

Geographical Distribution

Ergot infection of rye is still present in natural from in USSR, Spain, Portugal and Canada. However, because of control measures, the disease has been eliminated from most of the rye growing areas of Europe and America and as a result of increase in demand for pharmaceutical purposes, it is being cultivated parasitically on rye in Russia, Hungary, Czechoslovakia, Germany, Portugal and India.

ERGOT OF RYE (*Claviceps purpurea*)

Medicinal Uses

Ergot alkaloids have a wide spectrum of fascinating pharmacological activities and hence they have been the source of several drugs. The ergometrine derivatives are mostly used to stop haemorrhage after childbirth. Ergotamine is used against migraine. A derivative of ergotamine, dihydroergotamine is also used in migraine. Ergotoxine group of alkaloids (equal mixture of dihydrocornine, dihydrocrystine and dihydrocryptine) have been used for controlling essential hypertension and other peripheral circular disorders. A derivative of ergcryptine, called bromoergocrystine, is used against breast cancer and galactorrhoea. Additionally, a formulation of the dihydroderivatives of ergotoxine is used increasingly for the treatment of hypertension particularly in older patients.

Chemical Constituents

The main constituents of ergot are a group of indole alkaloids with a tetracyclic ring system called ergoline. More than two dozen alkaloids are known. However, the most important alkaloids, which are therapeutically important, are ergometrine, ergometrinine, ergotamine, ergotaminine, ergosine, ergosinine, ergocornine, ergocornine, ergocryptine, ergocryptinine, ergocrystine and ergocrystinine.

Soil and Climate

It grows best in medium and light sandy loam soils which are well-drained and rich in organic matter. Although, the plant prefers neutral pH, it can also be cultivated in salty soils up to pH 9. Temperate climates with mild summers and very cold winters are ideal for cultivation of rye for ergot production. However, it can also be cultivated as winter crop in subtropical areas like north Indian plains.

Improved Varieties

Strains ATCC 15 383, FI 275, OKI 56/1970, MUT 168, IC/39/20, OKI 00022, DM 838, PRL 1578, etc.

AGRO-TECHNIQUE

Land Preparation

Before planting the field is ploughed and harrowed several times to produce a fine seed bed like wheat. In case of insufficient moisture, a preplanting Irrigation is given before preparation of field.

Mode of Propagation

Rye can be propagated by seeds.

Sowing

The ideal time for sowing, both in temperate as well as in subtropical areas is October-November. Early planting, especially during the month of October, is ideal for subtropical areas like north Indian plains, as delayed planting result into decrease in the yield of ergot. Rye seeds @ 40 kg/ha are sown in rows at a distance of 25-30 cm by a tractor-drawn drill. In small areas seeds can be planted behind tractors by manual seeding.

Manures and Fertilizers

For good yield, 60 kg/ha phosphorus, 40 kg/ha potash and 20 kg/ha nitrogen is applied as a basal dose at the time of planting. 40-60 kg/ha nitrogen is applied in 2 split doses at the time of tillering and boot stage, depending upon the fertility of the soil. In very fertile soils, the amount of nitrogen required is much less as compared to soils with poor fertility. Nitrogen increases the yield of ergot, while phosphorus hastens flowering. In soils, which are not deficient in phosphorus and potassium, only nitrogen application is sufficient to grow a good crop of rye.

Irrigation

In temperate areas where there is snow and rain during winter, no Irrigation is required. However, in substropical areas like India, 2-3 Irrigations are required during the growing season from November to March to get a good crop of ergot and to maintain humidity.

Interculture

Rye should be kept free from weeds by at least one weeding after 8 weeks of planting. However, in large plantations hand weeding is uneconomical and it is advisable to apply chemical weedicides like 2, 4-D at the rate of 2-2.5 kg/ha, 4 weeks after planting when the seedlings are in 3-leaf stage.

INOCULATION OF THE CROP

Inoculation of the crop consists of two steps: (1) production of ergot inoculum, and (2) inoculation of the rye ear heads.

Production of Inoculum

For production of ergot inoculum, a virulent strain of the fungus, which has been selected for high alkaloid content and containing the required alkaloid, is cultured on sterile steamed rye or wheat grain in milk bottles. Recently, production of inoculum in liquid shake cultures has been tried and found suitable in India.

Inoculation

A suspension of the spores of the fungus is made by suspending the infested grain in water and straining the suspension in cheese cloth to separate the grain from mycelium and spores. The suspension is diluted to give at least 14000-15000 spores per mil. In case of liquid culture, after the optimum growth of the fungus, the culture is diluted with water to give the required amount of conidia. The ideal time of inoculation of the crop is the time when more than 2% of the earth-heads are out of the boot. This happens generally in the months of January-February in subtropical areas like north Indian plains and April-May in temperate areas like Western Europe and Kashmir valley in India. Previously, the spore suspension was sprayed by a sprayer on the open rye flowers. However, this gives erratic results.

The method of injection was developed in Hungary and now it is a practice to inject the spore suspension in the ovaries of rye flowers through needles. The simplest inoculation machine which was developed in Hungary consists of a metal board containing steel needles and a metal pad along with a 10 mm foam rubber pad. The foam pad is dipped in the spore suspension and while inoculating the crop, the workers press a group of rye earth-heads between the two pads. This allows direct injection of the spore suspension into the ovaries of the unopened flowers. Great care should be taken when the inoculation is carried out when the flowers are not open. This inoculation allows infection of at least 20% of the flowers at the boot stage. Sometimes, another inoculation is carried out one week after the first inoculation. The artificial inoculation produces the first cycle of infection resulting into production of microconidia and honey dew, which are then carried to healthy plants by insects, wind and rain causing secondary infection.

Harvesting and Processing

Ergot is ready for picking when the sclerotia can be easily detached by finger. This generally happens during the month of April in subtropical areas of India and in June and July in temperate areas of Western Europe and Kashmir valley in India. In most of the countries ergot is picked by hand. Generally, two pickings are required and finally the crop is harvested. Remaining sclerotia are separated from the grain by sieving. In countries like USSR, Hungary and Germany, special electrically- operated machines have been developed for harvesting sclerotia. The sclerotia are dried in shade and stored in air tight containers. It is advisable to keep 250 g camphor per container to prevent insect attack. However, the ideal way of storage is to store it in nitrogen.

Production and Yield

The yield of 80-100 kg/ha can be achieved in commercial cultivation depending upon the crop and efficiency of inoculation. However, climatic conditions play an important role in the yield of ergot. The alkaloid yield in the fementers, as a percentage of the mycelial dry weight, may exceed 20%, whereas natural sclerotia contain less than 1%, or even less than 0.1%.

Marketing and Trade

It has estimated the current annual world production of ergot alkaloids at 8500 kg of peptide alkaloids and in excess of 15000 kg of lysergic acid. The price of lysergic acid is today between 4000-5000 $/kg. The demand are not likely to be lesser in future. Therefore, many efforts are needed to make available the increasing requirement for ergot alkaloids above all of peptide type alkaloids and partial synthetic derivatives. The Poland, Spain, Portugal, USSR, Hungary, Czechoslovakia and Switzerland are major producer of ergot in the world.

Economics of Cultivation

The market of Ergot alkaloids and their products are highly volatile. The economics (per ha) worked out here are subject to fluctuations, depending upon time and place. On the basis of the prevailing market conditions, a net profit of about Rs. 25,000-30000/ha.

GLORY LILY (*Gloriosa superba* L.)

Vernacular Names

San.–Langali, Agnisikha; *Hin.*–Karihari, Languli; *Kan.*–Karadi, Kannina gadde, Huliyuguru, Sivasakthi balli; *Mal.*–Mentonni, Ventoni; *Mar.*–Karianag, Nagkaria; *Tam.*–Kanvali poo, Kaandal, Kalappai, Kizhangu; *Tel.*–Kalappa, Gadda, Potti dumpa, Adavi nabhi; *Eng.*–Supper lily.

Introduction

The Colchicines content in *Gloriosa superba* (family-Liliaceae) varies from 0.15 to 0.3% in tubers, and in the seeds it ranges from 0.7 to 0.9%. Until recently, only the tubers are exploited but after discovering the fact that the seeds also have a high quantity of alkaloids, the crop is grown mainly for its seeds which are in great demand within the country and in the international market. Colchicine has been extracted from the corms of *Colchicum automnale,* growing wild in some parts of Europe. Recently, the supplies of colchicines from the conventional sources has not been sufficient to cope

with its increasing demand. Among the other
Indian plants, the seeds of *Iphigenia stellata*
contain 0.90%, the corms of *Colchicum luteum*
contain 0.25% and the seeds of *Gloriosa superba*
were found to contain 0.60% of colchicines. The
easy availability from both wild and cultivated
sources make the seeds of *Gloriosa superba* a
potential source of colchicines in India.

Botanical Description

A beautiful, herbaceous, tall, glabrous,
branching leaf tip climber, about 1-3 m tall. The
leaves are ovate, lanceolate, acuminate, the tips
spirally twisted to serve as tendrils. The flowers
are large, solitary or may form a lax-corymbose
inflorescence, twisted and crisped with six
recurved or reflexed petals, blossoming yellow
but changing to yellow-red and deep scarlet.
Fruits is large oblong capsule upto 5 cm long.
Seeds subglobose, tasta spongy, wing like.

Geographical Distribution

It is a native of tropical Asia and Africa.
The genus derives its name from the Latin word

GLORY LILY (*Gloriosa superba*)

glorious, referring to the flowers. It is found growing throughout tropical India, from the North-west
Himalayas to Assam and the Deccan peninsula, extending upto an elevation of 2a20 m. In Karnataka,
it is commonly found growing all young the Western Ghats. It is also found growing in Madagascar,
Sri Lanka, Indochina and on the adjacent islands. The area under this crop in India is around 2000 ha.

Medicinal Uses

The plant has been used in the Indian system of medicine since time immemorial. Its tubers are
reported to have been used as a tonic, antiperiodic, antihelminthic and also against snakebites and
scorpion stings. The drug is a gasterointestinal irritant and may cause vomiting and purging. It is
sometimes used for promoting labour pains and, conversely, also as an abortifacient. It is considered
useful in colic, chronic ulcers, piles and gonorrhoea. It is used in local applications against parasitic
skin diseases and as a cataplasm in urological pains.

The leaves, when applied in the form of a paste to the forehead and neck, are reported to cure
asthma in children. The leaf-juice is used against head lice. The medicinal properties of the drug are
due to the presence of alkaloids, chiefly colchicines and gloriosine. Colchicine is used in the treatment
of gout, a common disorder in the temperate parts of the world. Gout is caused by the deposition of
microcrystals of uric acid in the joints, which is attributed to a defective regulatory mechanism for
endogenous passive synthesis. It is said that colchicines interrupts the cycle of new crystal deposition
which seems to be essential for the continuance of acute gout.

Chemical Constituents

The tubers contain an alkaloids, chiefly colchicines ($C_{22}H_{25}O_6N$; 0.15-0.3%) and gloriosine
($C_{22}H_{25}O_6N$). In addition, these alkaloids are also used as polyploidizing agents in polyploid breeding
in crop research.

Soil and Climate

It prefers sandy loam soils on the acidic side to red and black loamy soils (pH 5.5-6.5), good amount of humus with good drainage, for its successful growth. It is a tropical crop and comes up well in warm, humid regions. Under natural conditions, it is found growing up to an elevation of 600 m from seal-level. An annual rainfall of about 300-400 cm & temperature 10-20°C, is ideally suited for this crop.

AGRO-TECHNIQUE

Land Preparation

The field should be ploughed and harrowed several times until it is brought to a find tilth. All the grass-stubble and roots should be removed. The field must be leveled properly and drainage arrangements made to avoid water-logging during the rains. The field is then divided into subplots of convenient sizes. About 15-20 t/ha of FYM or compost should be mixed well into the soil. One foot deep furrows are opened at a spacing of 45-60 cm.

Mode of Propagation

It can be propagated by seeds and V-shaped rhizomes.

Sowing

The dormant tubers start sprouting from the month of May. Best planting time of tubers are July-August for good growth and yield. About 2.5 to 3.0 t/ha of tubers are required for planting. In order to avoid rotting of the tubers before sprouting, only healthy tubers should be selected for planting. In addition, the selected tubers or tuber pieces should be treated with suitable fungicides, preferably Emisan-6 (methoxy ethyl mercury chloride) @ 0.8%. The treated tubers are planted at a depth of 6-8 cm, keeping a plant-to-plant distance of 30 to 45 cm, depending upon the type of soil. Closer spacing has been reported to favour cross-pollination, thereby improving the fruit set. The provision of some kind of support is necessary for successfully growth.

Manures and Fertilizers

Though it makes satisfactory progress with little manuring and fertilization, the addition of well-decayed manure, bone meal and fertilizers to the soil ensures a vigorous plant, stronger tubers and better flowering. Normally, a fertilizer dose of 100-120 kg nitrogen, 50-60 kg phosphorus and 75-80 kg potash/ha is required for a good crop. Of the nutrients, the whole phosphorus and potash and one third of nitrogen is applied as a basal dose and the remaining two-third of nitrogen should be given in the first six to eight weeks after planting.

Irrigation

Frequent Irrigation is required during the sprouting time to keep the surface soft, so that there is no hard-pan formation, in order to facilitate easy sprouting and emergence of the growing tip outside the soil. Irrigation should be withheld until after the flowering is over, to prevent rotting of the tubers. Excess watering is harmful to the plants and causes yellow or brown-coloured patches on the leaves, which fall off prematurely.

Interculture

In the initial stages, it requires frequent weeding to control the weeds which will otherwise complete with plants for moisture and nutrients and will restrict the growth of the plant. While weeding, utmost care should be taken to avoid any damage to the growing tip, as once damaged, it does not sprout again during the season. Chemical weed control is possible only when there is wide spacing between the rows and the plants themselves.

Pests and Diseases

This crop has few Pests and Diseases. However, some Pests and Diseases pose a serious threat when they get favourable conditions, causing severe damage. Hence, great care has to be taken to control them. Mainly two pests *viz.* Lily caterpillar (*Polytela gloriosae*) & Green caterpillar (*Plusia chaleites*) are damage the crop. These pests can be effectively controlled by spraying metacid at a concentration of 0.2% at fortnightly intervals. The tuber rot or basal stem rotting & wilting and leaf blight are the major diseases, affected the crop. These disease can be controlled by spraying 0.3% Dithane M-45 or drenching the soil with Bavistin @ 0.2% or Cuprassol near the root-zone of the plant has been observed to control the disease.

Harvesting and Storage

It is a crop of 170-180 days duration. When planted in June, it starts bearing flowers after 50-55 days and continues to flower and fruit till October. The fruit requires about 110-120 days from the set to reach maturity. The right stage of harvest is when this capsule starts turning light-green from dark-green. After picking, the capsules should be kept in the shade for 8 to 12 days to facilitate the capsules to pen up, displaying deep orange-yellow coloured seeds. The seeds and pericarp are separated manually and dried for a week in the shade, by spreading them uniformly over any clean, dry floor or any platform specially erected for the purpose. At the later stages, the seeds are moved to the sunlight for a week till they dry completely. The dried seeds are then packed in moisture-proof containers and stored until exported or extracted for the alkaloids.

Production and Yield

The yield of seeds differs greatly, depending upon the vigour and age of the plant which, in turn, depend on the size of the tuber. The yield in the initial year will be low, but it gradually increases in the subsequently years. After three years, from a well-managed field, we may get about 2.0-2.5 quintals/ ha of dried seeds.

Marketing and Trade

The Glory lily tuber today finds a national and international market. The consumption of its tubers is nearly 2.5 t in 1999-2000 by Natural Remedies (P) Ltd., Banglore. A number of drug companies, using colchicines as raw material, are interested in buying this crop grown on a contract basis. The current market price of tuberous roots and seeds are Rs. 20-25 and Rs. 700-1000 per kg respectively.

Economics of Cultivation

The market of Glory lily and their products are highly volatile. The economics (per ha) worked out here are subject to fluctuations, depending upon time and place.

Expenditure (in Rs.)	—	10,500.00
Returns (in Rs.)	—	80,000.00
Net Profit* (in Rs.)	**—**	**69,500.00**

*   Estimation of the net profit is analysed on the basis of grower/collectors price. The grower/collector price is always 25-40% less than wholesale market price due to various stakeholders *viz.* Agent, middle man, Brokers, Commission agents.

GUGGAL (*Commiphora wightii* (Arn.) Bhan.)
Syn. *C. mukul* (Hook ex Stocks) Engl.

Vernacular Names

San.–Gugguluh, Mahisaksah; *Hin.*–Gugal, Guggul; *Kan.*–Guggulu; *Mal.*–Gulggulu, Mahisaksagulgulu; *Mar.*–Gugglu; *Tam.*–Gukkulu, Mahisaksi, Kukkil; *Tel.*–Guggulu, Maisakshi; *Eng.*–Indian bedellium tree; *Trade*–Guggal.

Introduction

This plant is a source of Indian bedellium, a oleo-gum-resin obtained by incision of the bark. The guggul resin is a complex mixture of steroids, occurs in vascular or stalactitic pieces, pale-yellow, brown or dull-green in colour with a bitter aromatic taste and balsamic odour. In Sanskrit, 'Guggulu' means 'that which protects against diseases'. This small tree belongs to the family Burseraceae.

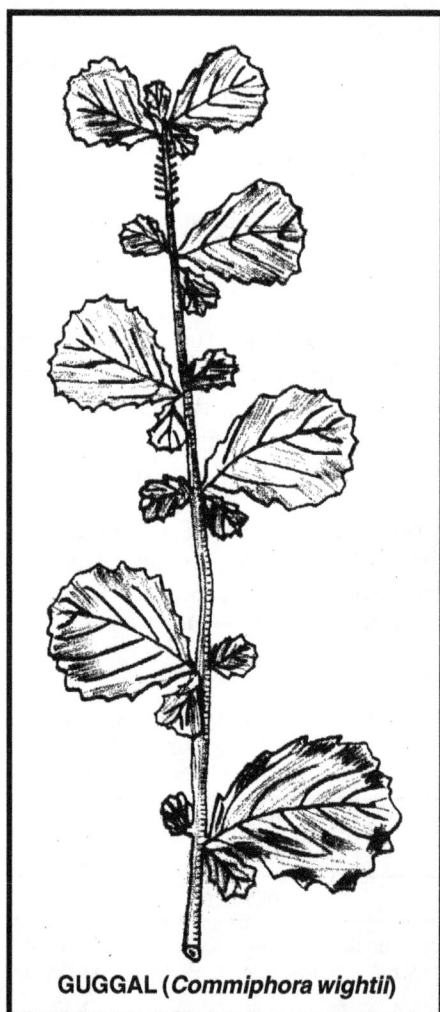

GUGGAL (*Commiphora wightii*)

Botanical Description

A small tree or shrub up to 3-4 m high, the branches are crooked, knotty, aromatic and end in sharp spines. The bark is papery and peels in strips from the older parts of the stem. The leaves are sessile, alternate or fascicled, 1-3 foliate, the leaflets are glabrous. The flowers appear in groups of 2 or 3 usually red and rarely pinkish-white in colour. Fruit ovoid drupe, red when ripe.

Geographical Distribution

Its origin in Africa and Asia and is widely distributed in the tropical regions of Africa, Madagascar, Asia, Australia, Pacific Islands, India, Bangladesh and Pakistan. But the plants are found growing naturally in arid zones of Western Indian. It is distributed in the states of Rajasthan, Tamil Nadu, Assam, Gujarat, Maharashtra and Karnataka.

Medicinal Uses

The resin is largely used as incense and as a fixative in perfumery and in medicine. In indigenous medicine, it is used as an astringent, antiseptic, anti-arthritic, anti-obesity, anti-inflammatory, anti-acne, anti-lipidaemic, stomachic, carminative and digestive. The oleo-resin causes an increase of leucocytes in the blood and stimulates phago-cytosis. It acts as a diaphoretic, expectorant and diuretic and is said to be a uterine stimulant and an emmenagogue as well. It is highly effective in the treatment of obesity, arthritis, indolent ulcers, weak and spongy gums, pyorrhea, alveolaris, chronic tonsillitis and pharangytis, ulcerated throat and chronic dyspepsia. Inhalation of the fumes of burnt guggul is recommended in hay fever, acute and chronic catarrh, chronic laryngitis, chronic bronchitis and phthisis. It is an ingredient of ointments for ulcers.

Chemical Constituents

Bark yield an oleo-gum-resin which pale-yellow, brown or dull-green in colour with a bitter aromatic taste and balsamic odour. The oleo-gum-resin is mixtures of resin (about 61%), gum (about 29.3%), volatile oil (about 1.45%) and occasionally combined with other substances, mainly myrcene, dimyrecene and polymyrecene. It has ketonic and non-ketonic fractions. The ethyl acetate extract of guggal has been named 'guggulipid."

Soil and Climate

It grows well sandy to silt loam soils, poor in organic matter but rich in several other minerals with moisture-retaining capacity. It prefers a warm, dry climate for a good yield of oleo-gum-resin.

Improved Varieties

'Marusudha'.

AGRO-TECHNIQUE

Land Preparation

The land is prepared well in advance of the rainy season by 2-3 ploughings and laid out into plots of convenient sizes. Pits of the size 0.5x0.5x0.5 m^3, are dug at a spacing of 3 to 4 m, in rows. The pits are filled with FYM and topsoil and mixed with Aldrin (5%), to prevent any damage to the plants due to termite attack.

Mode of Propagation

It can be propagated by seeds or vegetatively *viz.* stem-cuttings, air layering, etc

By Seeds

This is not a common method of propagating this plant, because the seed germination is slow and is also very poor (5%) due to the presence of the hard seed-coat. This achieves a good rate of germination, the seeds are mechanically scarified with sandpaper and are kept under running water for 24 hours. The seedlings may be raised in polythene bags during the kharif season and, after hardening, may be planted in the main field.

By Stem-cuttings

About 15-20 cm long and 10 mm thick, semi-hardwood cuttings are taken and the cut end is treated with IBA or NAA growth-regulator solutions and planted in well-prepared and manured nursery beds during the months of June and July. The beds are irrigated lightly after planting and regularly thereafter. The cuttings sprout in 10-15 days, they grow well and are ready for planting in the main field after 10 to 12 months, during the next rainy season. The percentage of rooting in stem-cuttings is around 80 to 94%.

Transplanting

The rooted cuttings are planted in the pits during the rainy season. As the plants grown, they are trained properly by cutting the side branches.

Manure and Fertilizers

The crop has not shown good response to fertilizers, except to low-level or nitrogen fertilization. Hence, urea or ammonium sulphate at 25/50 g/bush is given twice a year before Irrigation.

Irrigation

Providing light Irrigation during the summer season supports good growth of the plants.

Interculture

It is confined to one weeding and hoeing in the early stages of growth. But the soil has to be stirred around the bushes twice in a year. This practice of soil-stirring is found to increase the growth of plants.

Pruning

The guggal stems purned in May contain the maximum amount of guggal sterons (0.06%). This method saves the plant from drying and ultimate death, which was a result of the extraction procedure adopted earlier.

Pests and Diseases

Guggal plants are attacked by a leaf-eating caterpillar; white fly, which sucks the sap of leaves as a result of which the leaves turn yellow and finally drop off. The rooted cuttings also suffer from termite attacks in the early stages of growth. The crop is mainly suffering from leaf-spot and bacterial leaf-blight diseases.

Harvesting (Gum-tapping) and Storage

The plants attain normal height and girth after 8 to 10 years of growth, when they are ready for tapping. For tapping the gum, which is present in the balsam canals in the phloem, bark-deep, *i.e.*, a shallow, incision is made on the bark. If the incision is too deep, the plant either dies or yields little resin in the following years. While making the incision, a small quantity of guggal gum mixed with water may be applied to the incised place using a prick-chisel. The sharp end of the chisel is dipped in the guggal solution and an incision is make on the bark carefully. Usually the incision is made after November but before April. The resin in is collected at an interval of 10-15 days. Weather conditions influence the success of obtaining gum.

Production and Yield

The yield of gum-resin from each plant is about 700-900 g. The yield in the initial year will be very low, but it gradually increases in the subsequently years. After 5 years, from a well-managed field, we may get about 2.0 quintals/ha gum-resin.

Marketing and Trade

The unsustainable ways of harvesting and unrestricted marketing have to the reduction in population of guggul, high demand medicinal plants leading to sudden escalation in their prices in the market. The current market price of guggul is 150-200 per kg. According to estimation, the annual demand of guggal is nearly 1000 tonnes, while the capacity of the consumption of this drug is nearly 2300 t, used in the various preparation of Indian system of medicines. Total demand of Guggal by Herbal industries and ayurvedic drug producers of Maharashtra is nearly 4000 t annually. The oleo-resin of guggal is imported nearly 500-1000 t from Pakistan annually. The Natural Remedies (P) Ltd., Bangalore only consumed 1500 t in 1999-2000. The current market price of guggal is Rs. 1200-1500 per kg.

Economics of Cultivation

The market of Guggal and their products are highly volatile. The economics worked out here are subject to fluctuations, depending upon time and place. On the basis of the prevailing market conditions with gum-resin, a net profit of about Rs. 1,20,000/ha in a five-year-old crop, can be obtained and will be increased in subsequent year.

HENBANE (*Hyoscyamus niger* L.)

Vernacular Names

San.–Dipya, Parasikaya; *Hin.*–Khurasani ajavayan; *Beng.*–Khora-saniajowan; *Tel.*–Khurashanivamam; *Tam.*–Kurasaniyomam; *Kan.*–Khurasanivadaki; *Urdu*–Ajowan khurasani; *Eng.*–Black henbane, Common henbane.

Introduction

Henbane, comprising of the dried leaves and flowering tops of *Hyoscyamus niger* (black henbane) is an official drug of repute. "Hyoscyamus" finds its origin from Greek literature 'Hyos' meaning a 'Hog' and 'Kyamos' meaning a 'Bean'. The drug is valued because of the presence of alkaloid hyoscyamine, with smaller quantities of atropine and hyoscine (scopolamine). It is especially valuable as a sedative and narcotic in cases of maniacal excitement, sleeplessness and nervous depressions. It belongs to the family Solanaceae.

Botanical Description

An erect, viscidly hairy, biennial plant growing up to 160 cm in height. The leaves are simple, large, ovate or oblong, coarsely dentate and pinnately lobed. The flowers are yellowish, pale, sessile or subsessile, borne on terminal cymes. The fruit is a berry containing numerous minute seeds. The seeds are small, brown to black, marked with fine, conspicuous reticulations.

Geographical Distribution

It is a native of Europe, Central Asia, India and tropical America and is distributed throughout Europe from Portugal and Greece in the South on Norway and Finland in the North. In India, it is mainly found growing in temperate to subtropical regions and is grown largely in damp shady places of Kashmir and Uttaranchal at an altitude from 1500-3750 m. Its successful cultivation at Indore, recently, has indicated that it can also be grown in warmer regions as a winter crop.

HENBANE (*Hyoscyamus niger*)

Medicinal Uses

Leaves and flowering tops has anodyne, narcotic, sedative, antiseptic and mydriatic properties. The leaves are also smoked for intoxication. It is principally employed as a sedative in nervous affections and irritable conditions such as asthma and whooping cough. In veterinary practices used as a urinary sedative. Plant, especially the seeds, in large doses, produce poisonous effects similar to those of *Datura* poisoning, such as dry ness of tongue and mouth, giddiness and delirium.

Chemical Constituents

The herb and seed yield tropane alkaloids (0.03-0.06%), 90% of which are hyoscyamine and hyoscine and the rest are scopolamine, atropine, hyoscipirin, choline, fatty oil, mucilage, albumen and KNO_3. They are used as an anodyne, sedative, tranquillizer, antiseptic, antispasmodic, mydriatic, in asthma and whooping cough. Hyoscine is particularly useful in protection against shocks caused by accidents and loud noises.

Soil and Climate

It is a foliage and cool season crop, requires a level, well-drained fertile, sandy loamy soil (pH 7-8.5), rich in organic matter, for its optimum growth. Deep black and water-logged soils are not suitable for its cultivation. It requires optimum temperature 20-30⁰C with adequate moisture for good yield.

Improved Varieties

'Aela' is released by CIMAP, Lucknow .

AGRO-TECHNIQUE

Land Preparation

The field should be prepared to make good seed-beds, by light ploughing followed by two harrowing before the onset on the monsoon season.

Mode of Propagation

It can be propagated by seeds.

Direct Sowing

The best time for sowing the seeds directly in the field is early October in subtropical regions, while June-July is most appropriate in temperate regions. As henbane seeds are very small (2000 seeds/g) it is important to ensure that the soil is prepared to a fine tilth for sowing. Seeds must be sown not deeper than 2 to 2:5 cm below the soil for good germination. Soaking the seeds overnight and incubating them for 7-8 days at 25⁰C for sprouting before sowing, gives a better germination rate. In order to ensure the uniform distribution of seeds all over the field, the seeds should be mixed with dry, powdered earth or sand in a ratio of 1:10 before sowing. About 2-3 kg of seed would sufficient for an area of one hectare.

Nursery Raising

The best time for raising the nursery is September in the plains and April-May in temperate regions. The seeds are mixed with sand and sown by broadcasting in rows of 8 cm, in raised nursery beds. The seeds are then gently covered with a fine mixture of soil and FYM or sand, mulched and immediately watered. About 500 g seeds are required to raise enough seedlings to cover an area of one hectare. Germination takes place in about 8-10 days. At this time, the mulch is removed but watering and weeding are continued as usual. If the temperature is low, the seeds take a longer time to germinate and the growth of the seedlings is also very slow. The seedlings take about two months to reach a suitable stage for Transplanting. A foliar spray of 1% urea, when seedlings are about 20 days old, has been found to accelerate their growth.

Transplanting

The seedlings are transplanted at a spacing of 45 cm is rows and 15-30 cm from plant to plant in a well prepared field and followed up with light Irrigation. Since there is a heavy mortality in the transplanted seedlings, it is recommended that this crop should be grown on a commercial scale only by Direct Sowing.

Manures and Fertilizers

At the time of Land Preparation, well-decomposed FYM at the rate of 15-20 t/ha is added. In addition, a dose of 40-80 kg nitrogen, 30-40 kg phosphorus and 20-30 kg potash/ha, has been recommended to obtain the maximum dry-matter yield. The entire quantities of P and K along with 20 kg nitrogen is given as a basal dose. The remaining quantity of nitrogen is given as a top-dressing in two equal split doses during the growing season.

Irrigation

The transplanted crop is given the first Irrigation immediately after Transplanting. Subsequent Irrigations are given at 8-10 days intervals, until it is about a month old, and subsequently, the crop is irrigated at intervals of about 15-20 days.

Interculture

The crop raised by Direct Sowing should be thinned after a fortnight of germination and a spacing of 30 cm within the rows should be maintained. Usually, two weeding-cum-hoeing operations are necessary during the entire cropping season.

Pests and Diseases

No serious attack of insect Pests and Diseases are reported on henbane. However, the larva of the cotton bug attacks the leaves and capsule of this crop during July-August, causing heavy damage. Aphids also infest the crop. These two insects may be controlled by 2-3 sprayings (at 10-15 days' interval) of Endosulphon (2 ml/l of water) of Methyl Parathion of Metasystox (2 ml/l). A green mosaic caused by a virus has been reported to affect the crop in India. The nematodes *Meloidogyne incognita* and *M. javanica* cause root-knot disease.

Harvesting and Storage

It is a short duration winter crop. It matures within 90-100 days in the Indian plains. The leaves and flower tops of henbane are collected at the flowering stage (May) of the plant. The older leaves of the lower regions of the plant toughing the ground should be picked first and dried separately. The herb is sun-dried for 2-3 days followed by 6-7 days in the shade and about 80% moisture is lost, with constant raking over with sticks. For rapid drying artificial heat is given at a temperature of 40°C-50°C. After drying, the material is stored in gunny bags and kept in a cool, dry place for about two months before further processing.

Production and Yield

An average yield of 15-20 quintals of dry herb per hectare may be obtained.

Marketing and Trade

An estimation, the annual production of Henbane (50 Mts) *vis-a-vis* their 5 percentage consumption by Indian Ayurvedic pharmacies in 1999 (Source: ADMA, Mumbai). The current market price of henbane seed and leaves is Rs. 30-40 and 15-20 per kg respectively.

Economics of Cultivation

The market of Henbane and their products are highly volatile. The economics (per ha) worked out here are subject to fluctuations, depending upon time and place.

Expenditure (in Rs.)	—	6,700.00
Returns (in Rs.)	—	25,000.00
Net Profit* (in Rs.)	—	**18,300.00**

*   Estimation of the net profit is analysed on the basis of grower/collectors price. The grower/collector price is always 25-40% less than wholesale market price due to various stakeholders *viz.* Agent, middle man, Brokers, Commission agents.

HIMALAYAN GINSENG (*Panax pseudo-ginseng* Wall)

Vernacular Names

Eng.–False ginseng.

Introduction

Panax pseudo-ginseng (Family-Araliaceae) being analogous to the Korean Ginseng - *Panax ginseng* reputed as a health-food and 'Cure-all' vegetable drug has gained much attention of the medico-botanical researchers since last few years. Ginseng has been used in Chinese medicine for more than 4000 years. The earliest Chinese Herbal which refers use of this plant is "Shen-Nung Pen Tsao Ching". This appeared for the first time in written form in first century BC. There must have been a long previous verbal history. Chinese name of the plant has been translated in the western world as `Man-shaped Root' or `Man Essence.' This word is derived from the shape of the root, which resembles the form of man. It is also believed that the essence is really crystallized in human form. In old Chinese medicine the more the root resembles a human being, the more potent is its healing properties.

Because of widespread and indiscriminate collection of wild ginseng, the plant has been almost extinct, but because of great demand, collection of wild ginseng was practiced widely in Canada and USA. By 1750, because of depletion of forests and large-scale collection and supply of ginseng to China as well as the growing demand, supply of the drug was depleted.

Botanical Description

An erect, perennial herb with horizontal tuberiferous root stock. Leaves palmate long-petioled in whorl at the apex; leaflets 3-5, margins dentate, hairly on both sides. Inflorescence in terminal simple umbel; flowers small yellowish-green. Berry globose, slightly compressed, red or reddish-black when ripe, 2-seeded.

Geographical Distribution

It is found naturally in temperate forest of North East Himalayas especially in Sikkim from 3000-3600 m altitude. Presently, it is cultivated in some hilly areas of Sikkim.

Medicinal Uses

The Himalayan ginseng is prescribed as a tonic, stimulant and aphrodisiac. The drug is used in cases of neurasthenia, dyspepsia, palpitation and asthma. It is also used for controlling amnesia, headaches, convulsions, dysentery and cancer. It has been claimed that the drug is an adaptogen and enhances natural resistance/recuperative power of the body. Although, ginseng has not been adopted in modern medicine.

Chemical Constituents

The tuberous root contains triterpenoid saponins (saponins A,B.C,D), oleanolic acid, amino acids, leucine, valine and crysteine, Fe, Ca.

Soil and Climate

Ginseng thrives well in medium and light loam soils which are rich in organic matter. Heavy clay or very light soils are avoided for ginseng culture. Forest soils rich in organic matter and humus are ideal for cultivation.

It is a temperate plant and requires freezing temperatures at least for 3-4 months during the period when it is dormant, and cool summers (summer temperatures not exceeding 30⁰C) are ideal for ginseng culture. Chilling below 0⁰C is necessary for germination of seeds.

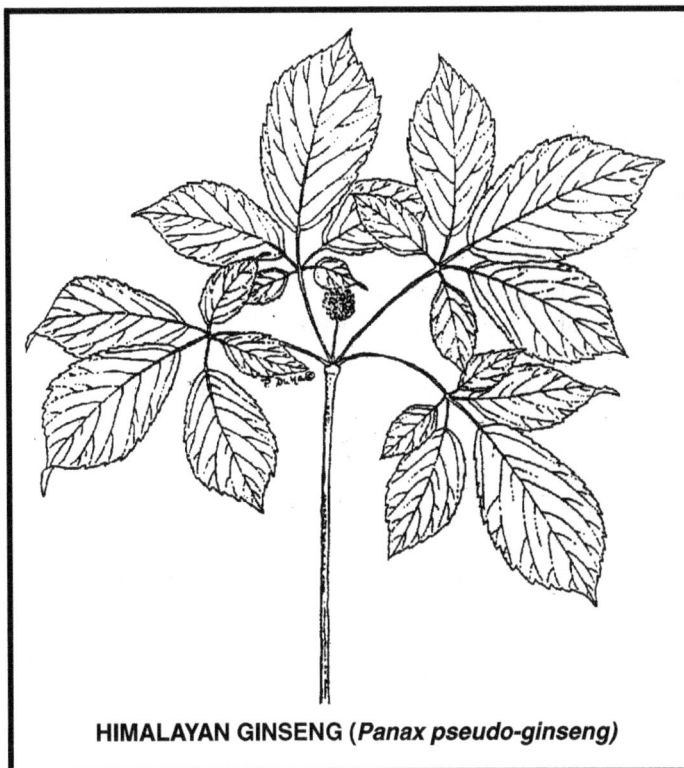

HIMALAYAN GINSENG (*Panax pseudo-ginseng*)

AGRO-TECHNIQUE

Land Preparation

The land should be ploughed and tilled several times during early fall to allow decomposition of organic matter and to avoid soil-borne diseases, pests and weeds. Beds should be 2 m wide with 46 cm walk-way between the beds.

Mode of Propagation

Ginseng can be propagated by seeds or vegetatively by roots.

Nursery Raising

In Ginseng, flowering takes place during May to June and fruiting from July to August. The seed ripen in September-October and is very small in size. The 4-year old mother plant provides the seed only once. The collected fruits are immediately stripped off their epicarps and mesocarps and the seeds are well washed with fresh water before drying it in the shade. The harvested seed, undergoes dormancy period and require further embryonic development. Before sowing the seed must be stratified to facilitate its germination as the seed coat is hard. In early November the stratified seeds are sown in the nursery beds in dotted lines. The germination of seed is over within 120-130 days.

The seedling gets prepared in early next April, when sprouts come out from the surface the beds are covered and protected from direct sun-light. The growth continues from the stage of germination throughout summer. In late September or early October the leaves and stem wilt off and so the covering of the stock with soil is necessary in order to develop the subterranean stem and to protect roots from the frost, in late March or in early April next year the seedlings are dug out and transplanted in the field.

Transplanting

One-year old seedlings can be planted at a distance of 12-15 cm in rows with plant to plant distance of 5-7 cm. Roots can be planted in fall or spring.

Silvicultural Management

One of the most important factors for successful ginseng cultivation is shade. The plant does not grow in full sunlight. Artificial shade is provided by wooden lathes, boards or propylene sheets supported with poles over the growing area. Boards and lathes can be supported on wooden poles. Shade cover should be stretched to form a canopy over the entire plantation area and it should be at least 2 m high.

Manure and Fertilizers

Normally, ginseng growers do not apply artificial fertilizers. Most of the ginseng is grown in fertile soils which are rich in organic matter, forest humus and compost which are applied to beds, provide enough nutrition. Extra manuring is not required.

Irrigation

Not much watering is required as the area remain moist. However, 1-2 Irrigation in a week is required during summers when there are no rains.

Interculture

The beds should be kept free from weeds by regular weeding/hoeing. Care is required in wooded planting, as bushes often sprout in between plants and there is excessive competition; such bushes should be removed. Generally two weedings in a year, one during May-June and other August/September.

Pest and Diseases

It is very susceptible to pests and precaution taken are necessary right from the seedling stage. Various diseases which caused serious problems in ginseng crop. The most important diseases are *Alternaria blight*, *Phytophthora blight*, anthracnose, grey mold, and Rhizoctonia stem rot. The disease can be controlled by regular spraying of Dithane Z-45, Maneb and Bordeaux mixture or by application of metalaxyl.

Harvesting and Storage

Harvesting should be started when the berries have ripened. The best time for harvesting in month of September. In case of good growth, it can be harvested after 4 years. In plantations under natural shade, harvesting is delayed to 6-8 years. Prior to harvesting top of the plant and mulch should be removed. Artificial shade should be removed and the roots dug out with spade. Mechanical diggers can also be used in bigger plantations. The roots are washed and dried in artificial drier at a temperature of 45⁰C. Roots should be packed in card-board boxes and stored in cool dry atmosphere.

Production and Yield

Nearly 1000 kg/ha dried tuberous has been obtained in well managed artificial shade plantations. In natural shade yield of 500-800 kg/ha dried roots are obtained.

Marketing and Trade

The unsustainable ways of harvesting and unrestricted marketing have to the reduction in population of ginseng, high demand medicinal plants leading to sudden escalation in their prices in the market. At present, 250-300 tonnes ginseng roots are produced in USA, 50-120 tonnes in Korea, 150-160 tonnes in Japan, and 300-400 tonnes in China. Total value of the product is approximately 600 million US dollars. An estimation, 24 t of other ginseng herbs was imported to India by Bulgaria, Morocco, Pakistan, Singapore and UAE in 1998-1999. On the other hand, the ginseng roots are exported from India and earned foreign revenue Rs. 802.45 lakh in 1993-1994 (*Source*: Exim bank Report, 1999).

Economics of Cultivation

It is a good source of income for the farmer living in the interior hilly areas where the farmers do not have much source of income from agricultural crops. But the actual Economics of Cultivation has not been done till now.

HIMALAYAN YEW (*Taxus wallichiana* (Zucc.) Pilger)
Syn. *Taxus baccata* Linn

Vernacular Names

San.–Dhaatripatra, Manduparni, Shakodarana; *Hin.*–Thuno, Birni, Zirnup birni; *Beng.*–Burmie, Bhirmie; *Kash.*–Birni, Poshil; *Kumaon*–Thaner, Thuner, Gullu; *Khasi*–Kseh Blei, Dingableh; *Trade*–Taxus.

Introduction

Himalayan Yew or common Yew has been in frontline focus recently due to its reckless exploitation. Initially, the tree was exploited through the extraction of the bark but later the demand shifted on to taxus needles which yield a greater amount of 'Taxine' the derivative obtained from Taxus leaves and which is a precursor to the drug 'Taxol'. Its IUCN status is Critically endangered -regionally. The "Biodiversity Hotspots Conservation Programme" (BHCP) of World Wide Fund for Nature-India (WWF-India) identified as the threatened species.

The taxus resembles the genus *Tsuga* and *Abies* and the distinction is made only through cones and the leaf colour which in case of the taxus is uniformly yellow green and lacks the conspicuous white stomata lines. Due to such resemblance adulteration of taxus with such species is quite likely.

Botanical Description

A slow growing long living evergreen tree, attains a height 15-20 meters and average girth of 1.5 m at breast height. Leaves distichously, linear, shining above, pale yellowish brown or rushy red below and more or less spirally arranged on erect shoots but appear mostly ranked on horizontal shoots. Bark reddish brown thin. Male and female strobili normally on different trees. Flowering occurs during March to may and needles are shed chiefly in may and June. Seeds ripen September to November of the same year.

Geographical Distribution

Taxus is widely distributed from Pakistan to South-West China mainly in Himalayas. In India, it occurs in Jammu and Kashmir, Himachal Pradesh, Uttranchal, Sikkim, West Bengal, Arunachal Pradesh, Meghalaya, Nagaland and Manipur mostly between 2300-3400 m in the evergreen and

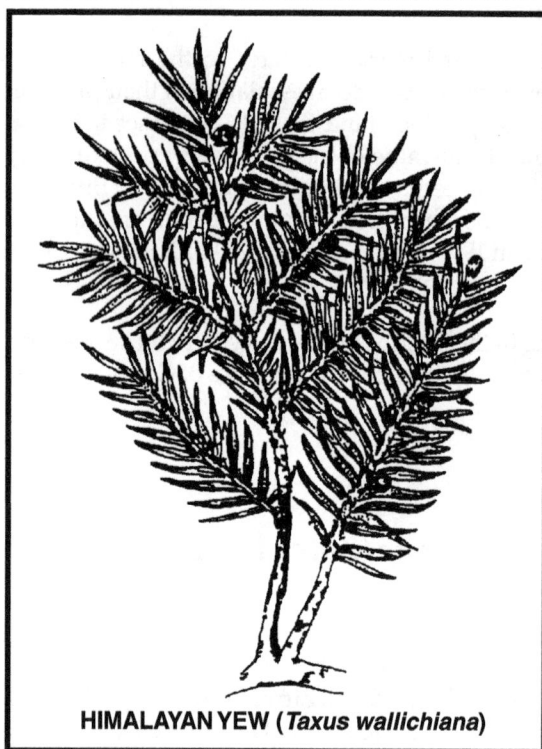

HIMALAYAN YEW (*Taxus wallichiana*)

confers forests. It is found associated with Kharshu Oak and Silver fir. In the Eastern Himalayas it accompanies Silver fir, Hemlock-spruce and Rhododendron. It is also reported from Nepal and Bhutan. Naturally, Taxus tree are found to occur sparsely in the moist shady mountain slopes with steep gradients.

Medicinal Uses

All the parts of the trees except it fleshy aril, are poisonous due to presence of an amorphous basic fraction named taxine & alkaloid taxol (a diterpene from inner bark) and is used for the treatment of 'ovarian' and 'breast cancer'. A medicinal tincture made from the young shoots has long been in the use for the treatment of headache, anorexia, rheumatism, cough, asthma, bronchitis, giddiness, feeble and falling, pulse, coldness of the extremities, diarrhoea etc.

The non poisonous and fleshy aril is eaten by the tribal. Extracts of the cone be added in cosmetics such as hair lotions, rinses, beauty and shaving creams and dentifrices.

Chemical Constituents

All the parts of the trees (except fleshy aril) contains an amorphous basic fraction named taxine (1.3%) & alkaloid taxol.

Soil and Climate

It can grow a variety of soil *viz.* Sandy, light brown lime soil and dry soil but flourish in wet loam drained soil slightly alkaline (pH 7.0-7.3). The Himalayan Yew is a conspicuous tree in the Himalayan forests, usually in moist and shady places under trees. It prefers moist temperate climate.

AGRO-TECHNIQUE

Land Preparation

The field should be prepared to make good seed-beds, by light ploughing followed by two harrowing before the onset on the monsoon season.

Mode of Propagation

It can be propagated by seeds and shoot cuttings.

Nursery Raising

Regeneration in Himalayan yew is very slow (only about 20-30%) and occurs through seeds which do not germinate until the second year. The best time for Nursery Raising is May-June. Seeds require shelter and moist shady areas for germination and do not survive in open areas. Major problem regarding the regeneration in wild is the grazing of the seedling by cattles.

The seedling gets prepared in early next April, when sprouts come out from the surface the beds are covered and protected from direct sun-light. The growth continues from the stage of germination throughout summer. In late September or early October the leaves and stem wilt off and so the covering of the stock with soil is necessary in order to develop the subterranean stem and to protect roots from the frost, in late May or in early June next year the seedlings are dug out and transplanted in the field.

Transplanting

Next year before onset of monsoon, the Taxus seedlings are planted in the field. Since the tree grows to a huge size a distance to 8 to 10 m. Before planting, it is necessary that planting place is appropriately marked, 1 m size pits are made well in mid summer, kept shady place. In each pit 3 to 4 baskets of FYM are mixed with dug soil and filled. Beginning of winter, the plants are planted in the center of these pits and staked properly.

Irrigation

Taxus tree is hardy in nature and stand very well against drought. However, the crop shall be benefited by giving two/three Irrigations at the time of full bloom and set.

Pest and Diseases

No Pests and Diseases has not been reported till now.

Harvesting and Storage

Earlier the bark of the tree was being extracted but later the demand for the Taxus leaf increases since the taxol content is higher in the leaf than in the bark. The growing season is from April to October and lopping during this season is banned. During the rest of the year leaves are collected through villagers. But snowfall during the winter season (Dec.-Feb.) hampers extraction. Sometimes chapped branches are added along with the leaves to increases the weight and the profit margin of the collectors.

Marketing and Trade

The unsustainable ways of harvesting and unrestricted marketing have to the reduction in population of Taxus, high demand medicinal plants leading to sudden escalation in their prices in the market. The cost of the 1.7 g taxol is nearly Rs. 3.0 lakh in international market. The current market price of dry needles/leaves and Taxus extract is nearly Rs. 40-50 and Rs. 400-450 per kg respectively.

Economics of Cultivation

The market of Taxus and their products (Taxsol) are highly volatile. The economics (per ha) worked out here are subject to fluctuations, depending upon time and place. On the basis of the prevailing market conditions with dry needles, a net profit of about Rs. 40000/ha in a five-year-old crop, can be obtained.

INDIAN ALOE (*Aloe barbadensis* Mill.)
Syn. *A. vera* (L.) Burm.

Vernacular Names

San.–Kumaree, Ghrit-kumari; *Hin.*–Ghikanvar, Kumari; *Beng.*–Ghirta-kumari; *Guj.*–Kunvar, kumarapathu; *Mar.*–Pivalaboel, Korphad; *Tel.*–Chinikala-banda, Mushambarum; *Tam.*–Chirukattalai,

Kariambolam; *Kan.*–Kathaligida; *Mal.*–Kattavala, Kattarvazha; *Punj.*–Musabbar; *Eng.*–Indian aloe; *Trade*–Aloe.

Introduction

Indian Aloe belongs to the old world and are indigenous to Eastern and Southern Africa, the Canary Islands and Spain. The species spread to the Mediterranean basin and reached the West Indies, India, China and other countries in the 16th century; and certain species are now cultivated for commercial purposes, especially in some of the West Indian islands of the north coast of South America. It is also cultivated throughout India. Aloe species, perennial succulents belonging to the family Liliaceae, are the source of the drug aloe. Aloe is obtained by cutting the leaves at their base and letting the yellow, bitter juice drain out. The water is evaporated off from the juice by heat, and the resulting light to dark-brown mass is the drug aloe.

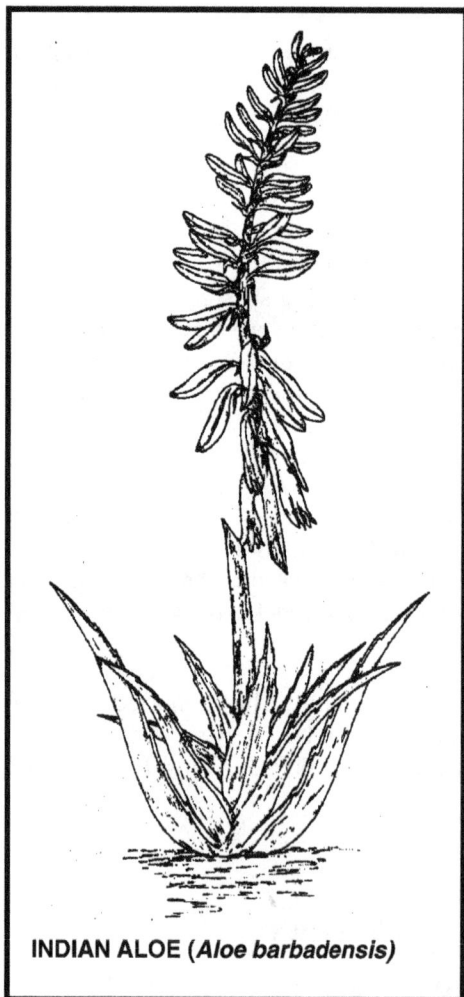

INDIAN ALOE (*Aloe barbadensis*)

Botanical Description

It is coarse-looking, perennial, shallow-rooted (multiple tuberous roots) plant with a short stem, 30-60 cm high. Leaves fleshy (about 60 cm long, 10 cm broad and 1.5 to 2 cm thick), are densely crowded, strongly cuticularized, and have a spiny margin with thin walled tubular cells. Flowers vary from yellow to rich orange in colour and are arranged in auxiliary spikes and blooming in the month of July-September.

Geographical Distribution

It is naturalized in Indian especially in the hot dry valleys of north-western Himalayas and throughout the central India. It is found in hotter provinces in India. Many of the form of this species are naturalized in India and semi-arid regions and dry westward valleys of the Himalaya. It is cultivated throughout India in many varieties some of which run wild as on the coast of Mumbai, Gujarat and South India.

Medicinal Uses

It used as bitter, purgative, stomachic, alternative, aphrodisiac, anthelmintic, emmenagogue, cathartic and blood purifier. Leaf pulp used to treat liver disorder, rheumatism, skin disorders, intestinal worms. It is used for evacuation relief in the presence of anal fissures after voctonal operations. Fresh aloe gel is well known for its domestic medicinal value. For this reason, *Aloe vera* is also called burn, first-aid or medicine plant.

Fresh juice of the leaves is cathartic and cooling. It is used in eye troubles and spleen and liver ailments. Aloin is also used in anti-obesity preparations. Aloe gel and sometimes the drug aloe are used as moisturizer, emollient or would-healer in various cosmetic and pharmaceutical formulations. Extracts of aloe or aloin are used in sunscreens, X-ray burns, dermatitis, cutaneous leishmaniasis and other cosmetic preparations. As a food, aloe extracts are used as a flavouring ingredient primarily in alcoholic and non-alcoholic beverages and in candy to impart a bitter note. It is dissolved in spirit it is used as a hair-dye to stimulate hair growth. The juice of the leaves mixed with a little opium and applied to the forehead relieves headache.

A plaster of the leaf juice is said to be the folk remedy for tumors, used also for condylomata, warts, and other abnormal skin growths, and for cancer or tumors of the tip, anus, breast, Larynx, liver, nose, prepuce, stomach, uterus etc.

Chemical Constituents

Aloe contains cathartic anthraglycoides as its active principles; these are mostly C-glucosides, notably barrvaloin, which is a glucoside of aloe emodin. The concentrations of these glucosides vary with the types of aloe ranging from 4.5 to 25% of aloin. Other constituents present include aloesin and its agycone aloesone (a chromone), free anthraquinones (*e.g.* aloe-emodin) and resins.

Soil and Climate

Aloe is a hardy plant and grows on a variety of soils. It does well in the sandy, sandy coastal to loamy soils of the plains with a pH of upto 8.5. However, water-logged conditions and problematic soils do not suit its cultivation. The plant has a wide adaptability and can be seen growing throughout the length and breadth of the country. It is well growing in warm, humid or dry climate with 150-200 cm to 35-40 cm rainfall annual.

Improved Varieties

In trials conducted at the National Bureau of Plant Genetic Resources, Delhi, the Aloe accessions IC 111271, 1C 111280, IC 111269, IC 111273, IC 111279 AND IC 111267 are reported to contain high aloin contents; and the accessions, IC 111267, IC 1112666, IC 111280, IC 111272 and IC 111277, high contents.

AGRO-TECHNIQUE

Land Preparation

Before cultivation, the land should be ploughed twice and the field should be cleaned thoroughly of weeds. If required, small canals may be prepared for drainage. About 25 t/ha of cowdung manure is also added. If the land is already been under some tree crops the Land Preparation will consists of only digging the small pits for planting the suckers.

Mode of Propagation

it can be propagated by root-suckers or rhizome-cuttings.

Nursery Raising

The aloe plant can be raised in the nursery from the seeds, suckers and stem cuttings. By seeds, however does not ensure the progeny having the same qualities as that of parent plants since they have a tendency to hybridize under natural conditions. Cutting having at least two nodes, from mature stems are planted in well prepared nursery beds which are filled with more sand and a little compost. The beds are kept moist. After some weeks the germination starts. The germinated cuttings are transferred to seedling bed. Nursery beds are partially shaded. The seedling should be protected from stray animals, harmful insects and certain fungi.

Transplanting

Indian Aloe can be planted throughout the year and the farmer can choose time when he is least occupied in the farm. However, from the physiological and growth habit point of view beginning of rainy season (May-June) or at the end of rainy season (August-September) is the best time to plant aloe suckers. When germinated cutting are 3-5 leaf stage they are planted out into the field. The plants are planted at a spacing of 60 × 30 cm or 60 × 45 cm. About 20-25 thousands of seedling are required for one hectare. About 15-18 cm long root-suckers or rhizome-cuttings are planted in such a way that two-thirds portion of the root-sucker or rhizome-cutting should be under the ground.

Fertilizer and Manure

It is a newly domesticated crop and its full production technology including manorial requirement is yet to be worked out. And adhoc basis application of a mixture of 150 kg/ha of nitrogen (N), potassium (K) and phosphorus (P) is recommended. The fertilizers are applied in the soil near the root-system, after the plants are established.

Irrigation

Soon after planting, the land should be irrigated. During the crop period, Irrigation must be provided according to the moisture content of the soil. Generally, 4 to 5 Irrigations per year are sufficient. However, water should not be allowed to stagnate near the plant.

Interculture

Weeding may be required periodically depending on the intensity of the weed growth. The best is to pluck or cut the weeds and spread them as mulch between the lines of aloe. The plant may require some kind of pruning and clearing to guide its growth into more leaf production. Weeding may be done twice a year and the land should be kept weed-free.

Pests and Diseases

Normally, the plant is not affected by any pests or diseases of a serious nature however, overwatering causes rotting. So the soil mixture in the field should have good drainage facility. But recently, the leaf-spot disease caused by *Altarnaria alternata* and *Fusarium solani* has been reported from India.

Harvesting and Storage

The plants are harvested 8-12 months after planting. While harvesting, the plants can be removed manually or with the help of a tractor-drawn disc harrow or cultivator. The broken rhizome part left in the soil throws out new sprouts in spring for raising the succeeding crop.

Production and Yield

An aloe plantation gives a commercial yield from the second year up to the age of five years, after which it needs replanting. The yield of the crop on a fresh-weight basis will be around 15-20 Tonnes/ha.

Marketing and Trade

The whole plant, juice extracted from the leaves and formulations are sold in the market. This herb market is growing day by day due to their demand in the global cosmetic market. Nearly 95-98%

supply of this herb has been fulfilled by wild resource (forest) of Maharashtra and Tamil Nadu and Rajasthan. According to estimation, the consumption of aloe is nearly 500-1000 tonnes annually by various pharmaceutical industries. On the other hand, the annual requirement *Aloe vera* only in Kerala is nearly 5120 Mts in 1995 (*Source*: AFC study on Medicinal Plants farm Project, 1995). In 1999-2000, the total annual consumption of Aloe leaves in Natural Remedies (P) Ltd. is nearly 15 tonnes. The current market price of fresh leaves of aloe is Rs. 2-3 per Kg.

Economics of Cultivation

The market of Aloe and their products are highly volatile. The economics (per ha) worked out here are subject to fluctuations, depending upon time and place.

Expenditure (in Rs.)	—	3,500.00
Returns (in Rs.)	—	20,000.00
Net Profit* (in Rs.)	**—**	**16,500.00**

*   Estimation of the net profit is analysed on the basis of grower/collectors price. The grower/collector price is always 25-40% less than wholesale market price due to various stakeholders *viz.* Agent, middle man, Brokers, Commission agents.

INDIAN LONG PEPPER (*Piper longum* L.)

Vernacular Names

San.–Pippali, Magadh, Kana; *Hin.*–Pipli; *Beng.*–Piplamor; *Guj.*–Pipli, Piplamul; *Kan.*–Pippali; *Mal.*–Thippali; *Mar.*–Pimpli; *Tam.*–Thippili, Pippili; *Tel.*–Pippallu; *Eng.*–Indian long pepper.

Introduction

Long pepper is obtained from the unripe spikes, belonging to the family Piperaceae. The spikes are sweetish, cooling, pungent due to presence of piperine and piplatin alkaloids. On steam distillation, it yields an essential oil with spicy odour resembling that of pepper and ginger oil.

Botanical Description

A slender, climbing, undershrub, creeping and rooting below. The young shoots are downy; the leaves are 5-9 cm long, 5 cm wide, ovate, cordate with broad rounded lobes at the base, sub-acute, entire, glabrous. The plant bears unisexual flowers in solitary, erect spikes during or just after the rainy season. The berries are ovoid, yellowish orange, sunk in thick rachis.

Geographical Distribution

Long pepper is a native of the Indo-Malaya region. It is found growing wild in the tropical rainforests of India, Nepal, Indonesia, Malaysia, Sri Lanka, Timor and the Philippines. Indian long pepper is mostly derived from the wild parts, but is also grown in small area in the Khasi hills, the lower hills of West Bengal, Eastern Uttar Pradesh, Madhya Pradesh, Maharashtra, Kerala, Karnataka and Tamil Nadu. It occurs wild in the forests of Andhra Pradesh and in the Andaman and Nicobar Islands as well.

Medicinal Uses

The root is pungent, and has heating, stomachic, laxative, antihelmintic, carminative properties, improves the appetite; is useful in bronchitis, abdominal pains, diseases of the spleen, and tumours. The unripe fruit is sweetish, cooling and useful in biliousness. The ripe fruit is sweet, pungent, a stomachic, aphrodisiac, alternative, laxative, anti-diarroeic, antidysenteric, is useful in *vata* and *kapha*, asthma, bronchitis, abdominal complaints, fevers, leucoderma, urinary discharges, tumours, piles, diseases of the spleen, pains, inflammations, leprosy, insomnia, jaundice, hiccoughs, tubercular glands and reduces biliousness (in the Ayurvedic system of medicine).

The roots and fruits are used in palsy, gout and lumbago. The root has a bitter, hot and sharp taste. It is used as a carminative, a tonic to the liver, stomachic, emmenagogue, abortifacient, aphrodisiac, haemostatic, diuretic, digestive, and as a general tonic, useful in inflammation of the liver, pains in the joints, lumbago, snakebite, scorpion-sting and night-blindness (in the Unani system of medicine). In the Travancore region, an infusion of the root is prescribed after parturition, to help in the expulsion of the placenta. It appears to partake, in a minor degree, of the stimulant properties of the fruit.

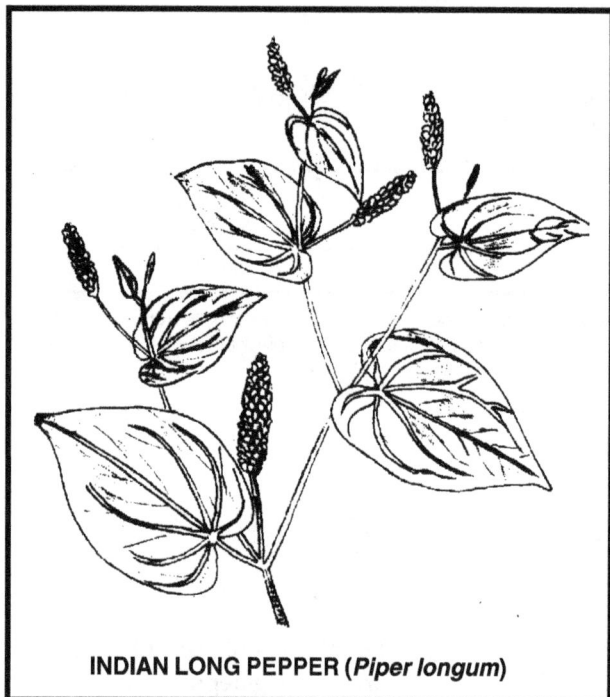

INDIAN LONG PEPPER (*Piper longum*)

Chemical Constituents

The spikes contain the alkaloids piperine (4-5%) and piplatin, piperolactum A, piperolactam B and piporadione. The roots contain the alkaloids, piper longuminine (0.2-0.25%) and piper longumine (0.02%) besides piperine. Besides, the dried spikes on steam distillation, yield 0.7% essential oil with a spicy odour resembling that of pepper and ginger oil.

Soil and Climate

It flourishes well in a rich, well-drained, loamy soil, limestone soil, laterite soils, rich in organic matter content with good moisture-holding, capacity are also suitable. It requires a hot, moist climate and an elevation between 330-3330 ft.

Improved Varieties

Gol Thippali (round ball type). Pipal Nonsori, Asali (true) and Suvali.

AGRO-TECHNIQUE

Land Preparation

The land should be ploughed 2-3 times and leveled properly. Then, raised beds of size 3 m × 2.5 m are prepared, in which pits are dug at a spacing of 60 m × 60 m the pits are filled with soil mixed with 100 g/pit of well-decomposed FYM or compost. In order to avoid water stagnation in the pits and beds, channels are made to drain excess rainwater.

Mode of Propagation

It can be propagated by seeds, suckers or cuttings.

Nursery Raising

The nursery can be raised during March and April. In the nursery, mealy bugs often attack the roots of the rooted cuttings. This can be avoided by mixing 10% DP with the potting mixture. Similarly, excessive Irrigation may lead to *Phytophthora*, wilt of rooted cuttings. Irrigation should be done on alternate days to avoid dampness and also to control the wilt disease.

Transplanting

The cuttings planted in nursery in March-April will be ready for planting in the main field by the end of May. The rooted cuttings or suckers are planted in the pits at the rate of 2/pit with the onset of the monsoon. The pits are gap-filled on month after planting.

Intercropping

Long pepper is planted as main intercrop with subabul, eucalyptus and coconut in different parts of the country.

Manure and Fertilizers

This crop needs heavy manure and fertilizers. The application of 20 tonnes farm yard manure (FYM) along with 50-60 kg nitrogen, 20-30 kg phosphorus and 70-75 kg potash/ha is required for optimum growth and yield of the crop. Of this, 50% nitrogen and the entire dose phosphorus and potash is given as a basal dose and the remaining nitrogen dose is given as a top-dressing.

Irrigation

The crop should be irrigated once a week if it is grown as a pure crop. In case the crop is grown as an intercrop with other crops, the Irrigation provided to the main crop is sufficient. The sprinkler system may be adopted for economizing Irrigation water. It is observed that summer Irrigation will induce continuous spike formation during the off-season.

Interculture

In the first year of planting, weeding is done when weed growth is noticed in the beds. After the application of FYM to the beds, earthing-up is done from the channels. Once the crop grows and covers the interspaces of the beds, it is difficult to do any Interculture operations.

Pests and Diseases

Mealy bugs infest the roots of plants and stunted the growth. Application of neem kernel extract at 0.25% concentration will control this insect. During the monsoon season, the fungus *Collitotrichum spp.* causes leaf, vine rotting, necrotic spots, blight on leaves during the warmer months. A spray of Bordeaux mixture at 1% concentration during May and 2 or 3 sprays subsequently, during the rainy season, is the suggested control measure.

Harvesting and Storage

The spikes are picked (in January) from the vines six months after planting. The spikes will be ready for harvest 2 months after their formation on the plants. The spikes should be picked when they are blackish-green and most pungent. If the spikes are not picked at this stage, there is a greater loss of pungency. The harvested spikes are dried in the sun for 4 to 5 days until they are perfectly dry. Besides the spike, the thick parts of the stem and roots which have medicinal value may also be harvested from 18 months after planting. While harvesting, the stems are cut close to the ground, the roots dug up, cleaned and heaped in the shade for a day, after which they are cut into 2.5 to 5 cm long pieces

The dried thicker parts of the stem and roots are called piplamool. There are three grades of piplamool. Grade 1 with thick roots and underground stem, it fetches a higher price than Grade II and III which consists of either roots, stems or fragments.

Production and Yield

The green to dry spike yield is around 350-400 kg/ha. The yield increases thereafter up to 3 years and it will be about 950-1000 kg/ha during the third year. After three years, the productivity of the vines decrease and they should be replanted. The average yield of roots is 400-500 kg/ha.

Marketing and Trade

The whole plant and seeds are sold in the market. This herb market is growing day by day due to their demand in the global herbal market. Nearly 60% supply of this herb has been fulfilled by cultivated resource of Maharashtra and Andhra Pradesh. According to estimation, the consumption of pepper is nearly 500-600 tonnes annually by various pharmaceutical industries. On the other hand, the annual requirement pepper only in Kerala is nearly 1000 Mts in 1995 (*Source*: AFC study on Medicinal Plants farm Project, 1995). An estimation, the demand of various parts of pepper (*viz*. infloresence, stem, rhizome) by herbal industries and ayurvedic drug producers only in Maharashtra is nearly 3000 tonnes annually. The current market price of long pepper is Rs. 80-100 per kg.

Economics of Cultivation

The market of long pepper and their products are highly volatile. The economics (per ha) worked out here are subject to fluctuations, depending upon time and place. (Economic life of plant is 6 years).

Expenditure (in Rs.)	—	31000.00 (in Ist year)	11,000.00 (in IInd year)	11,000.00 (in IIIrdyear)
Returns (in Rs.)	—	32000.00 (in Ist year)	48,000.00 (in IInd year)	56,000.00 (in IIIrdyear)
Net Profit* (in Rs.)	**—**	**1,000.00 (in Ist year)**	**37,000.00 (in IInd year)**	**45,000.00 (in IIIrd year)**

* Estimation of the net profit is analysed on the basis of grower/collectors price. The grower/collector price is always 25-40% less than wholesale market price due to various stakeholders *viz*. Agent, middle man, Brokers, Commission agents.

INDIAN TEJPAT (*Cinnamomum tamala* Nees & Eberm.)

Vernacular Names

San.–Tamalapatram; *Hin.*–Tejpat; *Kan.*–Kadu lavanga patte; *Mal.*–Paccila; *Tam.*–Tamalapattirii; *Tel.*–Talisa patri, Adavi-lavanga patri; *Eng.*–Indian cassia lignea, Cassia cinnamon.

Introduction

The bark is traded locally as Indian cassia bark or Indian Cassia Lignea, is collected from the trees growing at the foot of Sikkim Himalayas. There are regular plantations of in Khasi and Jaintia Hills, Garo Hills, Mikir Hills, manipur and Arunachal . Fresh and dried leaves, as tejpat leaves, are commonly sold as spice in markets and an oil is obtained by hydro or steam distillations of fresh, wilted or dried leaves. In Kashmir, leaves are used as a substitute for pan or betel leaves. The *C. tamala* bark is coarser than *C. verum* bark, although processing the ture cinnamom odour, and is a used as a substitute and adulterant for true cinnamon (*Cinnamomum cassia*).

Botanical Description

Two types of *C. tamala* are known; one whose leaves yield oil high in eugenol, the other in cinnamaldehyde. **(i).** *Cinnamomum tamala* (eugenol type) is an evergreen primarily forest tree generally below 8 m, with central trunk to 45-50 cm DBH, and mucilaginous bark. The leaves are large, dark green, generally lanceolate, acute, three ribbed, on short, stout petiole. The flowers yellowish, purplish drupe containing one small brown seeds. The tree is very long-lived, when only leaves are harvested, cultivated trees over 100 years is known in northern India. The eugenol type occurs generally over natural range of the species in India and wild trees make a substantial contribution to the total of harvested leaves and bark. The tree is also widely cultivated in large and small plantations to provide dried leaves as spice, but leaf-oil production is less common. **(ii).** *Cinnamomum tamala* (cinnamic type) is generally slightly smaller than the eugenol type, to 7.5 m, with the central trunk 30-40 cm DBH when mature. The leaves are opposite, glabrous, green to dark green, three ribbed, ovate-oblong, acute, on short petiole. Main flowering is in May and the fruits ripen in June-July.

Geographical Distribution

It native to the subtropical Himalayas from Kashmir to Assam and Bangladash. It occurs mostly in the north-eastern region of India, distributed largely in the Himalayas at 250 to 700 m above sea levels mainly in the foot hills of Sikkim Himalayas. There are regular plantations of in Khasi and Jaintia Hills, Garo Hills, Mikir Hills, Manipur and Arunachal Pradesh. The cinnamic type is more restricted in its range, although locally common in a specific area. It is especially plentiful in hills around Nainital, where it is also cultivated on small farms.

Medicinal Uses

Leaves are carminative, and used as in colic, diarrhea and rheumatism. It is significantly reduce blood sugar level and help in release or manufacture of more insulin. Tejpat is mostly used as a flavouring agent, in some vegetarian agent and most non-vegetarian Indian preparations including mughlai dishes. Besides flavouring, it is also used as a clarifier with myrobolans-the plant chemicals used for tanning and dyeing leather.

Chemical Constituents

Leaves yield about 2% essential oil. Its contains mainly 80-85% eugenol while

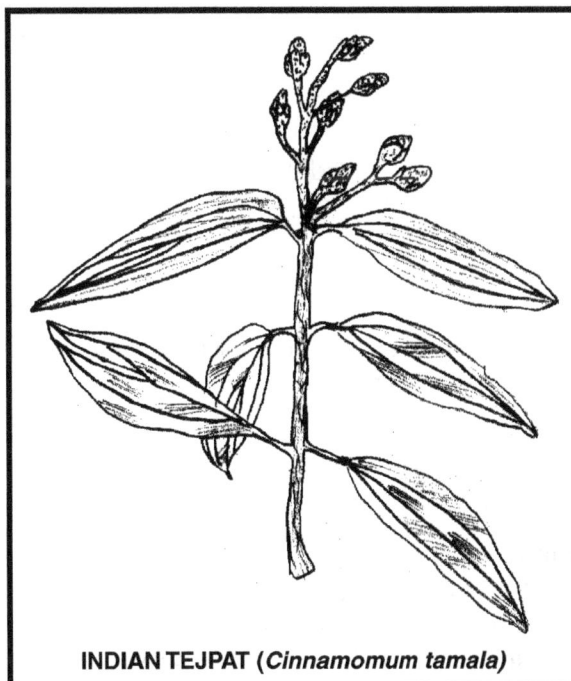

INDIAN TEJPAT (*Cinnamomum tamala*)

imported oil contains 75-80%. The essential oil from bark is pale yellow, and contains 70-85% cinnamic aldehyde. The trade is in cassia has declined appreciably with the advent of synthetic cinnamic aldehyde. The oil is often adulterant with cheap terpenes.

Soil and Climate

Suitable rich loam to poor laterite; warm humid climate with plenty of sunshine and rainfall (200-250 cm/year).

AGRO-TECHNIQUE

Land Preparation

The land is prepared well and brought to a fine tilth with the help of tractors and organic manure at the rate of 25 t/ha, to begin with, is mixed with the soil.

Mode of Propagation

It can be propagated by seeds or vegetatively *viz.* cuttings, layering, division of rootstocks.

Nursery Raising

Freshly harvested seeds was sown (June-July) for highest germination (94%). The seeds started germination after 7 days and completed in 20-25 days. Seeds should be sown thickly in rows 20 cm apart and covered with 2-3 cm of soil.

Transplanting

Nearly 4 months old seedlings may be transplanted into baskets, and planted out 4-5 months later. Plants are cut to within 15 cm of the ground, suitable sections of rootstocks removed with adhering soil and planted out immediately into prepared pit. The great advantage of this method is that stems are ready for cutting in 12-18 months compared to 2.5-3.0 years for seedlings. In India, a common spacing is 3 × 3 m with several seeds sown per site.

Fertilizers and Manures

The tejpat crop does not require heavy doses of manure and fertilizers. Organic manures @ 10-15 tonnes/ha should be applied along with the chemical fertilizers. Ammonium sulphate @ 60-80 kg/ha, when added to the soil shows positive results. The phosphorous fertilizer should be added to the field before sowing @ 40-50 kg/ha. Nearly 50 Kg of nitrogen should be added as a seed dose to the crop 1-2 year after sowing.

Irrigation

Seedlings are planted out just prior to the rainy season with supplementary watering, or just after the rains commence.

Interculture

Weeding three or four times annually is necessary during the first two years, thereafter twice per year should be sufficient.

Harvesting and Storage

The trees are not usually harvested for bark and the first harvest of leaves is at around ten years of age and continues annually until the tree dies up to 100 years. The leaves are harvested after the main monsoon rains have ceased (October-March) or during dry periods, as rain depresses oil content and aroma of leaves and thus their value as spice. The leaves are normally dried for 3-4 days before distilling and generally sold to a still owner for distillation; thus oil produced is very variable but acceptable for local purposes.

Production and Yield

Although exact statistics are not available, the total area covered is approximately 600 acres. A tree yield 10-20 kg dry leaves per year.

Marketing and Trade

The unsustainable ways of harvesting and unrestricted marketing have to the reduction in population of tejpat, high demand medicinal plants leading to sudden escalation in their prices in the market. The current market price of leaves is nearly Rs. 18-20 per kg.

Economics of Cultivation

The market of tejpat and their products are highly volatile. The economics (per ha) worked out here are subject to fluctuations, depending upon time and place. On the basis of the prevailing market conditions with dry leaves a net profit of about Rs. 20000/ha in a ten-year-old crop, can be obtained will be increased in subsequent years (Plant life span is 100 years).

IPECACUANHA (*Cephaelis ipecacuanha* (Brot.) A. Rich.)
Syn. *Psychotria ipecacuanha* Stokes.

Vernacular Names

Eng.–Ipecac.

Introduction

The name of the plant is the Portuguese form of the native world, *i-pe-kaa-guene,* which is said to mean ' road-side sick-making plant'. Ipecac is one of the oldest medicinal plants known to man. It belongs to family Rubiaceae. It is said to have been used by South American Indians before the beginning of recorded history. Dr. Adrian helveties of Paris is credited with the Introduction of ipecac to Europe. The early Brazilians appear to have applied the term ipecacunha to a number of roots with common emetic properties, until a Portugese scientist, Berhardino gomes, collected the authentic plants. These plants were later described by Flex de Aucllor Brotero, in 1803, who named it *Callicocca ipecacuanha.*

A Portuguese priest named Trislasu, traveling in Brazil in the later years of the 15th century, came to know of the curative properties of ipecac from the natives, which they called 'Igpecaya'. By the end of the 17th century a regular trade in this drug commenced, mainly through the efforts of the Dutch. Later, for fear of depletion of its supply from wild sources, many European countries became interested in its cultivation.

Botanical Description

A half-shrubby perennial 50-75 cm high, with a slender, cylindrical, corky, sinuate, underground stem (or false rhizome) from which extend several horizontal rots. When mature, the roots have a brick-red to dark-brown bark, smooth or more or less prominently transversely ridged or ribbed. The leaves are opposite near the top of plant, alternate below, dark-green on the upper surface and light-green below, rough, hairy with short, downy petioles. The flowers are white, funnel-shaped, borne in dense, globular clusters of 8-20 on a purplish-green, purple in colour.

Geographical Distribution

This plant is a native of the humid forests of Bolivia and Brazil. It has been grown on a commercial scale in Malaysia, and to a small extent in Burma, Brazil and India. It is cultivated in Mungpoo, Rongo, Munsong and Latpanchor (Darjeeling district) as a subsidiary crop in chinchona plantations and in the Nilgiris, especially near Kallar. It has been successfully raised at the Rungbee Chinchona Plantation in Sikkim. Jorhat and Srimargal in Assam, Sinduhans and Jalpaiguri in West Bengal and Bilaspur in Orissa are also considered to be suitable localities for its cultivation.

Medicinal Uses

The Pharmacopoeias contain a very large number of preparations of Ipecacuanha, most of which are standardized. Ipecac is an important drug being at once an emetic, a diaphoretic, one of the convenient known expetorants, and an excellent remedy for the treatment of amoebic dysentery. In large doses, its roots is emetic; in small doses, diaphoretic and expectorant, and in still smaller, stimulating to the stomach, intestines and liver, exciting appetite and facilitating digestion. It is also useful in the treatment of bronchitis and laryngitis. It is used in the form of syrup, powder, tinctures and lozenges. It has become one of the most prominent chemicals used in cancer research.

A mixture of ipecac powder and opium was prescribed as a sudontic at the onset of influenza, but this is no longer official. The syrup of ipeaac is often given in the early stages of acute bronchitis. It has been used successfully in the treatment of bilharziasis, guinea-worms and oriental sores. The chief adulterant of the drug is the aerial stem of the plant. It can be distinguished from the root by the pith composed of cells with lignified walls and by the surface scars.

IPECACUANHA (*Cephaelis ipecacuanha*)

Chemical Constituents

The root rich in non-phenolic compounds such as emetine in its stem (0.21-0.54%), leaves (0.58-1.22%) and roots (0.61-1.35%). Five major alkaloids have been isolated from the drug. Of them, emetine and cephaeline are the more important. Cephaeline is a phenolic alkaloid, too toxic for use, and is usually converted into emetine by methyline. Small amounts of psychotrine, methyl-psychotrine, emetamine, and traces of ipecamine and hydroipecamine have also been isolated.

Soil and Climate

It thrives best in forest areas in sandy loam soils, slightly acidic in nature (pH 6-6.5), rich in potash, magnesium and lime. It thrives well in a tropical mild, humid climate. An annual rainfall of 1200-300 cm well distributed throughout the year and a temperature 10⁰C-38⁰C, are essential for its successful cultivation. The plant is sensitive to frost, but can be grown successfully at altitudes of 350-850 m.

AGRO-TECHNIQUE

Land Preparation

Before cultivation, the land should be ploughed twice and the field should be cleaned thoroughly of weeds. If required, small canals may be prepared for drainage. About 25 t/ha of cow-dung manure is also added.

Mode of Propagation

It can be propagated by seeds or vegetatively by stem cutting.

Nursery Raising

When the seedling are to be raised for Transplanting, they should to be sown in well-prepared, raised nursery beds (1x3 m size) and are mixed with well-rotten leaf compost and sand. Seeds are small (1 g contains 50-55 seeds) and drilled or broadcasted in the beds and watered regularly. Ideal time for seed sowing is January to March, depending upon the local climatic conditions. It has been observed that treatment of seeds with lime water for 48 hours or with hydrogen peroxide for 96 hours improves germination. The germination of seeds is slow and erratic, it takes between 3 and 5 months. The small (up to 4 cm height), seedlings are transferred to another nursery-bed at a spacing of 5x5 cm. this is done during August-September, when the monsoon rains are over. The young plants are allowed to remain in this nursery till March-April of the following year.

Transplanting

Seedling are planted in field during June-July, after they are 2-2.5 months old, at a distance of 10 × 10 cm or 15 × 15 cm and allowed to remain there for the next 3 years, till they are dug out upon maturity. A bed of 15 × 1.5 m, proved to be the optimum for better survival, and higher yield with total alkaloids. The seed-raised plants are slow in growth. Growing the plants under the shade of plant like *Teprosia candida* and *Crotalaria anagryoides* and also inter-cropping with *Cinchona ledgeriana* have given encouraging results.

Fertilizers and Manures

The use of organic manures, leaf-mould and compost has been recommended to increase the quantity of nutrients in the soil and to improve drainage. There is not much information available on the nutritional requirement of ipecac. In an experiment conducted at Mungapoo, the highest root yield has been reported by the application of castor-cake @ 600 kg/; whereas, superphosphate in combination with nitrogenous fertilizers, augmented the formation of non-phenolic alkaloids. Considerable increase in the yield has been reported in West Bengal, when ammonimum sulphate was used in soil containing humus; the oil cake @ 200 g/bed and 100 g/bed super phosphate.

Irrigation

The nursery-beds require very careful Irrigation. Plants should be watered frequently in the initial stages. The frequency is gradually reduced and when the crop is well established, Irrigation once in about 15 days is sufficient during the dry period.

Interculture

In order to maintain the satisfactory development of roots about 2 weedings are necessary during the monsoon and one hoeing at the end of the growing season. Mulchign the nursery beds with coloured polythene sheets (25 gauge in thickness) has been reported to reduce the cost of weeding.

Pests and Diseases

The plants are affected by damping-off fungi like *Rhizoctonia spp.* In the nursery. As a control measure, the seeds should be treated with a suitable seed-dressing of fungicide before sowing. *Fusarium moniliforme* is reported to cause wilt disease, for which suitable control measures like field sanitation and drenching the plants with fungicidal solution like copper oxychloride (3 g/l) may be taken up. The other diseases reported on this crop are leaf-blight and rot caused by the fungi, *Alternaria alternata* and *Fusarium* spp., respectively.

Harvesting and Storage

Generally plant survives for 5-6 years. The plant starts flowering from the second year and fruits ripen 3 months later or so. The maximum production of root is in between third and fourth year. Therefore, harvesting should be done after the plant is 3 years old. It is reported that roots dug out when the plants have shed their leaves, are richer in total content of non-phenolic alkaloids. A light Irrigation should be given in advance to facilitate easy digging of roots. The roots are freed from the adhering soil, washed if necessary and thoroughly air-dried till they become brittle and are usually packed in gunny bags.

Production and Yield

At Rongo and Mungpoo in Darjeeling *c* 19-20 tonnes roots are produced during 1966-67. During 1982, about 80 ha were under ipecac cultivation at Mungpoo. The total area under this crop in India is about 100 ha. A 3-year old crop (4 years from seeds) yields the maximum amount of roots and also contains high alkaloid content. The Indian Pharmacopoeia requires that the roots contain not less than 2% of the total alkaloids, of which not less than 50% should be emetine in content. A 3-year-old plantation, on an average, may yield up to 6.5 t/ha of fresh roots, when managed properly.

ISABGOL (*Plantago ovata* Forsk)
(Syn. *P. isphagula* Roxb.)

Vernacular Names

San.–Snigdhabijah, Snigdhajirakah, Ishad gola; *Hin.*–Isabgol, Isabgul; *Kan.*–Isagfolu, Visamagolu; *Beng.*–Eshopgol; *Mar.*–Isabgola; *Guj.*–Isaghol, Ghoda jeera; *Mal.*–Snigdhajirakam, Ispaghal, Karkatasringi; *Tam.*–Iskolvirai, Isphogol; *Tel.*–Isapgalu vittulu; *Eng.*–Blond psyllium, Spogel seeds; *Trade*–Isabgol.

Introduction

Indian plantago is important for its seeds and husks which have been used in indigenous medicine for many centuries. Because of the superior quality and yield of its seeds, it has replaced the French psyllium, *P. psyllium* Linn, belongs to family Plantaginaceae, in the world market. In fact, France now depends on India for its requirements.

It derives its name from two Persian words, 'asp' and 'ghol' meaning a horse-ear, referring to its characteristic boat-shaped seeds. Plantago is from latin, meaning 'sole of the foot' and refers to the shape of the leaf, Psyllium is from the Greek, meaning 'flea,' in reference to the colour, size and shape of seeds (flea-seed) and 'ovata' refers to the ovate shape of the leaves. The husk of the seed is the economic part and is separated by a physical process.

Botanical Description

A stemless or short-stemmed, highly cross-pollinated, annual herb which attains a height of 30 to 40 cm, has alternate leaves, which clasp the stem, strap-like, recurved, tapering to a point, three-nerved, entire or toothed, coated with fine hairs. The flowers are white, minute, four-parted. The capsule is ovate, the top half lifting up when ripe, releasing the smooth, dull, ovate seeds, pinkish-grey brown or pinkish-white with a brown streak on the convex surface. The seeds are covered with a translucent membrane, known as the husk, which is odourless and tasteless. The husked seeds are dark-red and hard.

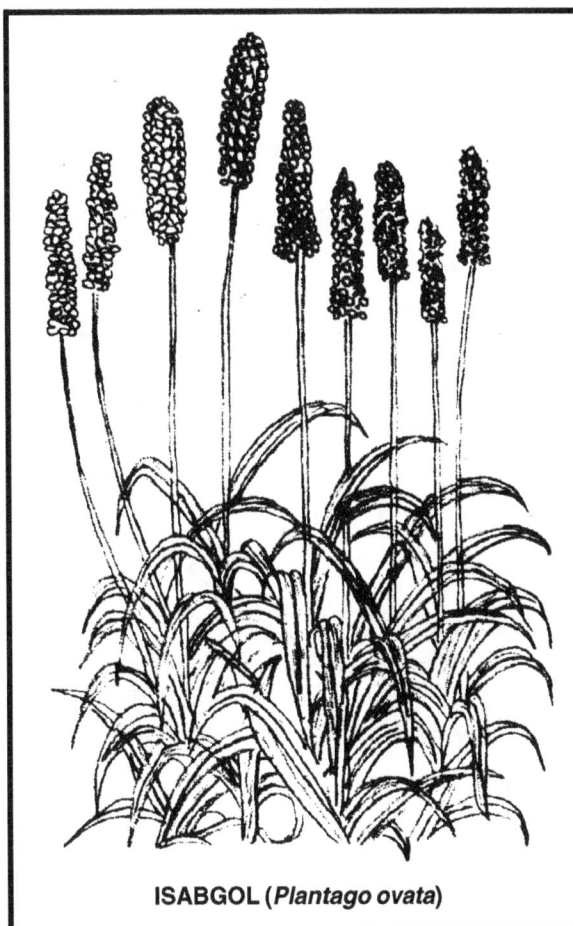

ISABGOL (*Plantago ovata*)

Geographical Distribution

The plant is indigenous to Persia and West Asia, extending up to the Sutlej, Sind and West Pakistan. The plant has also acclimatized well in Mexico and the Mediterranean regions. It has now been introduced in Southern Rajasthan, Punjab and, to a very small extent, in Maharashtra and Uttar Pradesh. The quality of husk and yield is comparable to the crop grown in Gujarat. In view of its sustained demand, there is still scope to increase the area and production and to intensify its cultivation.

Medicinal Uses

The husk has the property of absorbing and retaining water and, therefore, it works as an anti-diarrhoeal drug. It is beneficial in chronic dysenteries of amoebic and bacillary origin. It is also used for treating constipation and intestinal disorders because it works as calorie-free fibre food, promoting regular bowel movement. It is reported to have no adverse side-effects. The seed has also cooling and demulcent effects and is used in Ayurvedic, Unani and Allopathic system of medicine. The seeds are sweet, astringent, refrigerant, emollient, mucilaginous, diuretic, laxative, anti-inflammatory, anti-dysenteric, expectorant, aphrodisiac, roborant and tonic. The seeds and husks are used to cure inflammations of the mucous membranes of gastro-intestinal and genito-urinary tracts, duodenal ulcers, gonorrhea and piles. It can also be used as a cervical dilator for the termination of pregnancy. In addition to these Medicinal Uses, it has a place in dyeing, calico printing, in the ice-cream industry as a stabilizer, also in confectionery and cosmetic industries. The seed without the husk, which contain about 17-19% protein, is used as cattle-feed.

Chemical Constituents

The seed-husk contains a colloidal mucilage (30%), mainly consisting of dxylose, arabinose, galacturonic acid with rhamnose and galactose, etc. The seed (embryo) yield 14.7% of a linoleic acid-rich oil and small amounts of glycoside acubin and tannin.

Soil and Climate

It is an irrigated crop which grows well on light soils. Heavy soils and those with poor drainage are not conducive to the good growth of this crop. A silty-loam soil, having a soil (pH 4.7-7.7) with low nitrogen and moisture content, is reported to be ideal for the better growth. It thrives in warm temperate regions. It requires cool, dry weather and is sown during the winter months.

Improved Varieties

Gujarat Isabgol-1 (G1-1), Gujarat Isabgol-2 (G1-2), Haryana Isabgol (HI-5), Niharika, Jawahar Isabgol.

AGRO-TECHNIQUE

Land Preparation

The field must be free of weeds and clods and should have a fine tilth for good germination. About 10-15 t of FYM/ha is incorporated into the soil at the time of the last ploughing. The field should be divided into suitable plots of convenient size.

Mode of Propagation

It can be propagated by seeds.

Sowing

Sowing should be done in month of November-December. Fresh seeds from preceding crop should be shown for getting higher germination and optimum population. Before sowing, seeds pretreated with any mercurial seed-dresser (thiram) @ 3 g/kg seed, to protect the seedlings from a possible attack of damping-off. The seeds are small and light. Hence, the seed must be mixed with a sufficient quantity of fine sand or sieved FYM. The seeds are broadcast, because sowing in lines 30 cm apart. After broadcasting, the seeds are swept over lightly with a broom to cover them with some soil. The broom, however, should be swept in one direction only, to avoid deep burial of the seed and for uniform germination. The sowing should immediately be followed by Irrigation. Germination begins 4 days after sowing. About 5-7 kg of seeds are required for one hectare land.

Manure and Fertilizers

It does not require the application of heavy doses of fertilizers. A fertilizer dose consisting of 50 kg nitrogen, 25 kg phosphorus and 30 kg potash/ha, has given the maximum seed yield. The full dose of phosphorus and potash along with half of the nitrogen is given as a basal dose at the time of sowing itself, and the second split dose of nitrogen is applied as a top-dressing after one month of sowing. For other areas, a basal dose of 25 kg nitrogen and 25 kg phosphorus/ha is recommended. A top-dressing of 20-25 kg nitrogen/ha is given 35 days after sowing.

Irrigation

Immediately after sowing, a light Irrigation is essential. First, Irrigation should be given with a light flow of water. If the germination is poor, a second Irrigation should be given. Later on, Irrigations are given as and when required. The last Irrigation should be given at the time when the maximum number of spikes have reached the milk-stage. The crop requires totally 6-7 Irrigations for good productivity in medium sandy soils.

Interculture

The first weeding is done after about 20-25 days of sowing. Since the crop is sown by broadcasting, hand-weeding is a very costly operation. Ordinarily, 2-3 weedings within two months of sowing will help to keep the weeds under control.

Pests and Diseases

Sometimes, white grubs and termites damage the crop by cutting off the root. These can be controlled by broadcasting phorate 10G @ 10 kg/ha before the last ploughing. Aphids also attack this crop, and can be controlled by spraying 0.2% Dimethoarate. This crop is affected by a number of diseases but major one are leaf blight, damping-off disease and Wilt diseases. These are controlled by Bordeaux mixture @ 600-800 lit/ha or Copper oxychloride, Dithane M-45, Dithane Z-78 @ 2.0-2.5 g/l.

Increased yield can be obtained in this crop by spraying cycocel (CCC) at 25 mg/l twice at the seedling stage and just before the heading stage.

Harvesting and Storage

Blooming begins two months after sowing and the crop is ready for harvest in February-March (110-130 days after sowing). When mature, the crop turns yellowish in colour and the spikes turn brown in colour. The seeds are shed when the spikes are pressed even slightly. At the time of harvest, the atmosphere must be dry and there should be no moisture on the plant, otherwise harvesting will lead to considerable seed-shattering. The harvested plants are transported to the threshing yard, bundled in large cloth pieces and spread out. After 2 days, they are threshed with the help of a tractor or bullock. The plants are threshed and winnowed and the seeds sieved until they are clean. The seeds may be marketed whole or the husk may be sold separately. To remove the husk, the cleaned seeds are passed 6 to 7 times through stone-grinders, sieved and screened through several grades of mesh to sort according to fineness.

Production and Yield

Nearly 50,000 ha area is under cultivation and 45,000 to 50,000 tonnes seeds is produced every year in India. At present, the Isabgol crop has acquired the place of the 'dollar earner' crop of north Gujarat, where it is being cultivated on about 50000 ha with a production of 40000 tonnes. Experimental cultivation at Bangalore has been quite successful. Therefore, if Indian is to retain its monopoly in the production of this important foreign exchange earning commodity, there is an urgent need to evolve high-yielding strains and to examine suitable locations for growing this crop in other states of the country. It yields 9-10 quintals seeds/ha.

The processing is done in mills. About 25% of husk is obtained by weight basis. Husk is removed by grinding pressure and separated out by fans and sieve.

Marketing and Trade

About 75-80% of the total annual produce of north Gujarat is exported, which brings crores of rupees of foreign exchange annually. India continues to hold a monopoly in its production and trade in the world. Though India enjoys the monopoly in production and export of Isabgol, hardly 50% of the requirement of the USA is being met. The husk and peel are largely exported to USA, West Germany, UK and France. The total quantity of exported Isabgol husk to various countries is nearly 19277.66 tonnes and earn foeign revenue Rs. 200 Crores in 2000-2001 while nearly 1067.3 tonnes Isabgol seeds is also exported in 2002-2003 and earn revenue nearly Rs. 592.75 lakhs. The current market price of Isabgol seeds is Rs. 50-60 per kg.

Economics of Cultivation

The market of Isabgol and their products are highly volatile. The economics (per ha) worked out here are subject to fluctuations, depending upon time and place.

Expenditure (in Rs.)	—	7,500.00
Returns (in Rs.)	—	42,500.00
Net Profit* (in Rs.)	—	**35,000.00**

*   Estimation of the net profit is analysed on the basis of grower/collectors price. The grower/collector price is always 25-40% less than wholesale market price due to various stakeholders *viz.* Agent, middle man, Brokers, Commission agents.

JATAMANSI (*Nardostachys grandiflora* DC.)
Syn. *Nardostachys jatamansi* DC.

Vernacular Names

Hin.–Jatamansi; *Hima.*–Balchharh, Machhi, Mansi, Jatamansi; *Eng.*–Spikenar; *Guj.*–Jatamasi; *Tam.*–Jatamansi; *Beng.*–Jatamanshi; *Tel.*–Jalamansi; *Mar.*–Batacharea. .

Introduction

Jatamansi has been used since immemorial. In traditional system of medicine rhizome are used as laxative, diuretic, stomachic and heart tonic. It is reported to be useful in epilepsy, hysteria and mental disorders. Jatamansi is indigenous drug and is found in wild. The Jatamansi constitutes the sources of an important pharmacopoeal drug. WWF, IUCN, IDRC and other agencies are declared podophyllum as Critically Endangered (CR) due to uncontrolled exploitation, indiscriminate harvesting, loss of natural habitat and illegally for trade. Due to this threats, the natural availability of jatamansi from wild sources become uncertain. It belongs to family Valerianaceae.

Botanical Description

An aromatic, erect, perennial herb, 10-60 cm high with long, stout woody rhizomatic rootstock covered with tail like brown fibres left over from the withered leaves towards the stem, while the root continues to penetrate deep in the soil. Leaves are radical in nature, long, narrow and the flowers are creamy white, often rosy or pale pink in appearance arising in terminal corymbose cymes. Fruit is small, covered with minute hairs.

Geographical Distribution

The herb is naturally found in rock crevices and moist shady places of alpine Himalayas from Himachal Pradesh to Sikkim and Bhutan, at altitudes between 3000-5000 m. Globally, it occurs from Afghanistan to southern West China and Myanmar.

Medicinal Uses

The rhizome is considered as antiseptic, sedative, anti epileptic, hypotensive, cardiac, stomachic, diuretic, laxative, tonic and antidepressant properties. An infusion of the rhizome is reported is to be useful in hysteria, hypochondriasis, nervous disorders, mouth ulcers epilepsy, depression as a carminative and a tonic in the form of extract, infusion and tincture. Its essential oil is used as a tonic and stimulant in certain medicinal preparations. Prevalently used as a hair and heart tonic.

Chemical Constituents

The hairy roots contain essential oil (0.5%) having jatamansone, jatamansic acid, lupeol, jatamansinone, jatamansinol, jatamansin, etc. Rhizome also contains alkaloid-actinidin.

JATAMANSI (*Nardostachys grandiflora*)

Soil and Climate

The plant will grow in a variety to soils, but thrives best in a moist, rich and rather heavy loam. Some drought resistant forms also occur in chalky and limestone soils. It is a temperate plant and prefers very cold winter and mild summer temperatures for its good growth. The plant is often found to flourish in damp and shady places.

AGRO-TECHNIQUE

Land Preparation

The land is prepared well and brought to a fine tilth by ploughing it several times. Later, the field is divided into beds of convenient sizes and FYM is added at the rate of 15-20 t/ha.

Mode of Propagation

It can be propagated by seeds or cutting of root-stocks or rhizomes.

Nursery Raising

The nursery is raised in July-August or April-May. At first, the nursery-raised seedlings are transplanted at a spacing of 18-20 cm in rows 30 cm apart. When the seedlings are 6-8 weeks old and 10-15 cm tall, they are transplanted to their permanent sites. About 0.5 kg seeds per ha are required for sowing.

Planting

If the crop is raised through division of root-stocks or rhizomes, they are spaced at 30 cm × 30 cm. If the divisions are planted very early in autumn before the onset of frost. When seedlings are raised for growing the crop, the seedlings are transplanted at 20 cm × 20 cm spacing. The field is irrigated immediately after planting.

Manures and Fertilizers

It responds well to the application of manure and fertilizers. FYM may be incorporated into the soil along with 50 kg/ha each of phosphorus and potash at the time of Land Preparation, while 150 kg/ha nitrogen is applied in 3 split doses at 4-month intervals.

Irrigation

This crop requires 2-3 Irrigations during the dry period from June to September.

Interculture

The interspaces are harrowed to keep the soil loose and weed-free. A low ridge of soil is usually raised along the crop row to promote the formation of large root-stocks. The land is kept weed-free through regular weeding.

Pests and Diseases

Valerian is a hardy crop. No serious pests and diseases have been reported to attack this crop.

Harvesting and Storage

Although the plants can be harvested after 2 year of planting, it is advisable to harvest 3-year-old plants, since the yield of the underground parts would be higher and so also their essential oil content. The plants can be harvested when they become dormant during November, exhibiting yellowing of leaves and withering of plants, the rhizomes and roots are pulled out by hand or sometimes mechanically. The harvested material is cleaned of adhering soil by thorough washing. This is done by placing them in baskets or perforated boxes, which are then suspended under running water. They are occasionally stirred to hasten the cleaning. Then the rhizomes and roots are dried under the sun or in kilns and the longer rhizomes are longitudinally cut into pieces or into halves,

Production and Yield

Under favourable conditions, a yield of 9-10 quintals/ha of dried rhizomes is obtained from a 3-year-old crop.

Marketing and Trade

The unsustainable ways of harvesting and unrestricted marketing have to the reduction in population of jatamansi, high demand medicinal plants leading to sudden escalation in their prices in the market. Nearly 95-98% supply of this herb has been fulfilled by wild resource (forest) of Nepal, Assam, Himachal Pradesh and Uttranchal. According to estimation, the annual requirement of jatamansi roots by the herbal industries and crude drug producers is nearly 5000-6000 tonnes but the supply got reduced since 1995 after declared as engendered species and banned for export. On the other hand annual production of its roots *vis-a-vis* their percentage consumption by Indian ayurvedic pharmacies in 1999 is 200 Mts and 25% respectively. (CHEMEXCIL, Mumbai). The current price of jatamansi rhizome Rs. 90-160 per kg.

Economics of Cultivation

The market of jatamansi and their products are highly volatile. The economics (per ha) worked out here are subject to fluctuations, depending upon time and place.

Expenditure (in Rs.) (for 3 years)	—	1,20,000.00
Returns (in Rs.) (Roots plus seeds)	—	4,25,000.00
Net Profit* (in Rs.)	—	**3,05,000.00**

*   Estimation of the net profit is analysed on the basis of grower/collectors price. The grower/collector price is always 25-40% less than wholesale market price due to various stakeholders *viz.* Agent, middle man, Brokers, Commission agents.

KALMEGH (*Andrographis paniculata* (Burm.f.) Nees.)

Vernacular Names

San.–Bhunimba, Kirata; *Hin.*–Kalmegh, Kirayat, Mahatita; *Beng.*–Kalmagh; *Kan.*–Nelaberu, Vayu hullu; *Guj.*–Kariyatu; *Mal.*–Kiriyattu, Nelaveppu; *Tam.*–Nilavembu; *Tel.*–Neelaveemu; *Eng.*–Creat, Kariyat; *Trade*–Kalmegh.

Introduction

Kalmegh is indigenous to India and has been used in Indian systems of medicines since time immemorial. The plant is also known as 'Rice bitters' in West Indies and 'King of bitters' or 'Chiretta' in England. The fresh and dried leaves of kalmegh and the juice extracted from the herb are official drugs in Indian pharmacopoeia. Sometimes, it is substituted for *Swertia chirayita* for its better principles. The whole herb is the source of several diterpenoids, of which the bitter water-soluble lactone andrographolide is important and is distributed all over the plant body in different proportions.

The common ayurvedic drugs are 'Kalmeghasva' and 'Kalmeghnamayas Haub', the main constituents of which are *A. paniculata* besides trikatu (*Piper longum, P. nigrum* and *Zingiber officinalis*), *Cyperus scariosus,* Triphala, *Embelia ribes, Plumbago zeylanica* and Lauh Bhasm (organo-metallic salt of iron).

Botanical Description

An annual, erect, herb or shrub recumbent reach upto 1.0 m high. The branches are sharply quadrangular, often narrowly winged towards the apical region. The leaves are petiolate, lanceolate, oblong-attenuate at both ends, glabrous and acute. The flowers are small, solitary, in panicles with a rose-coloured corolla which is hairy externally. The seeds are numerous, yellowish-brown and glabrous.

Geographical Distribution

It grows wild as an undershrub in tropical, moist, deciduous forests. The plant is found in the plains throughout India and Sri Lanka. It is also reported from certain parts of China, Thailand and Bangladesh. In India, it is distributed in the States of Andhra Pradesh, Assam, Bihar, Karnataka, Kerala, Madhya Pradesh and West Bengal.

Medicinal Uses

Kalmegh forms the principle ingredient of a household medicine called 'Alui', extensively used in West Bengal for general debility and for certain forms of dyspepsia amongst adults and infants. The expressed juice of the leaves is prescribed with cardamom, cloves and cinnamon in the form of globules to infants for relief from bowel complaints, irregular stools and loss of appetite.

In India, the entire plant is used to treat snakebite. The hot water extract of the whole plant is used for acute jaundice, where the powder is mixed with garlic and 2 g of this mixture is given orally with buttermilk for four days and also as febrifuge and as an anti-dysenteric agent; while the extract of the dried leaf is used to treat stomach worms. The hot water extract of the dried leaf and stem is used as a stomachic, febrifuge, and tonic, while the fresh leaf-juice along with the leaf-juice of *Azadirachta indica* and *Tinospora cordifolia* are used to cure cholera. The roots are used as antipyretic, alternative and cholagogue agents. The decoction of the plant is used in some cases of anemia. In Thailand, the decoction of the dried leaf is used against high blood pressure. While in Bangladesh, the hot water extracts of leaf and stem are prescribed as a powerful tonic.

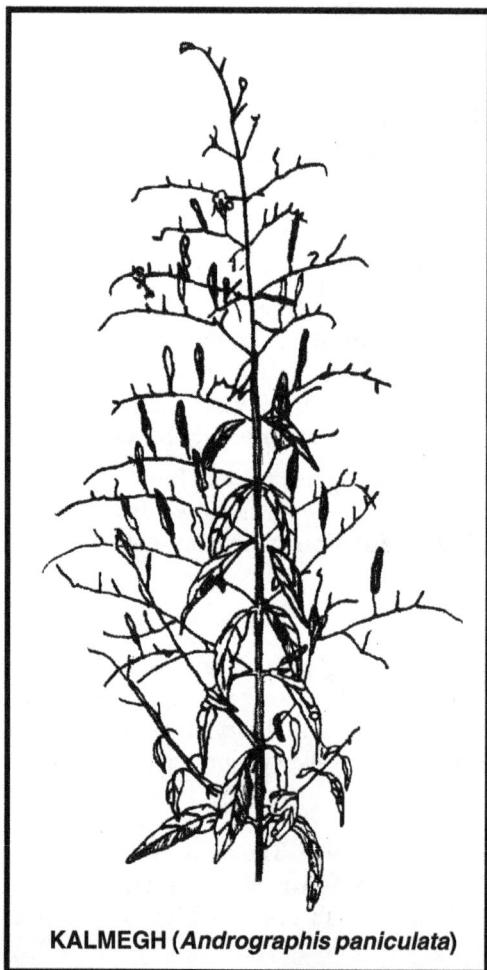

KALMEGH (*Andrographis paniculata*)

Chemical Constituents

The leaves contain a diterpene lactone called andrographolide (2.5%) while the stem contains a smaller amount (2.0%) of this active principle. In addition, the plant is an important source of the flavonoids, sesquiterpenes, phenylpropanoids. The roots contain the flavonoids andrographin, panicotin, aplegenin-4', 7-dimethyl ether, mono-o-methyl wightin and hydroxy-7,8,2', 3'-trimethoxy flavone and beta-sitosterol.

Soil and Climate

It is a hardy plant and can be grown on a variety of soils from clay to sandy loam soils but sandy loamy soil rich in organic matter is good for its growth and yield. It is tropical and subtropical crop. However, cooler climates, moist, shady places with well distributed rainfall are ideal.

AGRO-TECHNIQUE

Land Preparation

The land is prepared well by repeated ploughing and planking and brought to a fine tilth. FYM @25 t/ha is mixed well into the soil at the time of the last ploughing. The land is laid out into plots of convenient sizes along with Irrigation channels for easy management.

Mode of Propagation

It can be propagated by seeds or cuttings.

Direct Sowing

For Direct Sowing, furrows are made at a distance of 30 cm and about 3-4 seeds are sown in each spot maintaining a distance of 15 cm between two plants. About 3-5 kg seeds per hectare are required for Direct Sowing in the field.

Nursery Raising

For raising the seedlings, the seeds are sown in polythene bags or nursery beds filled with a mixture of soil, sand and organic manure in 1:1:1 proportion. The seedlings can also be raised in nursery beds of 3x1.5x0.15 m size. The nursery beds are mixed with well rotted, powdered cattle manure @ 20 kg/m². The best season for sowing is during the month May-June when it has the maximum growth and gives the highest yield. About 400 g/ha of seeds are required for sowing. The seeds are sown in lines 5 cm apart in the nursery beds. The beds or polythene bags are watered daily. The seeds germinate in about 8-10 days and the seedlings will be ready for Transplanting into the main field when they are 45-50 days old.

Transplanting

The seedlings are transplanted at a spacing of 30x30 cm or 30x45 cm. The first or second week of June is the best time for Transplanting. Studies have shown that in order to get the maximum herb yield, a spacing of 15x15 cm is the best.

Manures and Fertilizers

Kalmegh responds very well to the application of manures and fertilizers. Besides the application of about 25 tonnes of FYM, a fertilizer dose of 75 kg nitrogen, 7 kg phosphorus and 50 kg potash is required for a crop of one hectare. Of this, 50% nitrogen and the entire dose of phosphorus and potash is given as basal dose and the remaining 50% nitrogen is used as a topdressing 30 days after sowing or Transplanting.

Irrigation

It is mainly grows as a rainy season crop, in the areas with well distributed rainfall hardly any Irrigation is given. However, in case of long dry periods, the crop is irrigated initially at 3-4 days' interval and later the interval is increased to one week, depending upon the weather conditions.

Interculture

Kalmegh is a small herbaceous shrub and, therefore, the competition with weeds during its initial stages of growth should be avoided by keeping the land free from weeds. The first weeding is done after about 20-30 days of planting. After this, one or two weedings are done after 60 days of sowing to keep the weeds under control during the initial stages of crop growth. Later, weeding follows every harvest.

Pests and Diseases

It is a hardy plant and is not attacked by any Pests and Diseases of a serious nature.

Harvesting and Storage

The crop is ready for the first harvest after about 90 to 120 days of sowing when the plants start flowering. At this stage they are harvested by cutting the plants at the base, leaving about 10-15 cm of the stem for regeneration. The regenerated plants are ready for harvest in about 60 days of the previous harvest. Totally, 2 to 3 harvests can be made in a year. As a perennial crop, one in August and second in November/December. Crop remains dormant during winter. At the flower initiation, active principle andrographolide is high in leaves. Sometimes, the whole plant is also harvested after about 6 months. After harvesting, the plants are dried under shade for 3-4 days before storage.

Production and Yield

The average yield is about 2 to 2.5 t/ha of dry herb.

Marketing and Trade

According to estimation, the demand of kalmegh by herbal industries and ayurvedic drug producer of Maharshtra is nearly 16-18 tonnes in 2001-2002. The Natural Remedies (P) Ltd., Banglore is required nearly 11.5 t Kalmegh as crude drug in 1999-2000. Annual production of kalmegh is nearly 5000 Mts *Vis-a-Vis* their 20% consumption by Indian ayurvedic pharmaceutical industries in 1999 (*Source*-ADMA, Mumbai). The current market price of herbage is Rs. 15-20 per kg.

Economics of Cultivation

The market of Kalmegh and their products are highly volatile. The economics (per ha) worked out here are subject to fluctuations, depending upon time and place.

Expenditure (in Rs.)	—	12,500.00
Returns (in Rs.)	—	54,000.00
Net Profit* (in Rs.)	—	**41,500.00**

*   Estimation of the net profit is analysed on the basis of grower/collectors price. The grower/collector price is always 25-40% less than wholesale market price due to various stakeholders *viz.* Agent, middle man, Brokers, Commission agents.

KUTH (*Saussurea costus* (Falc.) Lipsch.
Syn: *S. lappa* Clarks

Vernacular Names

San.–Kusttah; *Bang.*–Pachau kur; *Guj.*–Upaletakur, *Hin.*–Kuth, Kur, Kostha; *Kan.*–Kosth; *Kash.*–Postkhai; *Mal.*–Kottam, Simakkattam; *Mar.*–Kushta, Kosht uplet; *Punj.*–Kut kooth kuddh; *Tam.*–Kastam, Gostham; *Tel.*–Changala, Kustam; *Eng.*–Kuth, Costus.

Introduction

It is indigenous to the north western regions of the Himalayan Mountains (Kashmir and Hazara). It has lately been introduced also to Chamba and Tehri. It is locally called "Kuth" and grows abundantly in the Kishengang valley of Kashmir, and in the higher elevations of the Chenab valley. Large quantities of costus root are exported from Kashmir to China and Japan for use in Temples, and to Europe for distillation purposes. Prior to World War II, about 2 million pounds of costus roots were said to be produced yearly, but only a small part of this quantity went to Europe for extraction of the essential oil. Its belongs to Astraceae family.

The Kashmir State Government had a monopoly over this drug as it was a main source of state revenue, and its extraction was directly supervised by the State Government. It was valued as the (then) present day narcotic and the state army used to be deployed for its protection against illicit extraction. But due to the lack of control and over-exploitation, this herb is now on the verge of extinction.

Botanical Description

A tall, sturdy, herbaceous perennial upto 2 m high; stem is robust, upper parts pubescent. Leaves are rough above, glabrous beneath and are irregularly toothed, 15 to 30 cms long. Flowers are bluish purple or almost black, borne on rounded flower heads; few flowers heads are clustered together. Fruits are small, curved, compressed and of brown appearance.

Geographical Distribution

It is distributed in temperate regions of Asia, Europe and America. It is native to India & found both as wild and cultivated. In Kashmir it occurs wild on moist, open slopes in Kishangang & Chenab valley. Its cultivation is done in Lahul spite, and Himachal Pradesh. It is also found in Uttaranchal hills at an altitude of 2500-3500 masl.

Medicinal Uses

In action it is anodyne, anti-arthritic, antiseptic, aphrodisiac, astringent, carminative, digestive, deodorant, disinfectant, diuretic, emmenagogue, expectorant, febrifuge, narcotic, spasmodic, stimulant, stomachic, tonic and vermifuge. It is also useful in asthma, bronchitis, colic, cough, dental trouble, diarrhoea and dysentery, fever flatulence, headache, hiccough and Hysteria. Its roots also used in the preparation of rich perfumes and hair oils. It is also an insecticide and insect repellant. Pieces of the root placed in the folds of woolen not only protect them from insects bit also impart them its characteristic perfume. For perfumed sachets these roots have the added advantage of use in gold braids and embroidery also. It is also used as an incense.

KUTH (*Saussurea costus*)

The costus oil is also used in high grade perfumes of heavy oriental type. It blends with sansl, vertiver, patchouli, rose and violet imparting to them unique and alloying tonalities that are difficult to trace. Because of its very strong and tenacious odour, costus oil is dosed most carefully unless the effects become unpleasant. The oil is expensive and not always readily available.

Chemical Constituents

The root contains a volatile oil (1.5-2.5%) and an alkaloid saussurine (0.05%), resin (6%) and tannins. The essential oil contains phellandrene (0.4%) camphene (0.4%), a-terpene alcohol (0.2%), aplotayene (20%), costene (6%), a-costene (6%), Costus acid (14%), costol (7%), costus lactone (11%), and dihydrocostus lactone (15%). Kushtin has been isolated from the roots.

Soil and Climate

A deep porous soil rich in organic matter is preferred. Plants growing on such soils develop long & thick roots. Heavy soils or soils with little fear of water stagnation should be avoided. It thrives in shady, moist places beneath birch and dwarf willow. Kuth needs cool and humid climate are found at an altitude of 2600-3200m. However, areas receiving blast of monsoon are to be avoided. Cultivation even at 2400 m altitude has been successfully done in Kashmir and Lahul valleys.

AGRO-TECHNIQUE

Land Preparation

The field should be prepared to make good seed-beds, by light ploughing followed by two harrowing before the onset on the monsoon season.

Mode of Propagation

It can be propagated by seeds or root cuttings

Sowing

The seeds are collected towards the ends of September to mid of October for raising seedling in nursery. For seedling and root cuttings raised in small nursery is done in the beginning of rainy season. Experiments on the cultivation of *Saussurea lappa* in the Garhwal districts were carried out and best results obtained by sowing seeds in nursery beds (instead of directly in the forest) and by transferring nursery plants to intensively prepared and manured beds. At the onset of monsooon, the one year old seedling are transplanted in the pit sat a spacing of 30 cm × 30 cm and are irrigated lightly.

Fertilizers and Manures

Well decomposed farmyard manure should be applied to nursery beds and in the field during land preparation.

Irrigation

4-6 irrigations during May-September. The land is irrigated when seeds are sprouting.

Interculture

The crop is kept free of weeds by regular weeding, so as to provide maximum nourishment to the crop.

Pests and Diseases

Since the cultivation of costus root is done on limited scale, the plant is relatively free from insect and pest attack. Also, no systematic attempt has been made so far to study the pest and disease problem.

Harvesting and Storage

The best time for collecting the roots was found to be the end of October or early November, when they are well matured. At this period of the year, the moisture content of the roots is the lowest. After harvesting, the roots are immediately cut into short lengths of about 5-10 cm and cleaned of earth before being placed on shelves for drying. Thus the collected roots are partially dried on the spot before the close of the season in early November, after which they are to be transported down to a drying "godown" at a lower altitude for further drying.

In Kashmir, the roots are dug up soon after fructification in September and October, cut into pieces 7-10 cm long, gently roasted and kept in the go-downs for complete drying.

Production and Yield

Until late seventies the production of oil and resionoid estimated to 100 kg. But it has crossed 500 kg in the world and the main producing country is India. Dry root yield varies form 2.0 to 2.5 t/ha. Root yield upto 4 t/ha is also reported. On an average, yield of oil ranges from 0.3 to 1.5%.

Marketing and Trade

The unsustainable ways of harvesting and unrestricted marketing have to the reduction in population of Kuth, high demand medicinal plants leading to sudden escalation in their prices in the market. Kuth roots has a great demand in Buddhist countries. In 1999-2000, the total annual consumption of kuth roots in Natural Remedies (P) Ltd., Bamglore is nearly 6.5 tonnes. Nearly 95-98% supply of this herb has been fulfilled by wild resource (forest) of Himachal Pradesh, Kashmir Uttaranchal. According to survey report, the annual requirement of kuth roots by the herbal industries and crude drug producers is nearly 500-600 tonnes but the supply got reduced since 1995 after declared as engendered species and banned for export.

Large quantities of its roots have regularly been exported for a long time through the ports of Kolkata and Mumbai to China. The main importing countries include Hong Kong, France, Singapore, Thailand, Veitnam, Japan and Sri Lanka. Prior to 1962-63 China was the main exporter of the costus root. The prices of root varies from to Rs. 50-70 per kg and oil varies from Rs. 10,000 - 12,000 per kg.

Economics of Cultivation

The market of Kuth and their products are highly volatile. The economics (per ha) worked out here are subject to fluctuations, depending upon time and place.

Expenditure (in Rs.)	—	10,000.00
Returns (in Rs.)	—	1,00,000.00
Net Profit* (in Rs.)	**—**	**90,000.00**

*   Estimation of the net profit is analysed on the basis of grower/collectors price. The grower/collector price is always 25-40% less than wholesale market price due to various stakeholders *viz.* Agent, middle man, Brokers, Commission agents.

LIQUORICE (*Glycyrrhiza glabra* L.)

Vernacular Names

San.–Uashti-madhu, Maduka; *Hin.*–Mulhatti, Jethi-medh; *Beng.*–Jashtimadhu, Jasishbomodhu; *Tel.*–Yashtimadhukam, Atimadhuramu; *Tam.*–Atimaduram; *Kan.*–Yashti madhuka, Atimad-hura; *Mal.*– Iratimadhuram; *Mar.*–Jeshtamadh; *Guj.*–Jethi madhu; *Eng.*–Liquorice, *Sweet wood*; *Trade*–Mulhati.

Introduction

It is a hardy perennial herb and undershrub, belonging to the family Papilionaceae. As its roots are very sweet, it is popularly called liquorice. In roots are very sweet, it is popularly called liquorice. In Greek, glycyrrhiza means 'Sweet Root'. It has been known in pharmacy for thousands of years as an important medicinal plant. In Hindu medicine, it is a principal drug of the Susruta in China also it was considered as an important drug. The Egyptians, Greek and Rotans utilized it for medicinal purposes and, liquorice maintains it place of importance, even in present day medical and pharmaceutical sciences.

Botanical Description

A perennial herb or undershrub attaining a height of 1-2 m with long, cylindrical, burrowing rootstock and horizontal creeping stolons. The leaves alternate, pinnate with 9-17 ovate, yellowish-green, leaf-lets about 2.5-5.0 cm long. The flowers are borne in erect, auxiliary, long spikes, usually 10-15 cm long, lavender to purple in colour and 1.25 cm long. The seed pods are maroon, 3 cm long, oblong, pointed, flattened and contain 2-4 kidney-shaped seeds.

Geographical Distribution

It is distributed in the subtropical and warm temperate regions, particularly in Western China, parts of Asia Minor, Persia, the Asian Republics of the erstwhile USSR and Afghanistan. It is also distributed in the warmer parts of the Mediterranean region of North Africa, Spain, Italy, Yugoslavia, Greece, Syria, Hungary and South Russia. It is cultivated in Italy, France, Russia, Germany, Spain and China. Although the origin of liquorice is attributed to regions from Southern Europe to Pakistan and Northern India, none of the liquorice yielding species occur in India. In an effort to curb import, substitution, attempts have been made to cultivate liquorice in several places in India, notably Baramulla, Srinagar, Jammu, Dehradun, Delhi, Anand, Indore and Bangalore and it has been found to come up successfully in quite a few places.

LIQUORICE (*Glycyrrhiza glabra*)

Medicinal Uses

The dried, peeled and unpeeled underground stems and roots constitute the drug. A few of the reputed uses of the drug are: demulcent, mildly laxative, expectorant and it is an important constituent of all cough and catarrh syrups, throat lozenges and pastilles. Mustong, a preparation in which one of the constituents is *G. glabra*, improves sexual inadequacy. It is also known for its analgesic, anti-inflammatory and anticonvulsive properties. Recent investigations have shown its growth and inactivation of virus particles. Large quantities of liquorice are used in confectionery and in tobacco blending. The extract is reported to lend sparkle and aroma to beer.

Chemical Constituents

Its extract contains about 5-20% saponin like glycoside glycyrrhizin, a characteristic sweet in nature. The other constituents are glucose (3.8%), sucrose (6.5%), mannite, starch, asparagines, resins, volatile oil (0.03-0.06%) and the colouring matter isoliquoricin. Glycyrrhizin ($C_{42}H_{62}O_{16}$) occurs as Ca or K salts of glycyrrhizic acid. On acid hydrolysis, it yields, glycyrrhizic acid $C_{30}H_{46}O_4$, a triterpenoid aglycon) and manuronic acid.

Soil and Climate

The plant requires loamy soils of light texture (pH 6.0-8.2) for high biological and economic productivity. Deep moist soils, particularly on the banks of rivers, subject to periodical inundation, are also suitable. It prefers a warm and dry, subtropical climate with well-defined winters. Heavy rains or frost affects its growth adversely. The plant thrives in locations receiving 50-100 cm rainfall annually, supported with Irrigation.

Improved Varieties

EC-111236, EC-21950 and Haryana Mulhati No. 1 are released by the Haryana Agriculture University, Hissar.

AGRO-TECHNIQUE

Land Preparation

Before cultivation, the land should be ploughed twice and the field should be cleaned thoroughly of weeds. If required, small canals may be prepared for drainage. About 25 t/ha of cowdung manure is also added.

Mode of Propagation

It can be propagated by seeds, shoot cuttings, root cutting and by tissue culture.

Seed Biology

Though the plants can be grown from seeds, this is seldom practiced because of the germination problem. It was found that treatment of seeds with 0.0025 to 0.005% of succinic acid, combined with scarification (in sand), increased laboratory and field germination. The treated seeds are drill-sown 15 cm apart in raised nursery beds (1x3) that are well prepared and manured. The germination starts after 15 days and is completed within a month. The seedlings are picked out when they are 10-15 cm high and planted in the main field.

Tissue Culture

Rapid clonal multiplication of *Glycyrrhiza glabra*, using auxiliary bud culture, has been reported. The technique has also been standardized for Transplanting test-tube grown plantlets to the field.

Propagation by Cutting

The traditional method of perpetuation is by utilizing the cuttings prepared from the old crown of the lifted roots, cut into pieces of 10-15 cm length. The plants are also produced from runners or underground stems which are prepared into cuttings of 10 cm long, each with 2 buds. These cuttings are kept in moist sphagnum moss for about 8-10 days and afterwards, when their buds start sprouting, they are planted in the main field.

Transplanting

The crop is usually grown continuously on the same land. Uprooting of the plant is done in October-November and this disturbs the soil to a depth of 80 to 100 cm. The soil is then leveled and planting ridges 60 cm wide and 30 cm apart are marked out. The rows are enriched with FYM and the soil from the intervening spaces is swept up to cover the manure to form rounded ridges whose summits are about 37 cm along the furrows. The planting is done in March or early April. The sets are planted in groups of 3, 30 cm apart in rows. The central planting material of every alternate group is an old crown which is placed at the top of the ridge with the two other sets (root-cuttings) on either side of the ridge approximately 30 cm apart. The crown are covered with 5 to 7 cm of wet soil. Dry conditions at planting time and for the next few months give the conditions required for a good crop. If cold weather prevails in May or June, 0 to 40% of the sets may fail to grow.

Manures and Fertilizers

In order to obtain good growth and development of the roots, the field must be liberally manured. Compost of FYM @ 10-15 t/ha is enough to meet the requirements. However, there is difficulty in getting organic manure in such large quantities, hence, a fertilizer dose at the rate of 40:40:20 kg/ha per year is recommended for application in the form of urea, super phosphate and muriate of potash. Of these, a full dose of super phosphate and potash are applied as a basal dose at the time of planting. Whereas, nitrogen is applied in three split doses: i.e., at planting, after six months and one year of growth, respectively. Since the crop remains in the field for 3-4 years, every year the same dose has to be reapplied.

Irrigation

Regular Irrigation is necessary until the cuttings put down roots and establish themselves. Later, the watering may be done at intervals of 8-10 days for 5-6 months and afterwards as and when required. Once the plants are established, they need no special attention as the crop is fairly drought resistant.

Intercropping

The crop occupies the land for a period of 4 to 5 years, and the growth is slow during the first 2 years. Thus, intercropping by planting carrots, potatoes or cabbage between the ridges, with the additional inputs required by the intercrop, is recommended. The field may be hoed periodically to keep down the weeds.

Pests and Diseases

Pests

Termites cause a lot of damage to young plants during the dry period. Aldrex at the rate of 25-30 kg/ha can be given during the last ploughing/harrowing to control termites. Cotton ash weevil (Myllocerus udecimpunctatus) is reported to cause injury to leaves and this insect can be controlled by spraying Metasystox (1.5 ml/l).

Diseases

The crop is found to be infected by root-rot disease. The roots and stolons of the affected plants turn soft and pulpy; 5 kg/ha Brassicol (or drenching the soil with 6:3:100 Bordeaux mixture) is applied during February to control the disease effectively. *Rhizoctonia bataticola, Sclerotium* spp.; and *Fusarium spp.* Causes root-rot, collar-rot and wilt diseases. Soil treatment/aerial spray of Bavistin or Benlate (0.05%) is reported to control these diseases.

Cercospora cavarae is reported to cause leaf-spot disease. To control the disease, spray 0.4% Bavistin. However, one protective spray should be given one-and-a half to two months after planting. The other fungi which are reported to cause leaf-spot diseases are *alternaria tenuis, Septoria glycyrrhizae* of Phyllosticta glycyrrhizae. *Uromyces glycyrrhizae* causes leaf-rust. Leaf-blight is another serious infection which can be controlled by foliar spraying of Blitox (0.2%) 3 to 4 times, at an interval of 5 days.

Harvesting and Storage

The crop is ready for uprooting about 3 to 4 years after planting and just before the plants have borne fruit. The plants are lifted in autumn (November-December) after the rains. A trench about 60 cm deep is first dug at the side of the ridges then, by working inwards, the soil is loosened from the roots so that they can be pulled out easily. The aerial parts are cut and removed. The broken root-parts raising the succeeding crop, only the gaps need to be filled with rooted cuttings.

Roots and underground stems are cut into 15 to 20 cm-long pieces, 1 to 2 cm in diameter and are dried alternately in the shade and sun, this may take several months. The drying process reduces the weight to 50% and the moisture from 50-60%. Artificial drying can also be done at 30-40ºC by using mechanical driers.

Production and Yield

Presently, India is importing the drug on a considerable scale (5000-10000 t) from Iran, Iraq, Afghanistan, Burma and Singapore. The yield depends upon the soil fertility, variety, the climate and cultivation practices. An yield of 5 t/ha of roots, plus 14-20 t/ha of trimmings is considered satisfactory.

Marketing and Trade

The unsustainable ways of harvesting and unrestricted marketing have to the reduction in population of liquorice, high demand medicinal plants leading to sudden escalation in their prices in the market. Pakistan imports nearly 105 tonnes of pulverized *Glycyrrhiza* annually from China spending foreign exchange equivalent to Rs. 2.5 lakhs and its demand in the world is unlimited. The roots of Liquorice is imported nearly 150-200 Mts annually from Western European countries, USA, Japan, Afghanistan, Iran and UAE. The current market price of liquorice roots is Rs. 40-60 per kg.

Economics of Cultivation

The market of Liquorice roots and their products are highly volatile. The economics (per ha) worked out here are subject to fluctuations, depending upon time and place.

Expenditure (in Rs.)	—	25,800.00
Returns (in Rs.)	—	2,80,000.00
Net Profit* (in Rs.)	**—**	**2,54,200.00**

* Estimation of the net profit is analysed on the basis of grower/collectors price. The grower/collector price is always 25-40% less than wholesale market price due to various stakeholders *viz.* Agent, middle man, Brokers, Commission agents.

MUSK MALLOW (*Abelmoschus moschatus* Medic)

Vernacular Names

San.–Latakasturika; *Hin.*–Mushdana, Latakasturi; *Tam.*–Kattu kashturi; *Mal.*–Vattilai kasturi, Kattukasturi; *Beng.*–Muskdana; *Mar.*–Musk bhindi; *Tel.*–Kasturibendavittulu; *Kan.*–Kadukasturi; *Eng.*–Musk mallow; *Trade*–Muskdana.

Introduction

Muskdana belongs to family Malvaceae, popularly known as musk mallow yields musk-like scented seeds. Its seeds yield an essential oil and gives a strong flowery musky brandy-like odour of remarkable tenacity because of the presence of ambretrolide, a macrocyclic lactone in the seed coat. Ambrette seeds are exported to Canada, France, Nepal, Spain, UAE and United Kingdom to the extent of about 116 quintals in a year because of its diversified uses.

Botanical Description

An erect, hirsute or hispid annual herb attends a height upto 180 cm. Stem possesses long deflexed hairs. Leaves simple, polymorphous, cordate, the lower ovate, acute or rowdish angled, upper palmately 3-7 lobed, divided nearly to the base. Flowers are large usually solitary, axillary, mucilaginous hairy, yellow with purple centre. Seeds are scent sub-reniform greenish brown in colour with numerous raised brown striate.

Geographical Distribution

It is native of India and grows throughout tropical regions of the country. It is found all over the Deccan and Karnataka in the hilly regions and at the foot hills of Himalayas. It has been introduced in tarai area of Kumaon (Uttaranchal) and Punjab as a Kharif season crop. It is also cultivated as an irrigated crop in U.P., Andhra Pradesh, Karnataka and Gujarat.

Medicinal Uses

In medicine, seeds are used as a tonic, aphrodisiac, antiseptic, diuretic, stomachic, demulcent and carminative. They allay thirst, check vomiting and cure diseases due to *kapha* and *vata* and are useful in treating intestinal disorders, dyspepsia, urinary discharge, nervous debility, hysteria and skin diseases like itch and leucoderma. The leaves and roots of the plants are recommended for the cure of gonorrhea.

MUSK MALLOW (*Abelmoschus moschatus*)

Aqueous and raw infusions of seeds are used for intestinal worms snake bite rheumatism, flu, asthma. Its seeds absorbent capacity inactivates snake venom. The pulverized roots are used to poultice boils and swellings. Its seeds are mixed with coffee as a flavoring agent. Besides seeds are also used to protect woollen garments against moth and imparting a musty odour to sachets hair powder, pan masala and agarbatti etc.

Chemical Constituents

Leaves, flowers and seeds of the plant contain beta-sitosterol and its glucoside, campesterol, stigmasterol, cholesterol, ergosterol. The seeds yield an essential oil (0.3%). The oil contains sesquiterpene alcohol arnesol, lactone ambrettolide (musk-like odour 0.3%) and ambrettolic acid.

Soil and Climate

It prefers well-drained fertile rich loam to sandy loam soil. Clayey and sandy soil, water-logged condition are not good for its proper growth. It requires tropical, worm and humid climate. It can be grown up to 1000 m altitude. It required sufficient rainfall during growth but comparatively dry weather during flowering and fruiting stage. The optimum temperature for its vegetative growth ranges between 20^0C and 28^0C.

AGRO-TECHNIQUE

Land Preparation

Before cultivation, the land should be ploughed twice and the field should be cleaned thoroughly of weeds. The field is prepared well, divided into convenient sized plots and thoroughly mixed with organic manure.

Mode of Propagation

It can be propagated by seeds.

Sowing

The seeds are sown in early June for better growth. Presowing Irrigation is beneficial for uniform germination. Usually 5 kg seeds are sufficient to sow in one hectare land under irrigated conditions. Seeds are sown about 2 cm deep in rows. The germination percentage of seed is about 85%, provided proper soil moisture is available. Presowing treatment of seeds with fungicide is preferred in water for a date prior to sowing is also beneficial in order to ease their germination. It begins to germinate within 4 days of sowing. The germination is complete within 15 days. After germination, only healthy seedlings is retained by pulling out the thin and weak seedlings.

Manures and Fertilizers

The crop responds well to the application of manure and fertilizers. FYM at the rate of about 8-10 t/ha should be added at the time of Land Preparation. In addition, a dose of 120 kg nitrogen, 100 kg phosphorus and 75 kg potash/ha, is recommended for obtaining a good yield. Nitrogen when applied in spilt doses during growth stage under sub-tropical climate gives good response in respect of seed yield per hectare.

Irrigation

If monsoon rains are regular, usually no Irrigation is given till December, subsequently depending upon rains 3-4 Irrigation at one month interval is given to meet the water requirement of crop.

Interculture

During early stage of crop growth, weeding cum hoeing is done at month intervals for 3 months. Thereafter, crop becomes tall and do not allow the seasonal annual weeds to grow for competition with the growth of crop.

Pests and Diseases

This crop is subjected to many Pests and Diseases from the time of germination till the fruiting stage. These pests of major significance included aphids, pink bollworms, leaf-worms which affect the seed yield. The foliar sprays of dimethoate, endosulphan, malathion, reduces the risks of aphids and thrips etc. Similarly, diseases of muskdana cause reduction in yield by affecting germination, killing the plants, lowering the productivity among the diseases. It can be controlled by repeated sprays of 0.5% Dithane Z-78, at one month interval.

Harvesting and Storage

The crop is shown as Kharif season crop in June and begins to flower in October and the fruit setting is almost simultaneous which continues till April. The crop is matured within 6-7 months. As soon as they attend blackish colour the mature fruits are plucked (before disease of pods) and seeds are taken out manually as otherwise seeds fall off.

Production and Yield

The average yield of seeds ranges between 8 to 10 quintals/ha. However, yield of seeds per hectare is more in Tarai of Uttaranchal and it ranges between 12 to 15 quintals/ha.

Marketing and Trade

In 1999-2000, India exported 172.75 tonnes of ambrette, valued Rs. 47.52 lakh. But that is quite less than the 95-96 (624.97 tonnes earned 83.13 lakh Rupees) although there is a rising/increasing trend. Muskdana seeds are imported in very less quantity and also irregularly. In 1997-98 only 0.94 tonnes worth of Rs. 0.69 lakh while in 1993-94, nearly 14.19 tonnes (Rs. 2.33 lakh) were imported. The current price of its seeds is nearly Rs. 100-150 per kg.

Economics of Cultivation

The market of Musk mallow and their products are highly volatile. The economics (per ha) worked out here are subject to fluctuations, depending upon time and place.

Expenditure (in Rs.)	—	8,000.00
Returns (in Rs.)	—	65,000.00
Net Profit* (in Rs.)	**—**	**57,000.00**

* Estimation of the net profit is analysed on the basis of grower/collectors price. The grower/collector price is always 25-40% less than wholesale market price due to various stakeholders *viz.* Agent, middle man, Brokers, Commission agents.

NEEM (*Azadirachta indica* A. Juss)
(Syn: *Melia indica* Brand; *M. Azadirachta* Linn.),

Vernacular Names

San.–Nimbah, Prabhadrah; *Hin.*–Nim, Neem, Nimb; *Kan.*–Turakabevu, Huccabevu, Cikkabevu; *Mal.*–Veppu, Aryaveppu, Aruveppu, Kaippanveppu; *Tam.*–Vempu, Veppu; *Tel.*–Kondavepa, Turakavepa; *Eng.*–Neem tree, Margosa tree, Indian lilac.

Introduction

Neem, is one of the most valuable and yet the least exploited of tropical trees, belonging to the family Meliaceae. It is commonly known as neem tree, margosa tree, Indian lilac, white cedar, paradise tree, chinaberry or crack jack in various places.

The neem tree has great potential for agriculture, industrial and commercial exploitation because of its multiple uses such as firewood, timber and in pharmaceutical and entomological preparations.

Every part of the tree from its roots, trunk, bark, leaves, flowers, fruits, seeds, sap and gum are known to have some use and have a place in the traditional folklore and medicine (especially in the Ayurvedic and Unani systems of medicine), so much so it almost plays the role of the village dispensary in this land.

Botanical Description

A large glabrous tree, 7 to 20 m high, with a straight trunk. The branches spread widely and form oval crown. The bark of the trunk is 1.2 cm thick, furrowed longitudinally and obliquely, dark grey outside and reddish-brown inside. The sapwood is yellowish-white, the heartwood is red or brown, especially the inner part. The leaves are alternate, imparipinnate, the leaflets are 9 to 13 in number, nearly opposite, shortly petiolate, 2.5 to 7.5 cm long. The inflorescene is an axillary panicle. The flowers are white, with a strong smell of honey, especially at night pentamerous, on short slender, pedicles with short, scattered hairs. The fruit is a drupe, ovoid-oblong, smooth, greenish-yellow when ripe, one or sometimes two-celled. The flowers appear from March-May and the fruit ripen during June-August.

Geographical Distribution

Neem is believed to be a native of upper Burma and possibly of the Siwalik. Deccan and other parts of South India. It is widely distributed in tropical Africa, Australia, the Andaman's, Pakistan, Sri Lanka, Myanmar, Malaysia, Indonesia, Bangladesh, Thailand, Middle East, Sudan, Niger, Fiji, Mauritius, Central and South America, the Carribean, Puerto Rico, the Virgin Islands, Haiti and other places. The largest known plantation of nearly 50000 trees is at the Arafat Plains, en route to Mecca in Saudi Arabia, for providing shade to the Haj pilgrims. In India, it is cultivated all over. It grows wild in the Tamil Nadu and Karnataka. The tree is well suited for afforestation. According to one report in 1991, India has approximately 14 million neem trees and has an annual potential to produce 0.1 million tonnes of neem oil and 0.15 million tonnes of neem-cake from 0.45 million metric tonnes of seeds.

NEEM (*Azadirachta indica*)

Medicinal Uses

Its leaves are used to cure many diseases of the bladder, kidney, eyes, and skin, for the explusion of intestinal worms, and with honey for curing jaundice, boils and urticaria. The dry flowers are considered as a stimulant, tonic and stomachic. The flowers are useful in some cases of a tonic dyspepsia and general debility. The bitter oil extracted from the flowers is useful in skin diseases. The bark is a bitter tonic, astringent and antiperiodic. It is usually prescribed in the form of a tincture or an infusion. The bark preparations are useful in fever, thirst, nausea, vomiting, skin diseases, scorpion and snakebites. The bark is also used as a vermifuge, the aqueous extract is reported to cause immobilization of human and bovine spermatozoa. The oil of the bark possesses a marked antibacterial spectrum against Gram-ve and Gram+ve bacteria. The fresh, tender twigs are used to clean the teeth, particularly in cases of pyorrhea.

The pericarp of the fruit is rich is carbohydrates and constitutes a promising substrate in methane gas generation. The oil displays a simulative, antiseptic, an alternative effect when used for massage of the body and is said to possess anti-inflammatory properties. At the same time, it is an active remedy against boils and ulcers. The oil has proved to be remarkably effective in infections like ringworms, itches, erysipalis, urticaria and scrofula. In veterinary surgery, the oil is used for foul sores and it has shown antidiabetic activity in rats. Medicinally, the oil is employed as an antithelminthic, used in bleeding gums, pyorrhea and asthma. It is used externally as a stimulant in convulsions and collapse from fevers, cholera and fits. The oil finds use in the Siddha system of medicine also. It has antifertility and anti-bacterial effects.

Several pharmaceutical preparations like liquors, emulsions, ointments, liniments, as well as medicinal cosmetics like creams, lotions, soaps shampoos, hair-tonics, toothpastes and gargles have been prepared with nimbidin and with the salts of the acid derived from it. Of the salts, sodium nimbitinate has so far been used for injection purposes. Nimbidin is reported to be efficacious in the treatment of skin diseases such as warts, extensive and persistent dandruff, furunculosis, arsenical dermatitis, scabies and saborrhoeic dermatitis. Nimbidin is also known to exhibit pronounced anticholinergic, diuretic, anthistamine, antinicotinic, anti-ulcer and moderately sedative effect in experimental animals. The oil is used in manufacturing a good number of insecticides as well.

Neem-cake is obtained after expressing the oil from the seeds. The cake represents 80% by weight, of the whole seed. The gum exudes from the bark in the form of clear, bright, ambercoloured tears or fragments. It is found mixed with gumghatti. The gum is prized medicinally as a stimulant, demulscent and tonic. It is useful in catarrhal and other affections which are accompanied by great weakness.

A sweet liquid, neem toddy, is occasionally obtained as an exudation from the upper parts of the trees or by exposing a healthy-looking root, cutting through it and placing a vessel beneath to receive the exuding liquor. It has the unpleasant odour of neem and contains sugar (6.5%) and albuminous and gummy matter (6.5%). The toddy or fermented sap of the tree is useful as a refrigerant, nutrient and alternative tonic. It is of great service in some chronic and long-standing cases of leprosy and other skin diseases, tuberculosis, tonic dyspepsia and general debility. The centrations is important in relation to the therapeutic value attributed to the exudates. Various neem products, either in combination with each other or in combination with other plant products, are also used or cure many disease.

Chemical Constituents

The leaves contain minerals like calcium, magnesium, iron and phosphorus; protein; vitamin A; crude fibre; carbohydrates; fat and alkaloids like quercetin, nimbosterol (b-sitosterol), azadirachtin, nimboflavone and nomicinol. The flowers are reported to contain nimbosterol, kaempferol, quercetin, myrcetin, nimbosterin, nimocetin, fatty acids and a highly pungent essential oil. The trunk bark contains a resin, nimbin, nimbiin, nimbidin, essential oil, tannins, the bitter principle margosine, desa-acetyl nimbin. The heartwood yield nimatone, tannins, inorganic calcium, potassium and iron salt.

The seeds contain calcium, magnesium, phosphorus, an amorphous morgosopicrin, diacetylnimbin, thionemone, azadirachtin, salanin, vilasimin, azadiracthtol, nimbine, nimbanidol, nomolicinol, salannotide, meliacin-4-epinimbin, nimlin and desacetyl nimbin. The seed-kernel contains about 45% of fixed oil called neem oil. The oil is non-drying, bitter and greenish-yellow to brown in colour and has a strong, characteristic, garlic-like odour. The bitter principles of the oil are nimbidin, nimbin and nimbidol. The cake contains oil (91.8%), nitrogen (6%), magnesium (0.88-1.3%), calcium (0.47-1.02%), sulphur (1.6%), proteins (37.5%), carbohydrates (25.6%), fibre (5.8%) and ash (5.4%). Besides, it also contains 110 ppm boron and tannic acid. The cake finds extensive use in corp production as an organic manure. The gum contains water (13-15%), ash (3%), galactose (12%), pentosans (26%), some albumins and oxidase. The gum is a non-amphotric, polyelectrolyte and behaves like a lyophilic colloid. The neem sap (toddy) exuded by the roots has been reported to contain glucose, sucrose, gums and protein. Its as contains iron, aluminium and calcium.

Soil and Climate

It grows well on moist, dry, stony, clayey or shallow soils, including moderately saline and alkaline soils, compact clays and lateritic crusts. It can tolerate a soil pH of up to 10. It is very hardy and thrives best in parts where normal rainfall ranges between 480 to 1000 mm and maximum temperatures as high as 48°C.

Improved Varieties

Local provenances or clones.

AGRO-TECHNIQUE

Land Preparation

The field should be ploughed and harrowed several times, leveled properly and drainage channels should be made. Since yams have a high requirement of organic matter for good tuber formation, a sufficient quantity (20-25 t/ha) of FYM is incorporated at the time of Land Preparation.

Mode of Propagation

It can be propagated by seeds and vegetatively *viz.* root, shoot cutting.

Direct Sowing

Direct Sowing is easy and can be successfully done by dibbling, broadcast, sowing in lines or in trenches, depending on the site conditions. It is reported that, in Rajasthan, neem seeds has been successfully dibbled under Euphorbia bushes in Ajmer district by placing 3-5 seeds in small pits and covering them with soil. Maharashtra, neem is sown with babul. While in the Bellary district of Karnataka, sowing on mounds in rows 2.7 m apart on black soil has proved successful.

Nursery Raising

The seeds are collected from June-August, which coincides with the rainy season. The seeds remain viable for a short period of 3 to 5 weeks only. Hence, they should be sown immediately or within 2-3 weeks of collection. It is reported that the seeds obtained 10-12 weeks after flowering possess the maximum germination capacity and longevity. The seeds can remain viable up to 2 months when stored in airtight tins, in pots with wood-ash and in gunny bags. The seeds are depulped before sowing.

The nursery is raised in beds in rows 15 cm apart with the seeds 2.5 cm apart in row. The beds are sparingly watered after sowing. The seedlings are transplanted from when they are about 2 months till 1-year-old. It is preferable to plant the seedlings when they are 7-10 cm in height. The nursery is protected against frost by providing screens. The seedlings can be raised in baskets and polythene bags also.

Transplanting

For planting the nursery-raised seedlings, the seedlings may be planted inside thorny bushes to provide initial protection from cattle damage. Neem can also be planted as a roadside avenue tree. Stumps from 2-year-old plants have been reported to give a better survival and faster growth rate than those from 1-year-old plants. The stumps are prepared with 2.5 cm of shoot and 23 cm of root pieces and should be planted at the onset of the rains.

Manures and Fertilizers

In neem, for a 2-3 year-old crop, the recommended fertilizers are 30-40 kg nitrogen, 65-70 kg phosphorus and 60-70 kg potash in split doses at an interval of 2-3 month from the time of planting in the first year and the same quantity in the second year. In the third year, only nitrogen at 30 kg/ha has to be applied in two split doses.

Irrigation

The crop needs Irrigation frequently in the initial stage during summer months. An interval of 4 to 5 days in summer and 7 to 10 days in winter is desirable for the proper growth and development of this crop and for economic yields.

Interculture

The competition of grasses should be avoided by proper and timely weeding because woody weeds compete with neem for moisture and nutrition.

Intercrops

During first 3 year of plant development, millets as mix-crops, and sesamum, cotton, sunhemp, groundnut, soyabean and sorghum as Intercrops has been successfully cultivated.

Pests and Diseases

In India, 38 insect species are reported on neem which include seed and flower insects, defoliators, root feeders, stem-and shoot-borer, sap-suckering insects and mites, stem-and shoot-borer, sap-sucking insects and mites. Apart from these insects, an unidentified species of mollusca also causes 15-20% damage to the seedlings in nurseries. To control the sap-suckers, including thrips, a systemic insecticide such as Dimethoate (0.01-0.2%) may be used. The larvae of the shoot-borer, L. koenigiana, can be controlled by 0.01-0.02% of Monocrotophos or Dimethoate. A mixture of 0.2% Dimethoate and vipul (2 ml/l) is also recommended. Against defoliator, 0.1-0.2%. Folithion: and for termite control, mixing Aldrex in the soil before planting, are recommended. The silvicultural measures and use of enemy complexes are also suggested. The plant is not attacked by any disease of a serious nature.

Harvesting and Storage

The flowering starts in February-March and ends in May-June, while shed their leaves during February-March. The fruits/berries can be collected from June till August. Fruiting starts generally sixth year. The fruits can be collected by giving jerk to branches so that fall on the ground. Collected fruits should be put in water to remove the stickiness on the skin and must be sun-dried initial and then in the shade, and stored in airtight gunny bags or tin containers. It must be ensured that the seeds do not contain moisture, which may otherwise invite fungal attack and lead to aflatoxin contamination.

Production and Yield

The potential of neem seeds in Indian is nearly 540,000 Mts every year. But only 25% of the total production of neem seeds is being collected. The 5 year old tree produces an estimated 35-40 kg of leaves and 20-25 kg kernels. As the tree grows yield vary from 37-100 kg kernels and about 300-400 kg leaves, depending upon rainfall and soil conditions. It also produce about 50 kg/tree of berries. The kernel yields 45% of fixed oil.

Marketing and Trade

The annual consumption of pesticides in India is worth Rs. 2000 crore. In next decade the neem based pesticides are expected to cross the 10% mark of the total pesticide consumed in India and it would be Rs. 150 crore as on today basis. The total resources of neem available in the world market would be worth SD $ 750 million. The neem cake and oil has vast potential for their export. It is very small amount considering the total world market of pesticides. Therefore, co-ordinated promote neem based pesticide formulations so that they can be able to fetch higher return in Indian as well as International markets. The current market price neem seeds kernels and dried leaves Rs. 3-4 and 8-10 per kg respectively.

Economics of Cultivation

The market of Neem and their products are highly volatile. The economics (per ha) worked out here are subject to fluctuations, depending upon time and place. On the basis of the prevailing market conditions with kernels, a net profit of about Rs. 30000/ha in a five-year-old crop, can be obtained. It will be increases in subsequent years.

OPIUM POPPY (*Papaver somniferum* Linn.)

Vernacular Names

San.–Ahifen, Chosa, Khasa, Khas-khash; *Hin.*–Afim, Afyun, Kash-kash, Post; *Beng.*–Pasto; *Tel.*–Abhini, Gasalu, Kasakasa; *Tam.*–Abini, Gashagasha, Kasakasa, Postaka; *Kan.*–Afim, Biligasgase, Khasakhasi; *Mal.*–Afium, Kshakhasa; *Eng.*–Opium poppy, White poppy; *Trade*–Opium.

Introduction

Opium poppy is an annual herb belonging to the family Papavaraceae. The commercial product 'Opium' is the air-dried concrete milky latex or exudation (inspissated juice) obtained by incising the unripe seed capsules (head) of opium poppy. A crop raised properly, in one hectare area, fetches about 1 lakh rupees. But its cultivation has to be done under the strict control of the Narcotics Department and it cannot be cultivated everywhere. It can be grown only in those areas specified by the Government of India. In India, this plant is mainly cultivated for its latex (opium) and the seeds come as a by-product. These seeds are quite a rich source of fatty oil and protein and, in many countries of Europe, employed as a major source of cooking oil.

The control and regulation of all aspects of the cultivation of poppy and wholesale trade has been under the purview of the Government of the India since 1773. After the partitioning of the Indian subcontinent in 1947, the essentials of the present system of narcotics control, which derived from the system instituted by the British in the 1857 Opium Act, were retained with few changes. The Government of India is able to effectively license farmers to grow opium. This is highly organized so as to allow elimination of those growers whose yield of opium is poor.

The 1953, the United Nation's Opium Conference Protocol (still in effect) for limiting and regulating the cultivation of opium poppy plants asserts that Bulgaria, Greece, India, Iran, Turkey, The USSR, Egypt, Czechoslovakia, Poland, Germany, Holland, China, Japan, Argentina, Spain, Hungary, Portugal and Yugoslavia are the countries that may legally produce opium. Other countries might grow it for other purposes.

Botanical Description

A small, erect scarcely branched (towards the top) herb attaining a height of 120 cm. Leaves large, alternative, sessile, amphlexicaual, irregularly lobed and toothed and attractive brightly coloured (white, pink, purple, red and variegated) flowers. The main shoot and branches terminate into large, oblong to globose capsules, filled with small white, flat seeds. The seeds are small kidney-shaped, whitish-yellow, grey-brown, reddish-brown or black, and rich in oil.

OPIUM POPPY (*Papaver somniferum*)

Geographical Distribution

It is supposed to have originated in the western Mediterranean region and from there it has spread through the Balkan Peninsula to Asia Minor and India. Since antiquity, its cultivation has been in vogue in Italy, Greece and Asia Minor. It was during the 15th century that the herb was introduced in India. First, it was cultivated along the sea coast and later penetrated into the interior of the peninsula. Its cultivation is restricted to about 25,000 ha in Madhya Pradesh, Rajasthan, and Uttar Pradesh. In Rajasthan, opium is grown in about 4068 hectares distributed in five districts.

Medicinal Uses

It is an important medicinal plant, the source of over 40 alkaloids including psychoactive agents, a great boon to psychiatry for the treatment of mental and nervious diseases and to medical research. The seeds are also reported to contain a high percentage of linoleic acid which lowers blood cholesterol in the human system. The alkaloids, morphine and codenine, are widely used as sedative to relieve pain and induce sleep, in addition to their use against cough. Opium is a very valuable but dangerous drug. It should be used in very limited quantities and under the strict supervision of a physician. The seed is also an important culinary item in India. It is extensively used in the preparation of native confectionery, pastries and bread. In some places, the young plants are also consumed as a leafy vegetable. This is one medicinal plants which is very lucrative.

Chemical Constituents

The seed capsules (latex) contain valuable alkaloids (on dry basis) like morphine (8-14%), codeine (2.5-3.5%), narcotine (5.5-7.5%) and papaverine (0.1-0.8%). Other minor alkaloids include aporeine, codamine, cryptopine, guoscopine, hydrocotarnine, laudanine, narcotoline, neopine, oxynarcotine and papaveramine.

Soil and Climate

The opium crop needs deep clay loam, highly fertile and well-drained soils (pH 6.0-7.5). Such soils, containing adequate organic matter, retain moisture and there is no need of Irrigation during lancing. The crop needs long cold season (20^0C) with adequate sunshine in the early season for a healthy vegetative growth; heavy rains after sowing cause loss in seed germination. Warm, dry weather with a temperature of 30-35^{0}C is required during the reproductive period. Cloudy weather, frost, hailstorms and high gusty winds, particularly during lancing, cause immense damage to the growing crop. Dry, warm weather conditions in February-March favour a good flow of latex and results in higher yields.

Improved Varieties

Talia, Ranghatak, Dhola Chota Gotia, MOP-3, MOP-16, Sharma, Shweta, BROP 1 (Botanical Research Opium Poppy-1) (NBRI-3), Kirtiman (NOP-4) Chetak (U.O. 285), Trishna (IC-42), Jawahar Aphim 16 (JA-16).

AGRO-TECHNIQUE

Land Preparation

Poppy seeds should be sown in very well-prepared soil. The field should be given 5-6 cross-ploughings, followed by planking. The land should be divided into small plots to facilitate Irrigation.

Mode of Propagation

It can be propagated by seeds.

Direct Sowing

The seeds are sown between late October and mid-November in different parts of the country. The seed rate is around 8 kg/ha when sown by broadcasting but when sown in lines at 30 cm apart, about 5-6 kg/ha of seeds are sufficient. Treating of seed @ 4-5% Thiram or Brassicol before sowing, protects the seeds from soil-borne diseases. After sowing, the seeds are covered by a thin layer of soil followed by a light Irrigation.

Manures and Fertilizers

The application of manures and fertilizers in opium poppy depends upon the texture of the soil. Organic matter content and the nature of earlier crop grown. Seed inoculation with azotobactor (W-5) reduces the chemical 'N' fertilizer requirement by 40 kg/ha and produces better latex and morphine yield and also lowers downey mildew infection. An application of 10-20 t of FYM is enough to maintain the soil in good physical condition and to support good initial growth. In addition, the field is applied with 30-50 kg phosphorus and, wherever potash is deficient, potash should be applied at the rate of 25-30 kg/ha. Both these fertilizers are placed 5-6 cm deep in the rows before sowing. The amount of nitrogen varies from 60 (10 t of FYM) to 90 kg (20 t of FYM/ha). It is applied in three equal split doses, the first dose is given at the time of sowing and the subsequent doses at 30-40 and 60-65 days after sowing. For correcting zinc deficiency, 12-30 kg/ha of zinc sulphate should added.

Irrigation

A light Irrigation is given immediately after sowing, if there is not enough moisture available in the soil. The first Irrigation is given 20-25 days after sowing and then frequent light Irrigations is necessary at an interval of 15-20 days. The crop needs Irrigation after second and fourth lancing for obtaining higher seed yield. A total of 8 to 12 Irrigations are required for this crop.

Interculture

All cultivation operations are done by hand. When the plants are 5 to 6 cm high with 3-4 leaves, *i.e.*, after 3-4 weeks of sowing, the first weeding is done. The intervals of the other weeding has to be adjusted locally. In all, about 4-5 weeding may be required during the whole cropping season.

Pests and Diseases

The Opium poppy is infested by a number of insect pests, but serious losses are recorded only occasionally. Important pests causing damage are root weevil, cutworms, aphidsand capsule borer. Most of these pests can be effectively controlled by dusting with 2% Carbaryl or 12% BHC @ 15-20 kg/ha. Aphids can be directly controlled by using methyl dimeton @ 0.025-0.05% and monocrotophos @ 0.05%.

This crop is susceptible to several diseases. Among the various diseases, downy mildew and powdery mildew can cause serious damage to the crop. Use of metalaxyl as seed dressing followed by 3 sprays during crop growth is highly effective in controlling such diseases. Sometime, spray of Dithane Z-78 or Dithane M-45 (2 g/l), beginning at the flowering stage and repeating it 2-3 times at an interval of 15 days also prevents the spread of the disease.

Lancing

After about 90 to 100 days of sowing, the plants which are waist-high begin to flower, *i.e.*, flowering will take place during first week of March, if the crop was sown during the second fortnight of November. Usually after 3 days of flowering, the petals fall off. The capsules mature at 110-120 days after sowing. Lancing of capsules at this stage exudes maximum latex. Usually each capsule is lanced 3 to 4 times. The raw opium is collected from surface with blunt edge of small iron scoop and kept in small plastic box or earthen pot.

Harvesting and Storage

The capsules, after the lancing operation and collection of opium latex, are allowed to dry on the plant itself. The drying process takes about 15 days after the lancing is completed. In India, the capsules are plucked by hand and the seeds are separated after breaking the capsules.

Production and Yield

The opium poppy is now produced for export is India which amounts to 1465 t annually and forms over 90% of the world production. In India, all the opium of commerce is now grown mainly in the states of Uttar Pradesh, Madhya Pradesh and Rajasthan covering an area of 18000 ha. On an average 40-50 kg/ha of crude opium and 400-500 kg/ha of seeds are obtained in India.

Marketing and Trade

Medicinal opium, opium alkaloids, semi-synthetic derivatives and their salts are manufactured in the Government Opium and Alkaloid Works undertaking at Ghazipur (U.P.) and Neemuch (M.P). India produces about 30,000 tonnes of lanced poppy capsules in a year. The lanced poppy capsules are exported from India for the manufacture of alkaloids. Poppy seeds are exported to European countries and Japan. The 80% of the world opium is produced in India is exported for the manufacture of medicinal opium and their alkaloids. On an average, India exported 1000 tonnes of opium at 900 consistence in a year, earning about Rs. 450 million in foreign exchange.

Nearly 92%, 4.5% and 2.5% of the morphine produced in the world is converted semi-synthetically to codeine, ethyl morphine and phocodine respectively. Only about 1% of the morphine produced is used as such. The India is the biggest producer of opium gum and its alkaloids which are being exported to other countries earning Rs. 300-400 million per year. The market price of poppy seeds is Rs. 100-150 per kg which goes up during festival days.

Economics of Cultivation

The market of opium poppy and their products are highly volatile. The economics (per ha) worked out here are subject to fluctuations, depending upon time and place.

Expenditure (in Rs.)	—	6,500.00
Returns (in Rs.)	—	35,000.00
Net Profit* (in Rs.)	—	**28,500.00**

* Estimation of the net profit is analysed on the basis of grower/collectors price. The grower/collector price is always 25-40% less than wholesale market price due to various stakeholders *viz.* Agent, middle man, Brokers, Commission agents.

PERIWINKLE (*Catharanthus roseus* (L.) G. Don)
Syn. *Vinca rosea* Linn.

Vernacular Names

San.–Nityakalyani, Shanthapuspa; *Hin.*–Sadabahar; *Beng.*–Nayantra, Gulferinghi; *Tel.*–Billa ganneru; *Tam.*–Sudakadu mallikai; *Mal.*–Ushamalai; *Mar.*–sadaphul; *Ori.*–Ainskati; *Guj.*–Barmasi; *Punj.*-Tattanjot; *Eng.*–Madagascar perwinkle, Rosy flowered Indian perwinkle; *Trade*–Sadabahar.

Introduction

Periwinkle is one of the few medicinal plants which has found mention in the folk medicinal literature as early as 2nd BC as diuretic, anti-dysenteric, anti-haemorrhagic and wound healing and is considered useful in the treatment of diabetes in Jamaica and India. It, however, gained commercial importance due to the anti-cancer activities in some of the alkaloids present in the plant, specially vinblastine and vincristine. The roots of the plant also contain an alkaloid raubasine (ajmalicine) which has anti-fibrillic and hypotensive properties and for this they have commercial demand. It belongs to the family Apocynaceae.

Botanical Description

An erect, perennial herb which grows up to 90 cm high. High stem pinkish-red, much branched. Leaves simple, opposite, obnate, glabrous on both sides, dark shining and petiolate. The inflorescence is racemose, with the flowers in axillary pairs. The flowers are rose-purple or white or white with a rose-purple spot in the center. Follicle cylindrical, narrow, slightly arched-recurved in pairs. Seeds are numerous, tiny blackish-browm in colour,

Geographical Distribution

The plant is a native of Madagascar and from there, it has spread to India, Indonesia, Indo-China, Philippines, South Africa, Israel, USA and other parts of the world. In India, it is being grown in Tamil Nadu, Karnataka, Andhra Pradesh, Madhya Pradesh, Gujarat and Assam in an area of about 3000 ha.

Medicinal Uses

The plant has been widely used as an abortificent, purgative, antidiabetic, diuretic, haemorrhagic, antimalarial, antidysentric and against skin diseases by the ancient people. Their alkaloids are well known for their hypotensive, antispasmodic properties and cancer therapy. Vincristine sulphate is being marked under the trade name ONCOVIN, which is used against acute leukemia, and vinblastine sulphate as VELBE to cure Hodgkin's disease and other lymphomas and choriocarcinomas. The purified alkaloids extracted from the leaves are effective in treating leukaemia and those from the roots are used to induce cerebrovascular dilation and for hypertension. In addition to the above, the alkaloids leurosidine, leurosovine and rovidine also possess anticancer properties, but they are not used clinically.

PERIWINKLE (Catharanthus roseus)

Chemical Constituents

Periwinkle contains more than 100 alkaloids, distributed through all the parts of the plant. The total alkaloid content in the leaf varies from 0.15 to 1.34%, of which the average content of vinblastine is 0.002%, while that vincristine is 0.005%. Other alkaloids are ajmalicine (raubasin), serpentine, reserpine .

Soil and Climate

The crop is quite hardy and grows well on a wide variety of soils, from deep sandy loam to loam soils of medium fertility, except those which are alkaline or water-logged. There is no specificity in its climatic requirements. It comes up well in tropical and subtropical areas. It can be successfully grown up to an elevation of 1300 m above sea-level. A well distributed rainfall of 100 cm of more is ideal for the crop.

Improved Varieties

(i) Rose-purple flowers, (ii) White flowers, and (iii) White flowers with a rose-purple spot in the center. The first type is being cultivated because of its higher alkaloid content. Recently, two white-flowered varieties named 'Nirmal' and 'Dhawal' have been released by the CIMAP, Lucknow.

AGRO-TECHNIQUE

Land Preparation

For this purpose, the land is ploughed twice and brought to a fine tilth. Weeds, stubble and pebbles are removed. The field is divided into plots of convenient size and the soil is mixed with the recommended doses of manures and fertilizers.

Mode of Propagation

It can be propagated by seeds or vegetatively through cuttings.

Direct Sowing

This method is best suited for large areas. Since the seeds are very small, for ease in handling and distribution, they are mixed with sand about 10 times their weight. The seeds had an initial viability of about 95 to 96%. The seeds @ 2.5 kg/ha are broadcasted at the onset of the monsoon in June-July, in lines spaced 30-45 cm apart and lightly covered. Germination takes place after about 7-8 days. After the germination is complete, the seedlings are thinned at a spacing of 30-40 cm within the row. The flowering starts 40-45 days after sowing.

Nursery Raising

The seeds are sown in well prepared, raised nursery beds in March-April by broadcasting them in rows spaced at 8-10 cm apart and about 1.5 cm deep. They are then covered with a light soil and leaf-mould mixture and watered to keep the bed moist. About 500 g of seeds will be enough to raise seedlings to cover a one hectare area.

Transplanting

In about 2 months time after germination, the seedlings are ready for Transplanting into the field. The seedlings are transplanted at a spacing of 45 cm × 30 cm apart in the field. With this spacing, one hectare requires a population density of about 74000 plants.

Vegetative Propagation

The raise the plants by this method, soft-wood cuttings obtained from the lateral roots have proved better than either the hard or semi-hard wood cuttings. Cuttings of about 10-15 cm length with a minimum of 5-6 nodes, have been reported as being ideal and result in about 90% rooting. Soaking the cuttings over-night in NAA solution of 25 or 50 ppm concentration has been found to further improved rooting to the extent of 96%. This method can be profitable used for multiplying the clones which have high alkaloid content.

Manure and Fertilizers

In areas where FYM is available, it is applied at the rate of 10-15 t/ha to obtain good growth and yield of alkaloid. If Irrigation is available, it is recommended to grow leguminous corps like sunhemp or horsegram and when they reach the flowering stage bury them inside the soil before sowing or Transplanting periwinkle. Green manure will act as a substitute for FYM and is useful in the areas where it is either difficult to procure or it is very expensive. The seeds of the green manure crop should preferably be treated with bacterial inoculants prior to sowing, to increase the development of root nodules which absorb atmospheric nitrogen and fix it in the soil. In case organic manure is not applied, it is advisable to apply a basal dose of 15-20 kg nitrogen, 25-30 kg phosphorus and 25-30 kg potash per hectare per year. In addition, a top-dressing with 20 kg nitrogen can be given in two equal split doses during the season.

Irrigation

In places where rainfall is evenly distributed throughout the year, the plants do not require any Irrigation. However, in areas where rainfall in restricted to a few months in a particular period, about 4-5 Irrigations will help the plants to give optimum yield.

Interculture

This crop requires two weedings in the initial stages of its growth. The first weeding may be done after about 60 days of sowing and the second at 120 days of sowing. Mulching the field with cut-grass or rice-straw will also minimize the weed growth. Application of the chemical weedicide Sinbar at 4-5 kg/ha as a pre-emergent spray is highly effective against all moncot weeds. Similarly, the application of a mixture of 2,4 D and Gramaxone @ 25 kg/ha to the soil before sowing keeps the weeds under control.

Pests and Diseases

The plant is sufficiently hardy and practically free from the attack of insect Pests and Diseases. However, the oleander hawk moth is reported on this crop. Occasionally, some plants have been found to suffer from the 'little leaf' disease, due to infection by mycoplasma resulting in stunted growth and resetting of the leaves of the plant. The disease can be effectively checked by uprooting and destroying the affected plants.

Recently, another disease 'dieback' or twig blight or top rot caused by *Pythium butleri*, *Phytophthora nicotianae*, *P. debaryanum*, *Alternaria tenuissima* and *Collectotrichum dematium* has been found to affect the crop during the monsoon in some parts of the country. The disease can be controlled by spraying Dithane Z-78 at an interval of 10-15 days.

Harvesting and Storage

The crop is harvested after 12 months of sowing. The plants are cut about 7.5 cm above the ground level and dried for the stems, leaves and seeds. If there is a demand for leaves, two leaf strippings - the first after 6 months and the second after 9 months of sowing - can be taken. A third leaf stripping is also obtained when whole plant is harvested. After the plant is harvested, it is dried in the shade. Its seed-pods dehisce and release the seeds with a light threshing, which can be used for the next sowing. Therefore, in order to obtain good seeds, it is advisable to collect them from mature pods two to three months before the harvest of the crop. The field is then copiously irrigated and when it reaches the proper condition for digging, it is ploughed and the roots are collected. The roots are later washed well and dried in the shade.

Production and Yield

Under irrigated conditions, about 4 t/ha of leaves, 1.5 t/ha of stem and 1.5 t/ha of roots, on an air-dried basis may be obtained. Whereas, the yield under rainfed condition may be about 2 t/ha of leaves and 0.75 t/ha each of stem and roots on an air-dried basis.

Marketing and Trade

USA is the world's largest user of this plant's raw material. A single firm which has the patent to manufacture Vinblastine and Vincristine sulphate have been consuming more than 1000 t of leaves of the plant annually. Most of it has been imported from Malagasy and the remaining from India and Mozambique. Hungary is also been one of the major consumer of its leaves. Nearly 1000 t of roots consumed by West Germany, Italy, Netherlands and UK. They extract ajmalicine (raubasin), serpentine, etc. The total demand from these countries is more than 1000 t of roots annually. According to an estimation, nearly 35-40 quintals of ajmalicine is isolated from natural sources by pharmaceutical industries the world over.

The cost of vinblastine sulphate about Rs. 5-6 million per kg and the cost of raubasine is about Rs. 30,000-40,000 per kg. Therefore, it is worthwhile to consider the possibility of production of these alkaloids in India and find suitable market for their export for earning good exchange. The current market price of roots and leaves Rs. 12-15 and 8-10 per kg respectively.

Economics of Cultivation

The market of Periwinkle and their products are highly volatile. The economics (per ha) worked out here are subject to fluctuations, depending upon time and place.

Expenditure (in Rs.)	—	8,500.00
Returns (in Rs.)	—	57,000.00
Net Profit* (in Rs.)	—	**48,500.00**

*   Estimation of the net profit is analysed on the basis of grower/collectors price. The grower/collector price is always 25-40% less than wholesale market price due to various stakeholders *viz.* Agent, middle man, Brokers, Commission agents.

PODOPHYLLUM (*Podophyllum hexandrum* Royle)
Syn *P. emodi* Hook. f. & Th.)

Vernacular Names

Hin.–Bakra chimaka, Bhavan bakra, Papra, Papri; *Beng.*–Papra; *Mar.*–Padwel, Patwel; *Guj.*–Venivel; *Punj.*–Bankakri, Papri; *Urdu*–Ban kakri; *Kash.*–Banwangan; *Eng.*–Mandarke.

Introduction

Podophyllum has been used since immemorial. However, use of the roots has somewhat changed. In traditional system of medicine roots are used as cholagogue, alterative, emetic, vermifuge and bitter tonic. It is reported to be useful in many skin diseases and tumerous growths, its use in curing canceroustissues now in experimental stages. Podophyllum is indigenous drug and is found in wild. The Podophyllum constitutes the sources of an important pharmacopoeal drug. WWF, IUCN, IDRC and other agencies are declared podophyllum as Critically Endangered (CR) due to uncontrolled exploitation, indiscriminate harvesting, loss of natural habitat and illegally for trade. Due to this threats, the natural availability of podophyllum from wild sources become uncertain.

Botanical Description

An erect, glabrous, succulent, perennial herb, 35-60 cm high, with creeping alternate, fleshy rhizome, bearing numerous roots, stem erect unbranched. Leaves 2 or 3 deflexed at first, peltate, orbicular - reniform, palmate, with lobed segments; flowers solitary, erect, rarely 2 , cup shaped, an oldong or elliptic berry, 2.5 - 5 cm diameter, orange or red or pinkish white, containing many seeds embedded in the pulp.

Geographical Distribution

The origin of *Podophyllum hexandrum* is the temperate zone of Himalaya. It is indigenous to India. It occur in sub-alpine and alpine meadows at an altitude of 3000-4500 m in the inner range of Himalayas from Kashmir to Sikkim. It has been found that *Podophyllum hexandrum* is generally found in association with *Rhododendron, Salix, Juniperus and Viburnum*.

Medicinal Uses

The rhizomes and roots of these plants constitute the drug. Dried rhizomes and roots contain a medicinal resin known as podophyllin. Podophyllin is an amorphous powder, vary in colour, with characteristic odour and bitter and acidic taste. The dried rhizomes and roots of *P. hexandrum* (Indian Podophyllum) are official in I.P. and form the source of medicinal resin. Podophyllum is utilized almost entirely for the preparation of Podophyllin commonly used as purgative.

Podophyllin is considered as cholagogue, alterative, emetic vermifuge and a bitter tonic. Kuth oil is very useful in the treatment of bronchial asthama, particularly of the vagotonic type. It is also used in veterinary medicines as a cathartic for dogs and cats. An ointment of Podophyllin is employed to remove warts in animals. Podophyllin can also be used like cokhicine in cytological work because it affects spindle formation and disperses chromosomes. Owing to its cytotoxic action it is used as a paint in the treatment of soft venereal and other wastes.

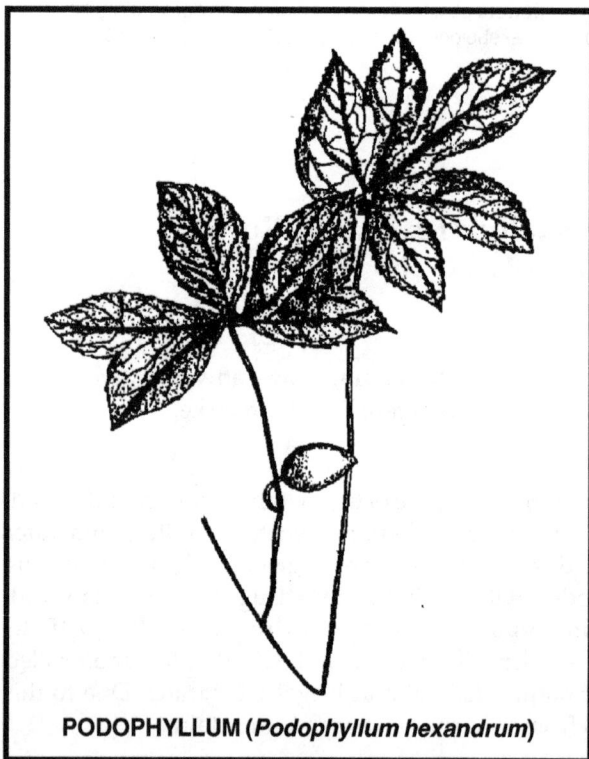

PODOPHYLLUM (*Podophyllum hexandrum*)

Chemical Constituents

Podophyllin from *P. hexandrum* contains podophyllotoxin as the major active constituent, in amounts ranging from 32-54%; a number of other related compounds and their glucosides have been isolated from the resin. It also contains quercetin (8%), kaempferol, astragalin (kaempferol-3-glucoside), an essential oil (3.7%), responsible for the odour of podophyllin, wax (8.6%) and mineral salts.

Soil and Climate

The plant flourishes well in soils rich in humus and decayed organic matter. The plant prefers fertile, medium or light medium loam, well drained soil for its cultivation. In the forests of Himalaya, *Podophyllum* is found growing as an under growth. The plant is generally associated with the species of *Rhododendron, Salix, Juniperus and Viburnum* and less frequently found in open Alpine meadows. The plant prefers temperate, moist and shady locations. For the good growth, it required Rainfall annually 80 - 1300 mm and optimum temperature 0ºC - 16ºC.

AGRO-TECHNIQUE

Land Preparation

The land is ploughed 2-3 times until a fine tilth. FYM is applied before 2nd or 3rd ploughing.

Mode of Propagation

It can be propagated by seeds and also vegetatively from rhizomes cuttings.

Seeds Biology

The seeds have a dormancy period of about one to two years. They may even fail to germinate if they are sown without the pulp. Presowing treatments like soaking the seeds in water or treating them with sulphuric acid have no favourable effect on germination. The seeds may be sown immediately after their collection. The percentage of germination is from 42 to 45%.

Direct Sowing

The seeds are sown with the pulp in June-July immediately after collection. They remain dormant for 9-10 months and showed germination only in the following spring, after the melting of snow. The seedlings are transplanted in the field when they are 6-10 cm tall. The crop raised for seedlings usually take more number of years to produce fan sized markable rhizomes. About 4-5 kg seeds per ha are required for direct sowing.

Vegetative Propagation

Rhizome cutting 1.0-2.5 cm in length are taken from the youngest tip portion. The rhizomes bearing leafy buds give the best results both in the percentage of sprouting and growth of sprouts. The rhizome cuttings may be planted from May to the beginning of July. However, planting at the end of June or the beginning of July gives the best results. However, the growth of plants propagated by rhizome cutting is also slow and the plant takes up 4-6 years to produce rhizomes suitable for exploitation.

Planting

The plants raised in the nursery is transplanted at a spacing of 30 × 30 cm. If the crop is to be raised by direct sowing of seeds, freshly extracted seeds at the rate of not less than 3 seeds should be dibbled at the same spacing to ensure in form stock. The seedlings may be thinned out when they are 5-7 cm high retaining only one plant per hill.

Manure and Fertilizers

It is desirable to give a dressing of well-rotted and sieved farmyard manure by the end of October or the beginning of November. A fertilizer mixture consisting of 20 kg N, 40 kg P and 40 kg K/ha should be applied at the time of planting. In addition, 60 kg/ha N should be applied in two split doses as a top-dressing during the growth period.

Irrigation

The plant may be irrigated during the dry period depending on the soil type and weather conditions. Generally 2-4 Irrigations are required.

Interculture

Weeding and hoeing may be done as and when necessary. Generally one or two weeding are required to be done during rains. Hoeing may be done with second weeding.

Pests and Diseases

No serious incidence of any insect pests or diseases are noticed on this crop. However, rhizome-rot disease is sometimes noticed, for which suitable control measures may be taken up, by treating the rhizomes with a suitable fungicide.

Harvesting and Storage

The development of rhizomes in this plant is very slow. The rhizomes are ready to be collected for the market only when they are 2 to 4 years old. The yield of rhizomes and roots is reported to be high if the crop is harvested 5 or 6 years after planting. In this plant, the underground rhizomes remain dormant during winter and produce aerial shoots in April or May. The shoots bear flowers and fruits are down in November. Usually, the rhizomes which bear 3-5 aerial shoots are considered suitable for harvesting. In spring or autumn, such plants are dug up and the rhizomes and roots are separated and cleaned. Rhizome lose more than 60% of their weight on drying. Then they are cut into longitudinal pieces and carefully dried. The rhizomes harvested in spring are reported to contain a higher resin content than those obtained in autumn, and also the freshly-harvested rhizomes contain large quantities of active principles which are lost on prolonged storage.

Production and Yield

The yield of dried rhizomes and roots varies from 1.5-2.0 t/ha.

Marketing and Trade

The market price of Podophyllum roots is Rs. 1000-1200 per kg.

Economics of Cultivation

The market of Podophyllum and their products are highly volatile. The economics (per ha) worked out here are subject to fluctuations, depending upon time and place.

Expenditure (in Rs.) (in 5 years)	—	4,50,000.00
Returns (in Rs.) [Roots and Planting material (i.e. seed)]	—	12,00,000.00
Net Profit* (in Rs.)	**—**	**7,50,000.00**

*   Estimation of the net profit is analysed on the basis of grower/collectors price. The grower/collector price is always 25-40% less than wholesale market price due to various stakeholders *viz.* Agent, middle man, Brokers, Commission agents.

PRIMROSE (*Oenothera lamarckiana* L.)

Vernacular Names

Eng.–Evening primrose

Introduction

Primrose commonly called the evening primrose, belongs to the family Onagraceae. The name

Oenothera is derived from the Greek word, oinos (wine) and thera (hunt or chase). Plants of this genera are native to North and South America.

Botanical Description

A biennial herb growing to 20 cm. Stems are reddish and the flower-buds hairy. The leaves are simple, alternate or opposite, entire or have a toothed margin, the stipules are absent or caduceus. The flowers are solitary, auxiliary or arranged in racemes, spikes or panicles.

Geographical Distribution

The genus Oenothera is distributed chiefly in temperate America, with some species occurring in the tropics. Some species have been introduced into Indian gardens, of which few have run wild.

Medicinal Uses

The flowers, leaves and stem possesses astringent and sedative properties. All three parts have been employed in the treatment of whooping for digestive problems and asthma, and used as a poultice to ease the discomfort of rheumatic disorders. The GLA-rich oil marketed principally as a health food that is used to treat various ailments such as cardiovascular disorders,

PRIMROSE (*Oenothera lamarckiana*)

metabolic disorders, dermatological conditions, diabetes, breast problems, premenstrual syndromes, actopic eczema, rheumatic arthritis and alcoholism.

Chemical Constituents

The seeds yield an oil (20-25%) containing gamma linolenic acid (GLA) (8-10%)

Soil and Climate

Although the crop can grow well on a wide range of soils, a fertile loam soil is considered ideal. It thrives on wasteground, especially in the dunes and sandy soil. However, the plant is sensitive to water-logging and good drainage is critical for its survival. It prefers a moderately cool climate for its successful growth. A soil temperature of 20⁰C is ideal for the yield and the quality of the oil

AGRO-TECHNIQUE

Land Preparation

The land is prepared well by repeated ploughing and planking and brought to a fine tilth. FYM @10-12 t/ha is mixed well into the soil at the time of the last ploughing. The land is laid out into plots of convenient sizes along with Irrigation channels for easy management.

Mode of Propagation

It can be propagated by seed and by cutting also.

Nursery Raising

For raising the seedlings, the seeds are sown in polythene bags or nursery beds filled with a mixture of soil, sand and organic manure in 1:1:1 proportion. The seedlings can also be raised in nursery beds of $3 \times 1.5 \times 0.15$ m size during the month of August. The nursery beds are mixed with well rotted, powdered cattle manure @ 20 kg/m². About 0.5 kg of seeds are enough to cover 1 ha area for Transplanting.

Transplanting

When the seedlings are about 60 days old they are transplanted into the main field on ridges, in such a way that a spacing of 90-100 cm between the rows and 30 cm between plants is maintained. A spacing of 60×30 cm is also recommended.

Manures and Fertilizers

As this is a newly domesticated crop, very little information on the nutrient requirement of this crop is available and there is a need to work it out locally. However, in the absence of any other recommendation, FYM @ 2 t/ha and fertilizer dose comprising of 40 kg nitrogen, 60 kg of super phosphate and 40 kg/ha of muriate of potash may be used to raise this crop.

Irrigation

It does not require much water. Initially, for 2-3 months, Irrigation is provided at 4 to 5 days' intervals. After this period, when the crop has developed the proper root system, the interval is increased to once in 10 days.

Interculture

Weeds are kept under control through regular weeding and hoeing, since there are no herbicides recommended for this crop. The crop must be kept free from weeds, because weed seeds are difficult to separate from the primrose seeds and they bring down the quality of the oil, making it unacceptable to the market.

Pests and Diseases

Cut-worms are reported to attack the plants during the early period of growth, damaging the foliage seriously and causing a setback. Tarnished plant bugs and aphids are the insects which appear in late June or early July, and continue feeding throughout the remainder of the season, retarding the growth of the plant. Spraying the crop with Carbaryl @ 4 g/l of water controls these pests. A variety of fungus *Botrytis* spp., may attack the new growth in early spring. Drenching the soil with 0.2% of Bavistin helps to control the disease.

Harvesting and Storage

The specific growth habit of the plant indicates variance of seed maturity on each plant. The maturity of pods is observed at the bottom and the top ones are the least mature. Physiologically mature seeds at the bottom are dark reddish-brown or black in colour with rough surfaces. The crop would normally be harvested in April-May. The crop is normally considered ready for harvesting when the bottom 5 to 6 seed pods are split and start shedding seeds. It must be harvested and allowed to dry for 5 to 7 days before threshing and winnowing. The seeds should contain not more than 10% moisture for safe storage.

Production and Yield

A good crop of primrose may yield about 8-10 quintals/ha of dried seeds.

Marketing and Trade

The current market price of seeds is Rs. 25-40 per kg.

Economics of Cultivation

The market of Primose and their products are highly volatile. The economics (per ha) worked out here are subject to fluctuations, depending upon time and place.

Expenditure (in Rs.)	—	6,500.00
Returns (in Rs.)	—	22,000.00
Net Profit* (in Rs.)	—	**15,500.00**

* Estimation of the net profit is analysed on the basis of grower/collectors price. The grower/collector price is always 25-40% less than wholesale market price due to various stakeholders *viz.* Agent, middle man, Brokers, Commission agents.

PYRETHRUM (*Chrysanthemum cinerariaefolium* (Trev) Vis.)
Syn. *Tanacetum cinerariaefolium* (Trev) Schultz-Bip.

Vernacular Names

Eng.–Pyrethrum; *Kan.*–Kriminaashoovu.

Introduction

The term pyrethrum is applied to dried flower heads of *Chrysanthemum cinerariaefolium* (family Asteraceae). Flowers are the source of pyrethrins, which are the most useful insecticides for the control of household insects and protection of food grains. The systematic cultivation of pyrethrum plant in India as a source of powerful insecticide was taken up by the Forest Department of Kashmir State in 1931.

Pyrethrum is produced mainly in developing countries and exported to the industrialized markets. During the last 20 years annual output has been rising on an average of 6.7% in terms of valve. The five main pyrethrum-producing countries are Japan, Kenya, Tanzania and Rwanda. Other countries producing Pyrethrum in smaller quantities include Bolivia, Brazil, Hungary, Indonesia, India, Papua New Guinea, Peru, Rhodesia, Taiwan, Zaire and USSR. At present, Kenya and neighbouring countries of Tanzania and Rwanda produce over 80% of the total pyrethrum produced in the world.

Botanical Description

A perennial bushy plant growing up to 1 m height. Leaves are alternate, petiole and divided into lobed and dentate segments. Heads are large, terminal, long peduncled or small and corymbose. Flowers are usually yellow, white or rosy. Achenes subterete or angled variously ribbed or winged.

Geographical Distribution

It is commercially cultivated in Kashmir and with largest acreage in south India especially in Tamil Nadu (Nilgiri hills), Andhra Pradesh etc.

Medicinal Uses

Pyrethrum is rich source of pyrethrins. The pyrethrins having insecticidal properties are much sought after due to its low mammalian toxicity, fast degradation and eco-friendly nature. Pyrethrum is effective in pediculosis and scabies. It is useful as an anthelminitic against *Ascaris lineata* and other intestinal parasites in veterinary practices. It is a contact poison highly toxic to insects. It is used either as powder or as spray to control agricultural and household pests.

Chemical Constituents

The alkaloids present in all parts of the plant, but mainly concentrated in the developing seeds of the achenes. The active constituents consists of 3 pairs of esters: Pyrethrin I, jasmolin I, pyrethrin II, cinerin II and jasmolin II and collectively referred to as Pyrethins. It varies from 0.04 to 2.00% with an average of 1.2%.

Soil and Climate

It grows best on fertile, deep and well-drained soil. But it is also reported to come up well in acid soils. It requires cool dry climate for better vegetative growth. Fall in night temperature triggers flowering. Thus flower production and pyrethrin content will be highest in March when the maximum and minimum temperature will be optimum. Well distributed rainfall of about 1000 mm is good for its growth and yield. Frosts during early tender stages of growth affect the crop performance adversely.

PYRETHRUM (*Chrysanthemum cinerariaefolium*)

Improved Varieties

The local selection C761, SL 7, SL 71564, strain 387 and Hansa.

AGRO-TECHNIQUE

Land Preparation

The land should be prepared well before Transplanting. Good soil tilth is obtained by 1-2 deep ploughing followed by harrowing and planking. The soil should be uniformly leveled before Transplanting to prevent water logging.

Mode of Propagation

It can be propagated by seeds or shoot cuttings.

Nursery Raising

The nursery is prepared in the spring season *i.e.* February to March. Soil should finely prepared to facilitate proper drainage, good seed germination and seedling growth. Seeds should be clean and free from pest damage. The raised beds of 15 cm in height and about 1.5-2.0 cm wide with furrows on either side will help proper drainage and aeration. The seeds will germinate in about 3 weeks. The growth hormone gibberellin (GA) gave the highest germination rate (76%) and the shortest germination time (14 days). Further nursery beds of pyrethrum inoculated with a mixed culture of Glomus faciculatum, G. caledonium and Gigaspora margarita resulted in increased seedling height, number of branches/seedling and roots/seedling, root length of seedling dry weight compared to untreated controls. When transplanted in the field inoculated plants showed a 7-10 day advancement of flowering. Care should be taken for proper watering and prophylactic measures against seedling rot and damping off by spraying fungicide. The seedlings will be ready for Transplanting in 8-10 weeks.

Transplanting

Transplanting in the main field may be started with the pre-monsoon in south India. While in central India Transplanting during first week of September recorded higher yield. Good soil moisture in the soil is very essential for ensuring better Transplanting and higher success rate. Seedlings should be carefully lifted from nursery without affecting the root system. They may be dipped in fungicide solution to safeguard against fungal diseases. Young seedlings are usually transplanted in 60 cm rows giving 30-45 cm intra-row spacing. Proper Irrigation after Transplanting to maintain soil moisture is necessary and at any point of time the transplanted seedling should not suffer for want of moisture before they are well established.

Manure and Fertilizers

The main field should be rich with organic matter and available nutrients. Application of about 10-15 t well decomposed organic manure and thorough mixing in the soil will be helpful to crop. Further application of about 20-30 kg nitrogen, 90-120 kg phosphorus and 15-30 kg potash/ha will be needed for harvesting good crop. Among sources application of rock phosphate 45 days before planting was at par with single super phosphate.

Irrigation

Although crop is grown in rainy season, due to lack of good rains, crop is made to suffer from soil moisture stress. When the soil moisture comes near to wilting point light Irrigation should be given through flooding or furrow Irrigation. Crop needs frequent Irrigation in the initial stage while in later stages fewer Irrigation will be sufficient. During summer crop needs frequent Irrigation as compared to rainy and winter seasons.

Interculture

It requires clean cultivation, proper weeding and hoeing to obtain good yield. The first hoeing can be done in autumn after the last crop is harvested, which should be followed by two weeding and hoeing in spring. For harvesting a good crop, weeding after every eight weeks is essential.

Pests and Diseases

The crop does not suffer from any severe insect pests or diseases once established in the main field. However, a few need based sprays against fungal diseases viz. seed rot, damping off and seedling rot will help better seedling and crop stand after Transplanting.

Harvesting and Storage

The well grown crop comes to flowering in about 7-9 months and thereafter the continuous flowering will be there. The right time of harvesting is important for getting higher yield and maximum pyrethrin content. The crop gives poor yield during the first year. The optimum yield are obtained only during second and third year. The pyrethrin content will be maximum when three quarters of the disc florets had opened. They are harvested at regular interval of 10-15 days or as and when the new flushes of flowers are mature and ready for harvest. In temperate areas like Kashmir, only one harvest is obtained in the end of June or first week of July. The flowers are cut at base on the stalk. Once the flowers are harvested they are sun dried for some time to reduce the moisture content in them. Bearing of fresh flower with out drying results in rotting and reduction in pyrethrin content.

The flowers should be dried immediately after picking. In India the flowers are dried in sun in thin layer. These should be turned frequently in order to avoid fermentation. The ideal method is to dry the flowers in hot air-dryers for larger areas.

Production and Yield

The yield of flowers depends on cultivation practices and climatic conditions. The average annual yield of pyrethrum is about 200-300 kg/ha in Kashmir and 400 kg/ha in Kodaikanal (Tamil Nadu).

Marketing and Trade

After world war-I, Japan become the principle exporter of pyrethrum to the world market and in 1935 its maximum output was 12,500 tonnes. In 1933, Keyna started production of pyrethrum on a large scale and was the second largest producer in the world in 1938, with an output of nearly 2,000 tonnes. In 1938, Brazil was entered the pyrethrum export market by 250 tonnes. The total world production of pyrethrum flowers is over 15,000 tonnes and America is its biggest consumer.

India has been importing the pyrethrum flowers from Japan and Kenya. In 1947, the Central Govt. Planning Committee was pointed out that the annual requirements of pyrethrum flowers was in the neighbourhood of 4,000-6,000 tonnes, and was likely to increase ten times in the course of next 15 years.

The increased awareness of the dangers of synthetic insecticides and the rapid building up of resistance in the insect population against synthetic insecticides are acting in favour of pyrethrins, which do not show any of these disadvantages. The demand of pyrethrum flowers is, therefore, resizing rapidly in the world market. The pyrethrum industry appears to have promising future.

QUINGHAO (*Artemisia anna* L.)

Vernacular Names

Sans.–Chauhara; *Hin.*–Kirmala; *Kan.*–Murni; *Guj.*–Chhuvaria ajmoda; *Mar.*–Kirmani ova; *Eng.*–Sweet wormwood.

Introduction

It is a strongly scented annual belonging to the family Asteracece. Quinghao is the Chinese name given to the wormwood or sweet wormwood, *Artemisia annua* L. The plant has been used in Chinese system of medicine for more than 2000 years for malarial fever and certain other diseases. A decoction of leaves in water was given to cure fever. As a result of studies carried out at the Chinese Institute of Traditional Medicine, it was found that the extract of the herb was feective in curing malaria. As a result of further research, the active component artemisinin (quinghaosu) was isolated in 1972. Further studies conducted by the Chinese have proved that the drug is effective against cholorquin-resistant strain of Plasmodium falciparum as well as against cerebral malaria. Two derivatives of artemisinin, namely, arteether and artesinate have been found to be very effective and clinical trials in China have shown that both of these derivatives are effective against chloroquin-resistant strains of malarial parasite as well as against cerebral malaria. The plant is indigenous to China, southern USSR, Turkey, Iran, Afghanistan and naturalized in USA. Its experimental cultivation has been carried out in USA and India.

Botanical Description

An annual, determinate, erect herb with a very slender and glabrous stem. The branches are deeply grooved,the leaves are broad, 3-pinnatisect or decompound, the segments are serrate or lobulate; the heads are 2 mm in diameter, subglobose, secured in very slender panicled racemes, the pedicle is pendulous, the involucre bracts are glabrous, the inner obicular is green and shiny with a scarious margin.

Geographical Distribution

The plant is indigenous to China as mentioned earlier. It is found in India, ascending to 5 000 ft. It also occurs in Afghanistan and Northeastern Asia, USA, UK and France. In India, it is cultivated in temperate as well as subtropical conditions in the Kashmir Valley and in the hills of Himachal Pradesh and Uttar Pradesh on a limited scale.

Medicinal Uses

The essential oil, called the 'Artemisia oil', is used in perfumery, cosmetics and in dermatology. The artemisizi oil has some specific antimycotic and antimicrobial action. It is useful in the treatment of skin moulds. Artemisinin, the antimalarial drug, has been recently introduced into the Indian market. It is very useful in cerebral malaria as well. Artimisinin is adminstered-as an intramuscular injection and is also reported to have an inhibition rate of 100% against human gastric cancer cell-line SGC-7901 at 50 fg/ml.

Chemical Constituents

The main constituent of the plant is artemisinin, which varies from 0.05417% with an average of 0.1%. Other constituents include the sweet smelling essential oil, artimisilene, arteannuin, artemusinic acid, phytene-1, 2-diol, pinene, cineole, 1-camphor. scopoletin, scopolin, arteannuin—,quercetagetin-6, 7,3',4'-tetramethyl ether, n-nonacosane, n-pentacosane, —amyrin acetate and qinghaosu.

QUINGHAO (*Artemisia anna*)

Soil and Climate

The crop is adapted to a wide range of soil types from sandy loam to laom, which are free from water-logging, but a well-drained, light-loam, rich in organic matter, is reported to be the best suited. It is a short-day, temperate plant, which requires a cold winter and a moderate summer. It can also be cultivated in subtropical areas as a winter crop.

AGRO-TECHNIQUE

Land Preparation

The land is ploughed several times to produce a fine tilth before the seedlings are transplanted. For convenient management, it is laid out into beds, depending upon the local conditions, after applying the required dose of manures and fertilizers.

Mode of Propagation

The crop can be propagated by seeds.

Nursery Raising

Nursery-beds of convenient sizes are prepared and well-decomposed FYM compost at the rate of 10 kg per bed is applied. Further, 250-500 g of seeds (sufficient to raise seedlings for I ha) mixed with sand are spread uniformly over the nursery-beds and covered with a thin layer of soil or sand. The beds are kept moist by frequent watering, with the help of a sprinkler or a rose-can. The seeds can be sown in the nursery during September-October for the late rainy season crop and during December for the summer crop. The seeds germinate in about 5-8 days. The seedlings would be ready for Transplanting after 6-8 weeks.

Transplanting

The beds are irrigated a day prior to Transplanting. The seedlings withstand Transplanting very well. Transplanting is done at a spacing of 30-40 cm or 60 cm between rows, and 45-60 cm between the plants. Gap-filling should be done within 8-10 days of planting. It is advisable to plant two seedlings per hill at the time of Transplanting.

Manure and Fertilizers

The crop responds well to fertilizers. A fertilizer dose of 60-80 kg nitrogen, 40-60 kg phosphorus and 60 kg potash/ha may be applied for a good yield. The full dose of phosphorus and potash and two-third of nitrogen is incorporated into the soil at the time of Land Preparation. The remaining one-third of nitrogen is applied in two equal split doses at 30 and 60 days after Transplanting. A basal application of Borax at the rate of 8 kg/ha is recommended.

Interculture

The Crop requires 2-3 weedings and hoeings during the growth period. The herbicides, Metachor and Chloramben (2.2 kg/ha) as pre-emergent treatment and Trifluralin (6.5 kg/ha) as post-emergent treatment have been found useful for weed control.

Irrigation

The field is irrigated frequently for establishment of the crop. Once established, only 3-4 Irrigations are required in all.

Pest and Diseases

The crop is affected by ants and other pests. The ant menace can be minimized by mixing about 10 kg/ha of Carbaryl dust into the soil at the time of Land Preparation. Damping-off disease can be overcome by adjusting the planting time, so that bright weather prevails during the first few days in the early stages of establishment.

Harvesting and Storage

The duration of this crop is about 4.5-5.0 months. The crop is harvested as soon as the flower initiation takes place. Delayed harvest or harvesting before the flowering, result in a low yield of artemisinin as well as essential oil. The crop is cut 15-30 cm above the ground-level with a sickle and dried in the shade. After the drying process is over, the leaves and flowers are separated from the stalks.

Production and Yield

The fresh herbage yield is about 10-15 t/ha which, in turn, gives 30-40 kg/ha of essential oil. The essential oil percentage generally varies from 0.2-0.4% under Indian conditions. The essential oil content is found to be generally higher (0.34%) in the early sown crop (September) and it is known to decrease due to delay in sowing/planting. The essential oil content is highest in the inflorescence. Further, the artimisinin content is found to be highest, at the anthesis and declines after flowering. The yellow flowers are reported to contain 2-to 4- fold higher concentration of arteminsin as compared to the leaves.

RED SORREL (*Hibiscus sabdariffa L.*)

Vernacular Names

San.–Patwa; *Eng.*–Rozelle, Red sorrel; *Hin.*–Lal ambori, Patwa; *Tel.*–Yerra gogu; *Tam.*–Pulichai, Kerai; *Kan.*–Pulachakiri, Pundibija; *Mal.*–Polechi, Puichehai; *Mar.*–Lal ambadi, Patwa.

Introduction

Roselle (family-Malvaceae) is an important multipurpose annual under-shrub used in Unani and modern systems of medicine throughout the tropics and subtropics, the world over. Apart from its wide range of uses, roselle is becoming increasingly' important because of the growth in its export as a source of natural dye, obtained from the dried calyces. The calyces which are valued as the main items of trade contain acids, pectin, mucilage, crude protein, carbohydrates, ascorbic acid, anthocyanins and minerals.

Botanical Description

An erect, branched, glabrous undershrub, 0.5-1.8 m tall, with a strong tap-root, The leaves are alternate, the stipules subulate, deciduous, 5-8 mm long. The flowers are solitary, axillary, 3-5 cm long with a greyish yellow corolla and an orange staminal column. The fruit is a ovoid capsule, villous, closely invested with an enlarged, persistent fleshy calyx, dehiscing by 5 valves when ripe. The seeds are reniform, dark-brown, pilose, 4-6 mm long and are 22-34 per capsule.

Geographical Distribution

It is native of Tropical Africa. It is also being cultivated in warm countries, particularly in the Philippines, Malaysia, Indonesia, India, Cuba, Central America, California, Florida, Egypt, Sudan, Nigeria, West Indies and Sri Lanka. In India, it is cultivated in Punjab, Bihar, Uttar Pradesh, West Bengal, Assam, Orissa, Maharashtra, Karnataka, Andhra Pradesh and Tamil Nadu.

RED SORREL (*Hibiscus sabdariffa*)

Medicinal Uses

The dried calyces have acquired a considerable reputation in the international trade as a substitute for tea and coffee and as a popular soft drink. The fresh calyces are used as a vegetable or eaten as salad or used in confectioneries, in the preparation of soups, purries, chutneys, sauces, jams, jellies, squashes, tarts, tisanes and wines. The astringent herbal drink made from hibiscus, known as medicinal tea, is used as a cure for fever, cold, cough, pneumonia, and is said to possess diuretic, demulcent, emollient, laxative, antiseptic, antiscorbutic, antispasmodic and choleretic properties. The drug is also used to cure heart and nerve diseases, asthma and skin diseases. Besides, the seeds contain fatty oil, proteins, starch and lipids. The seed-meal is made into soup. Medicinally, the seeds are reported to have demulcent, diuretic and tonic properties and are recommended in cases of dyspepsia, and general debility, The oil of the seed is said to have antibacterial properties.

Chemical Constituents

The principal water-soluble acids present are citric acid, d-malic acid, tartaric acid and hibiscus acid. The fresh calyxes contain moisture (88.26%), acid (3.74-4%), pectin (3.19%), and others *viz.* crude protein, carbohydrates, calcium, phosphorus, iron, manganese, aluminium, sodium and potassium in traces.

Soil and Climate

For good crop, a loamy soil, sandy soils and heavy loam soils capable of retaining moisture with a good quantity of organic matter is ideal. It prefers a warm humid or dry climate with an even rainfall of 150 to 200 cm annually. It can also grow in drier rainfed areas as an irrigated crop, but it cannot withstand heavy continuous rains, water-logging, winter cold and frost, it is known to do well from locations at sea-level up to an elevation of 600-780 m.

AGRO-TECHNIQUE

Land Preparation

Before sowing, the field is thoroughly prepared by bringing the land to a fine tilth with repeated ploughings. It is then laid out into plots of convenient sizes by constructing bunds and channels.

Mode of Propagation

It can be propagated by seeds and shoot cuttings.

Sowing

As the seeds lose the viability rapidly, fresh seeds should be used for sowing. The crop is usually sown from April-May or October-November. After the preparation of the plots, the soil is incorporated with 5 t of well-decomposed FYM or compost. Ridges and furrows are made at a spacing of 75 cm along the rows and the rows are irrigated lightly immediately after sowing. About 2-3 kg seed is required to cover one hectare.

Manures and Fertilizers

It responds well to the application of manures and fertilizers. Well-decomposed FYM, at the rate of 5 t/ha, is incorporated into the soil -at the time of Land Preparation. Subsequently, a fertilizer dose of 50 kg nitrogen, 75 kg/ha each of phosphorus and potash and is given at the time of sowing. An additional quantity of 200 kg nitrogen is applied in two equal split doses as top-dressing at 30 days and 70 days after sowing.

Irrigation

After sowing, the plots are immediately provided with light Irrigation. Later, Irrigation is given at regular intervals till the seedlings emerge and are well established. In the crop sown in the rainy season except during some long dry spells, practically no Irrigation is required. However, the crop sown in October-November needs to be irrigated once in 7-8 days, depending on the weather conditions.

Interculture

Generally 2 weedings during the early period of growth are enough to suppress the weeds. As it has a good canopy in the later stages, the weeds do not get much of a chance to grow.

Pests and Diseases

The crop is very hardy and there is no report of any serious insect pest in red sorrel. However, some insect pests like capsule borer, flea beetle and the cotton strainer bug have been observed to affect the crop. Two sprays of 0.2% Quinolphos or 0.15% Monocrotophos will control the attack of the capsule borer and flea beetle. Some fungal diseases affected the growth and their yield mainly are stern-rot, root rot. These fungal diseases are controlled suitable fungicides. Treating the seeds with 0.2% Captan or 0.1% Calixin or Kitazin before sowing and then drenching the plants with 0.1% Carbendiazim or 0.2% Diafoltan or 0.3% Nlancozeb chemicals a fortnight after germination, controls the disease.

Harvesting and Storage

The crop is mature in 5-6 months. The plants come to the flowering stage after about 45-60 days of sowing and it extends over a period of 4 months. The calyxes grow rapidly and are ready for picking within 15-20 days of blooming. While harvesting, the whole fruits are plucked when they are tender, plump, fleshy, crisp and deep-red in colour. The calyx lobes gathered this way are placed fresh or dried in the shade for 12-15 days till a moisture level of 12% is reached. The harvesting season lasts for 2-3 months.

For seed production, the calyxes are picked while the capsule is still attached to the plant. After harvest, the plants are allowed to dry in the field for some time. They are then brought to the threshing yard and the seeds are separated. These seeds are cleaned by winnowing and packed in airtight containers till they are marketed.

Production and Yield

A good crop yields nearly 13-14 t/ha of fresh calyxes which in turn may yield 1.8-2 t/ha dried calyxes. The recovery of the water-soluble extracts containing the dye may be around 1.97% on a fresh-weight basis. In addition, a seed yield of 2.5-3.0 t/ha, which contains 16.2% fatty oil, is also obtained from this crop.

SAFED MUSLI (*Chlorophytum borivillianum*)

Vernacular Names

San.–Musali; *Hin.*–Safed musli, Kulai; *Mar.*–Kuli, Safed musli; *Guj.*–Gholi-musali, Janjaria; *Tam.*–Tanirureta; *Tel.*–Kushelli; *Mal.*–Shedeveli.

Introduction

Tubers are white, and hence it is called as safed mush. Mush tubers available in the markets are normally collected from tribals from hilly areas. Safed musli *is* an important Ayurvedic medicinal plant. The small seedlings of musli are found in forests during rainy season.

Botanical Description

Leaves look like that of garlic but slightly yellowish, and white flowers with 6 petals are arranged on the flowering stalk which emerge from the centre of the plant. About 2025 flowers on the flowering stalk appear in July. The seed is very small, black and enclosed in flowering boles. In one bole there are about 10-12 seeds.

SAFED MUSLI (*Chlorophytum borivillianum*)

Geographical Distribution

The plant is observed in the valley of Himalaya, Satpuda, Vindhya, Arawah and in hilly areas of Bihar and Assam. It is also found in the parts of Rajasthan, Gujarat in some parts of Maharashtra.

Medicinal Uses

It is useful on *itla* and *vata*. It *is* explained *as vrishya, dhatuvardhak, sheet, madhur, rasayana,* slightly bitte, healthier and energetic in Ayurveda. It is effective on *fatigueness, daaha* and in blood purification. It is useful in certain diseases like renal calculus, dhupani, sangrahani, leucorrhoea. It is lactating, energetic to heart.

Chemical Constituents

Tubers are of great medicinal value containing the steroid sapoginine (1 -2 %), proteins (10-12%) and calcium to sorrie extent With some water-soluble minerals. Tubers are fat free.

Soil and Climate

The loamy, gravelly soil with rich organic matter and well drained soil with pH ranging between 6.5 to 7.0 is suitable for growth. Good amount of soil moisture during growing season favours freshly root development. Dry weather is favourable for its growth. It can survive at low temperature. It can be grown under warm humid climatic conditions. Nearly 25-35°C temperature and 50 to 75 cm annually rainfall is suitable for its growth.

Improved Varieties

RC-2, RC-16, RC-36, RC-20, RC-23 and RC-37.

AGRO-TECHNIQUE

Land Preparation

The land should be ploughed and harrowed so that there should be proper aeration. Weeds like *Cypents rotundus, Cynodon dactylon* should be killed completely. After Land Preparation, the broad bed furrows 90 cm broad and 15 cm height should be prepared. Water channels should be prepared for each bed. Well decomposed FYM about 15-20 tonnes/ha should be mixed in the soil properly before planting, or the vermicompost can be used. Due to the use of organic manures the water-holding capacity of the soil increases and it becomes porous.

Mode of Propagation

It is propagated by seeds and tubers. But the germination percentage by seed is very less and it takes about 18 months for harvesting tubers. Hence, the sprouted tubers are economical for planting. It takes about 6 months for maturity in the rainy (kharif) season. Previous year tubers are used for cultivation.

Sowing

The bunch of tubers should be removed from soil in May and tubers are separated from the bunch in such a way that some portion of crown should remain attached to each tuber. The number of tubers varies from plant to plant and on an average 5-30 tubers/ plant. Tubers separated should be stored in gunny bags filled with mixture of fine sand and silt with moist condition under shade in June, tubers get sprouted naturally and these are used for planting.

Planting

Planting should be done on the onset of monsoon, preferably in June. The soil should be moist at the time of planting. The planting should be done at the spacing of 6-8 cm between tubers while distance between rows should not be less than 30 cm. Tubers should be placed deeply in the soil. Nearly 6-7 quintals tuber are required for 1 hactare land.

Irrigation

Irrigation should be given as and when required during growth period to avoid any water stress. Sprinklers may be used if available for Irrigation purpose.

Interculture

Weeding should be done by hand. As the plants of musli appear like that of grass, there is possibility of getting removed these musli plants along with grass. The field should be kept weed free.

Harvesting and Storage

Tubers are ready for harvesting after 6 months of planting. Harvesting is done in December-January, when the matured leaves drop down. This is the best indication for harvesting tubers. The tuber bunches are removed by digging the soil.

Bunches removed from soil should be thoroughly washed with water. Such clean bunches are dried in shade for 2 days. The skin on the tubers should be removed by pressing the tubers by holding between 2 fingers. Such peeled mush should be again washed with water and dried completely in shade. The hard dried musli is marketed.

Production and Yield

Fresh tubers 3.0-3.5 tonnes/ha or dried- tubers-0.6-0.65 tonnes/ha.

Marketing and Trade

The safed musli roots and formulations are sold in the market. This herb market is growing day by day due to their demand in the global herbal market. Nearly 35% supply of this herb has been fulfilled by wild resource (forest) of Maharashtra and M.P. while 45% from cultivated sources. The rest 20% imported from Asian countries. According to estimation, the consumption of safed musli is nearly 50-80 Mts annually by various Indian pharmaceutical industries. Presently, the high demandable medicinal plants leading to sudden escalation in their prices in the market day by day. The current market price of roots of safed musli is Rs. 700-1200 per kg depend on quality.

Economics of Cultivation

The market of safed musli and their products are highly volatile. The economics (per ha) worked out here are subject to fluctuations, depending upon time and place.

Expenditure (in Rs.)	—	2,25,000.00
Returns (in Rs.)	—	6,00,000.00
Net Profit* (in Rs.)	—	**3,75,000.00**

*   Estimation of the net profit is analysed on the basis of grower/collectors price. The grower/collector price is always 25-40% less than wholesale market price due to various stakeholders *viz.* Agent, middle man, Brokers, Commission agents.

SARPENTINE ROOT (*Rauvolfia serpentina* (L.) Benth. ex Kurz)
Syn. *Ophioxylon serpentinum* Linn.

Vernacular Names

San.–Sarpagandha; *Hin.*–Chota chand, Sarpagandha; *Kan.*–Sarpagandha, Sutranabhi; *Mal.*–Sarpaganthi, Amalpori, Tulunni; *Guj.*–Sarpagandho; *Tam.*–Saraganthi, Sivan amalpodi; *Tel.*–Patalaganthi, Sarpaganthi; *Ori.*–Patal garur, Sanochado; *Mar.*–Harki, Sapasanda; *Eng.*–Serpentine root; *Trade*–Sarpagandha.

Introduction

It is one of the most important native medicinal plants of India. The hypotensive properties of sarpagandha was first discovered by Chopra in 1953, but the attention of the Western countries was drawn to it only after the isolation and identification of its most active alkaloid 'Reseripine' by the Swiss scientists, Schiller and Muller, of CIB Pharmaceuticals in Switzerland, in 1952. After that, there was a great demand for its roots, leading to its indiscriminate uprooting from wild sources where it grew luxuriantly. Intense collection brought the plant to the verge of extension. So, in 1955, the government of India put a ban on the export of the raw drug and attempts to cultivate the plant were taken up at a number of places. About 30 alkaloids are known to exist in this plant. The commercial supplier, this raw material is often adultetered with stems of *Rauvolfia serpentina*, roots of other *Rauvolfia* species and roots of the *Clerodendrum* species.

Botanical Description

A perennial undershrub, growing to a height of 60-90 cm. Leaves simple, elliptic or lanceolate, glabrous, bright-green above and pale-green beneath, pointed and occurring in whorls of 3-5. The inflorescence is a many flowered corymb with white or pink flowers. The fruit is a drupe, 0.5 cm in diameter and shiny black when fully ripe. The fresh roots emit a characteristic acrid aroma and are very bitter in taste.

Geographical Distribution

It is indigenous to the moist, deciduous forests of south-East Asia including Burma, Bangladesh, Sir Lanka, Malaysia, the Andaman Islands and Indonesia. Most of the drug is obtained from wild sources in these countries. It is cultivated on a small scale in India and Bangladesh.

In India, it is found in shady, moist in the central region, *i.e.* between Sirmor and the Gorakhpur district of Uttar Pradesh. In Eastern region, it occurs on the forest margin of mixed deciduous forests in Bihar, North Bengal and

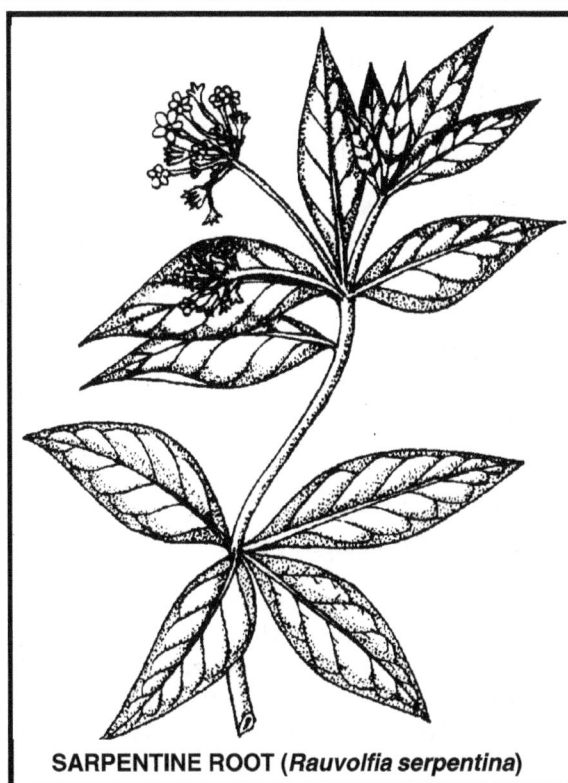

SARPENTINE ROOT (*Rauvolfia serpentina*)

Assam as well as in Khasi, Jaintia and Gharo Hills. In the Western Ghats. It occurs more frequently in Goa, Coorg, the North Kanara and Shimoga districts of Karnataka and Palghat, Calicut and Trichur in Kerala. In Orissa, Andhra Pradesh and Himachal Pradesh are areas comprising the catchments of the river Godavari are the richest. The plant is chiefly associated with Sal (*Shorea robusta*) forests as well as bamboo brakes.

Medicinal Uses

In India, the root of sarpagandha has a 400 year history of use in treatment of snakebite, insect stings, nervous disorders, mania and epilepsy; intractable skin disorders, such as psoriasis, excessive sweating and itching; gynaecological ointments for menopause, toxic goiter and in conditions such as angina pectoris and to promote uterine contraction in childbirth. The importance of the drug and the alkaloids obtained from it has been recognized by the allopathic system in the treatment of hypertension and as a sedative or tranquillizing agent.

Chemical Constituents

The root contains an alkaloids 'ophioxylin', an orange coloured crystalline principle, resin, starch and wax. The total alkaloid content varies from 1.7 to 3%. Five crystalline alkaloids isolated are ajmaline, ajmalicine, serpentine, serpen-tinine and yohimbine. Other constituents identified are phytosterol, oleic acid and unsaturated alcohols of formula $C_{25}H_{44}O_9$. The root-bark, which constitutes 40-60% of the whole root, is rich in alkaloids.

Soil and Climate

It grows in a wide of soils, from sandy alluvial loam to red lateritic loam or stiff dark loam but prefers clay or clayey loam with a large percentage of humus (pH 4.6-6.2). It flourishes in hot, humid with tropical or subtropical belt, having the benefit of monsoon rains. A climate with a temperature (10-30⁰C) and annually rainfall (250-500 cm) seems to be well suited for good growth and yield.

Improved Varieties

RS-1 is released by Jawaharalal Nehru Krishi Vishwa Vidyalay, College of Agriculture, Indore.

AGRO-TECHNIQUE

Land Preparation

The land is cleared of weeds and ploughed to a depth of 30 cm. Raised beds are made, which should contain one-third quantity of well-rooted FYM and two-thirds of fine soil. Well-rotted FYM at 25-30 t/ha is added during Land Preparation.

Mode of Propagation

It can be propagated by seeds or vegetatively *viz.* root-cuttings, root-stumps, stem-cuttings.

Seed Biology

Its seed germination is low percentage, but quite variable, ranging from 10-60%. Fresh seeds, collected from ripe fruits and immediately sown, show a higher percentage of germination (58-74%). The collection of mature seeds is usually done from September-February. Fruits mature between July-November. The viability of the seeds drops markedly with the increase in the interval of time between collection and sowing. Therefore, the collection of ripe fruits twice a week is necessary. The mature seeds are thoroughly dried in the sun and stored in dry places or in airtight containers, retained their viability for about 6 months. The germination rate of the seed also differs under varying agro-climatic conditions.

Nursery Raising

The nursery should ideally be located in partially shaded areas with Irrigation facilities. The time of sowing seeds in nursery very under different agro-climatic conditions. Sowing should be done in April end in Maharashtra and M.P., first week of May in West Bengal and during third week of May in Jammu. Seeds are sown in the April-May. The seeds should be soaked in water overnight and the light seeds which float can be discarded. The seeds can be treated with Thiram @ 3g/kg of seeds. About 5-6 kg of seeds will yield seedlings sufficient to plant one hectare. The germination is gradual and the growth of the seedlings is slow. Germination starts after 15-20 days and continues up to 40-50 days after sowing. The nursery should be kept moist throughout the germination period.

Transplanting

The seedlings of 40-50 days old which have 4-6 leaves are ready for Transplanting. The seedlings are carefully dug out and the tap-root should be cut. They are then dipped in a 0.1% solution of Emisan fungicide before planting, to protect them against soil-borne fungus causing damping-off disease. About 15 cm deep furrows are dug at a distance of 45 cm. The seedlings are transplanted into the furrows, by making holes large enough to receive the seedlings along with the accompanying clump of earth. A spacing of 30 cm between the plants should be maintained. The seedlings are buried up to the first pair of leaves and the soil around them is lightly pressed. Irrigation after Transplanting is essential, and should be continued at regular intervals until the seedlings are established.

Vegetative Propagation

By Root Cuttings

Large tap-roots with a few filiform lateral secondary rootlets are used. Cuttings of 2.5-5.0 cm length are planted in holes, at the beginning of the monsoon (June-July) and are almost completely covered with earth, leaving only 1 cm above the surface. Nearly 50% of the root-cuttings sprout in about a month. Trials have shown that under irrigated conditions, root-cuttings of about 0.25 cm diameter planted during March-June give a 50-80% success rate. About 100 kg of root-cuttings are required to plant 1 ha. The high percentage of success obtained by using root-cuttings makes it more preferable than propagation by seeds. However, the recovery of roots from them has been found to be not as high as in plants raised from seeds.

By Root Stumps

Propagation by using about 5 cm of root with a portion of the stem above the collar has also been attempted. This method gives about 90-95% success, sometimes even 100%. Such plants transplanted in May-July into irrigated fields become well established by the end of September. This method has its limitations, as only one plant can be raised from a single stump.

By Stem-cuttings

Stem cuttings taken from woody twigs have also been tried as a source of propagation. Hard-wood cuttings have been found better than soft-wood cuttings. Cuttings of 15-22 cm length, with three internodes are the most suitable. Nearly 60-100% of rooting is obtained by treating hard-wood cutting with beta-indole acetic acid solution of 30 ppm for 12 hrs and the treated cuttings root within 15 days. Stem cuttings planted in the nursery during the early monsoon (June) and kept moist until they sprout, give about 40-65% success rate. Such cuttings, though they start sprouting 3-4 days after planting, naturally strike roots mostly after about 75 days. Stem cuttings have been found less satisfactory than root-cuttings, since many of them do not root easily.

Manures and Fertilizers

The use of organic manure, leaf-mould and compost has been recommended to increase the quantity of nutrients in the soil and improve the damage. The plants respond better to chemical fertilizers than to organic manures. Nitrogenous fertilizers induce more vegetative growth, followed by organic manure. Nitrogen seems to have a stunting effect on the root. But the combination of nitrogen either with FYM or phosphates results in better root growth than nitrogen alone. Application of phosphates induces more growth of thick as well as thin roots. Since good manure is in short supply and uneconomical, artificial fertilizers should, therefore, be preferably used. It is better to apply 25-30 t of well-rotted FYM at the time of Land Preparation and 10-20 kg N, 50-60 kg P_2O_5, and 30-40 kg K_2O/ha as a basal dose. Later, two equal doses of N, each of 8-10 kg/ha in moist soil is given at 50 days and 170 days after planting.

Irrigation

The crop may be irrigated fortnightly in the hot dry season and about once a month in winter. The crop can be cultivated under rainfed conditions also, but the yield is considerably poorer.

Interculture

In order to maintain the satisfactory development of roots about 2 weedings are necessary during the monsoon and one hoeing at the end of the growing season (December). This may be done in large plantations using a tractor-drawn cultivator which is cheaper than manual labour. Hoeing by means of a tractor-drawn wheel-hoe, is the most economical.

Intercropping

Soyabean, onions and garlic.

Pests and Diseases

Insect pests

> ☞ Root knots appear as galls of various sizes, covering the root-system. Stunted growth of plants, etiolation and decrease in leaf-size are the symptoms in the aerial portion. The galls, on examination, show the presence of mites, various soil fungi and nematodes (*Hterodera* sp.). Generally, brown clay soils supporting the natural growth of *R. serpentina* are found to be infested with nematodes, while dark clay soils are free from them. Application of 25 kg of 3 G Carbofuran or 20 kg of 10 G. Phorate granules per hectare will control them.

> ☞ A pyralid caterpillar (*Glyphodes vertumnalis*) causes appreciable damage to the leaves. The caterpillars roll the leaf and feed on the green matter. Some other caterpillars (*Daphnis nerii, Deilophola nerii.*) are also reported to feed on tender leaves, causing defoliation of the plant. They can be controlled by spraying 0.2% Rogor.

> ☞ Cockchafer grubs (*Anomala polita*) attack the seedlings about 2 cm below the hypocotyl, resulting in their drying up. To control the attack of grubs, mix phosphate granules with the soil at the time of nursery preparation.

Diseases

> ☞ Leaf-spot caused by *Cercospora rauvolfiae* manifests as dark-brown coloured spots on the upper surface of the leaf and yellowish-brown on the lower surface. The affected leaves turn yellow, become dry and subsequently fall off, resulting in defoliation. To control this disease Dithane Z-78 or Dithane M-45 @ 0.2% is to be sprayed in early June, before the monsoons and repeated at monthly intervals until November.

> ☞ *Afternaria tenuis* attacks the leaves, resulting in minute, brownish or dark-coloured circular spots with a yellowish margin on the ventral side of the leaves; these spots enlarge to form prominent dark-brown circular lesions. The fungus also affects the flowers and fruits. The crop should be sprayed with 30 g Blitox in 10 litres of water, whenever the symptoms are seen.

> ☞ Mosaic is a common disease on this crop. The primary symptom comprises of vein-banding and the gradual yellowing of the leaves, later the leaves curl from the margin to the mid-rib on the abaxial surface and, ultimately, the leaves drop off. Strict selection of seedlings at the nursery stage helps in avoiding the disease.

The other diseases reported include target leaf-spot caused by *Coryneospora cassicolo* and *Felficularia filamentosa*, leaf-blotch caused by *Cercospora serpentina*, anthracnose caused by *Collitotrichum gloeosporoides*, die-back caused by *Collitotrichum denicrium*, powdery mildew caused by *Leviellula* taurica and fusarium wilt (*Fusarium oxysporum*). The root-knot nematode (*Meloido*gyne spp.) is also reported on this crop.

Harvesting and Storage

The roots of exploitable size are generally collected 2-3 years after planting, i.e. from 18 months onward. It is reported that roots dug out in winter (December), when the plants have shed their leaves, are richer in total content of alkaloids than the roots harvested in August. A light Irrigation should be given in advance to facilitate easy digging of roots. Care should be taken to keep the root-bark intact as the bark constitutes 40-56% of the whole root and has a higher alkaloid content. The roots are freed from the adhering soil, washed if necessary.

The roots are sun-dried till they retain a moisture content of 12%. The roots are stored away from sun-light. They are cut into small pieces (2-15 cm long and 3-20 mm in diameter). The roots are cylindrical and almost unbranched. The pieces of old root possess longitudinal ridges while those of young ones wrinkles and are usually packed in gunny bags.

Production and Yield

The yield of fresh roots per plant varies widely from 0.1-4.0 kg. The total yield of roots in the case of plants raised from seeds works out to about 11-12 tonnes/ha on an air-dried basis, as compared with 1-2 tonnes/ha in the case of plants raised from stem-cuttings, and 3-4 tonnes/ha in the case of plants raised from root-cuttings. A yield of 2.2 tonnes/ha of air-dried roots has been obtained from a 2-year-old plantation and 3.3 tonnes/ha from a 3-year-old plantation, under irrigated conditions.

Marketing and Trade

A major part of the commercial supply of this drug used in the USA and European countries originates from India, Pakistan, Sir Lanka, Myanmar (Burma) and Thailand, with India being the major supplier. The presently commercial supplies of the roots of rauvolfia are mostly from Uttar Pradesh, Bihar, Orissa, West Bengal, Assam, Andhra Pradesh, Tamil Nadu, Kerala, Karnataka and Maharashtra. There is also a great demand for the alkaloids as well as the raw drug in the international market.

The annual requirement of roots in the country for the manufactured sarpgandha extracts, simple and total alkaloids, have been estimated at about 650 tonnes against the present annual supply of 30 tonnes from all the sources. The world requirement of dried sarpagandha roots is around 2000 t/annum. Indian imported Sarpgandha roots from Mayanmar is nearly 28 t in 1997-1998 while exported nearly 4 t in 1993-94. According to estimation, the Annual production of Sarpgandha (800-1000 Mts) vis-a-vis their percentage consumption (20%) by Indian Ayurvedic Pharmacies (*Source:* ADMA, Mumbai). On the other hand, the annual requirement of Sarpgandha only in Kerala is nearly 21-25 Mts in 1995 (*Source:* AFC study on Medicinal Plants Farm Project, 1995). The current market price of its roots is Rs. 80-150 per kg.

Economics of Cultivation

The market of Sarpgandha and their products are highly volatile. The economics (per ha) worked out here are subject to fluctuations, depending upon time and place.

Expenditure (in Rs.)	—	22,000.00
Returns (in Rs.)	—	1,50,000.00
Net Profit* (in Rs.) (After 3 years)	**—**	**1,28,000.00**

* Estimation of the net profit is analysed on the basis of grower/collectors price. The grower/collector price is always 25-40% less than wholesale market price due to various stakeholders *viz.* Agent, middle man, Brokers, Commission agents.

SENNA (*Cassia angustifolia* Vehl.)

Vernacular Names

San.–Swarnpatri, Bhumiari, Bhupadmamarkundi; *Hin.*–Sanaya, Buikhakhasa; *Beng.*–Sannamakki, Son-pat; *Tel.*–Nela-tangedu, Cinemanaku; *Tam.*–Nila virai, Nilavakai; *Mal.*–Nila vaka; *Mar.*–Sonamukhi, Mulkacha; *Guj.*–Nat-ki-sana; *Beng.*–Sona mukhi; *Kan.*–Nelavarike; *Eng.*–India Ssenna, Tinnevelly senna.

Introduction

Seena is a small perennial under shrub, a native of Yemen, South Arabia. It was introduced into Tamil Nadu in the eighteenth century where it is grown as an annual crop of 5 to 7 months duration in 8,000 to 10,000 ha both under rainfed and irrigated conditions. The main advantage of the crop is that it does not require the application of any fertilizers and it is not devoured by insects, animals or birds. It is cultivated in South India in Tirunelveli, Madurai and Tiruchirapalli districts and Mysore also, where it is found to do well. Recently it has been introduced in Western part of India that is in Rajasthan.

Botanical Description

An erect, small perennial undershrub, upto I m in height. Leaves, are large, compound and pinnate while leaflets are bluish-green to pate-green in colour and emit a characteristic fetid smell when crushed. The flowers are bright yellow in colour, arranged in axillary, (on subterminal) erect, many-flowered racemes. The pods are slightly curved, 3.5 to 6.5 cm long and up to 1.5 cm broad, green in the beginning changing to greenish-brown to dark brown on maturity and drying, Each pod has 5 to 7 ovate, compressed, smooth, dark-brown seeds.

SENNA (*Cassia angustifolia*)

Geographical Distribution

Senna is indigenous to Somalia, Southern Arabia, part of Sindh (Pakistan) and the Kutch area of Gujarat. In India, it is cultivated in Gujarat, Maharashtra, Orissa, Karnataka, Andhra Pradesh, Kerala, Rajasthan and in Tamil Nadu it is cultivated in the district, of Tirunelveli, Parnanathapurani, Madurai, Salem and Tiruchirapalli.

Medicinal Uses

They are used in the form of decoctions, powders, confections and many other household preparations. The leaves/pods as such or in powder form do not lose potency easily. *Senna is* used in medicine as a cathartic. It is especially useful in habitual constipation. It increases the peristaltic movement of the colon. The pods have the tendency to gripe caused by senna may be obtained by combining it with an aromatic or a saline laxative.

It is reported to be a safe and effective drug against habitual constipation, abdominal disorders, leprosy, skin diseases, leucoderma, jaundice, dyspepsia, cough, bronchitis, typhoid fever, anaemia and tumors.

Chemical Constituents

The chief laxative principles of senna are two glycosides, *viz.*, sennoside A and sennoside B. The Indian sennosides varies from 1.5-3.0%. It also contains beta-sterol (0.33%) and flavanols-kaemferol, kaempferin and isorhamnetin.

Soil and Climate

It can thrive on a variety of soils, but prefers well drained red loams, alluvial loams, clayey rice-fields (pH 7-8.5). It is a sun-loving. An average rainfall of 25-40 cm annually is ideal for good yield.

Improved Varieties

ALFT-2, Sona, etc.

AGRO-TECHNIQUE

Land Preparation

The land is prepared deep and exposed to the sun for 110-115 days to dry out the roots of perennial weeds, followed by two cross ploughings, harrowing and leveling, FYM is incorporated into the soil at the time of the final cross-ploughing. Then the land is laid out into plots of convenient sizes with Irrigation channels.

Mode of Propagation

It can be propagated by seeds.

Direct Sowing

The seeds are broadcast or, more preferably, drilled at a distance of 30 cm in lines made at 30 cm apart at 1.5 to 2.5 cm depth on well-prepared land. It can be grown as an early summer (Feb-March) or a winter (Oct-Nov) crop but in North Indian conditions, it is reported to be the ideal time from June to October. The germination commences on the third day and is completed within a fortnight. Before sowing, the field should be perfectly levelled, lest it hamper uniform seed germination. It is found that treating the seeds with Thiram, Captan or Agroson @ 2.5 g/kg protects the seedlings from damping-off and seedling blight diseases which are common occurrences. The seeds can also be dibbled on the inner sides of the ridges opened at 45 cm distance, maintaining a plant-to-plant spacing of 30 cm. Only about 5 kg of seeds are required for this method of sowing.

Manures and Fertilizers

Application of 5-10 t/ha of well-rotted FYM at the time of sowing. In generally, 50-100 kg of nitrogen, 20-50 kg of phosphorus and 30 kg of potash/ha, depending upon the growth and number of pickings are required for good growth. Of these, the entire dose of phosphorus and potash and 50% of nitrogen should be applied at the time of sowing, and the remaining 50% of nitrogen is to be applied 90 days after sowing.

Defloration

Defloration practice (removal of flower bud) delays reproductive activity and reduces pod bearing in the second pinking. However, defloration at 45 to 65 days results in higher foliage and sennosides yield in leaves.

Intercropping

Cotton, sesamum, chillies, brinjal, gram, ginger, okra and tomatoes. Under the different crop sequences like senna-mustard and senna-coriander rotations gave higher profits.

Irrigation

Senna can be economically grown under rainfed conditions. In most years, the crop needs no Irrigations except during prolonged drought. However, when it is grown as a semi-irrigated crop, the yield increases considerably. In all, about 5-8 light Irrigations are enough to raise a good crop of senna as heavy Irrigations are injurious to the crop.

Interculture

When the plants begin to grow, once or twice interculturing is given after which the rows close up. The first weeding-cum-hoeing is done at 25-30 days, a second at 75-80 days and a third at 110 days, to keep the soil free from weeds. The use of Teeflan herbicide as a pre-emergent spray at the rate of 4 kg/ha has been reported not only to increase the yield, but also the anthraquinone content.

Pests and Diseases

Pests

The larvae of several leaf-eating caterpillars feed on the green senna leaves. Usually, the spraying of Carbaryl (4 g/1) periodically in the growing season controls the infestation. The white butterfly (Catopsilia pyranthae) attack on this crop can be minimized by sowing the crop in March-April, instead of June-July in North India. Another pest, a pod borer is also reported to attack the pods and can be controlled by spraying Endosulphan (0,05%) or Carbaryl (0.25%) at an interval of 10-15 days.

Diseases

- ☞ The leaf-spot disease is the most serious disease and causes severe damage to the crop. The infested leaves drop off, resulting in poor yield. The disease is caused by *Alternaria alternato, Phyllosticta spp.* and *Cercospora spp.* The sennoside content in the leaves is inversely proportional to the intensity of the disease. The spraying of 0.15% Dithane M-45 at fortnightly intervals, 3 times in a period of 5-6 weeks, has been found to control-the disease effectively.

- ☞ Damping-off of seedlings is the most devastating disease, which is caused by *Rhizoctonia bototicola*. The disease spreads rapidly if the fields are affected by stagnating moisture at this stage, Treatment with Thiram or Captan at 2.5 g/kg of seeds protects the growing seedlings. At a later stage its other physiological form, called *Macrophomina phaseoli* develops; it causes dry rot in the crop and kills the plant. The fields can be drenched with 0.2% Brassicol or 0.5 to 1.0%. Rhizoctol, but this only gives partial control.

- ☞ The leaf-blight disease is caused by *Phyllosticta spp.* usually occurring at the later stages of growth, in September-October. Cloudy days and humid weather conditions favour the spread of the disease. The leaf-spot disease initially appears as small, brown, irregularly scattered lesions on leaves which grow bigger in size and turn black in colour at a later stage. The spraying of 0.15% Dithane M-45 at fortnightly intervals can control the disease.

Harvesting and Storage

The senna plants produce foliage containing higher sennosides between 50-90 days of sowing. The first picking of the foliage crop should be done at 50-70 days age. The picking of leaves is done by hand so that most of the growing tops are removed at harvest; this also induces more branching which, otherwise, reduces the foliage growth considerably. A second picking is taken at 90-100 days and the third picking between 130-150 days. When the entire plants are removed, so that the harvested material includes both leaves and pods together.

The harvested crop should be spread in a thin layer in an open field to reduce its moisture. Further drying of the produce is done in well-ventilated- drying sheds. It takes 10-12 days to dry completely. The dry leaves and pods should have a light-green to greenish-yellow colour. The pods are threshed during drying to remove the seeds. The produce is baled under hydraulic pressure and wrapped in gunny bags for export.

Production and Yield

A good irrigated crop of senna, under efficient management conditions, produces 15-20 q/ha of dry leaves and 7-10 q/ha of pods. The yield under rainfed conditions is about 10 q/ha of dry leaves and 4 q/ha of pods.

Marketing and Trade

India is the largest producer and exporter of senna leaves, pods and total sennosoides concentrations to the world market. A major part of them produce is exported in the form of leaves, pods and sennoside concentrates, though several pharmaceutical houses utilize it for the manufacture of calcium sennoside granules, tablets and syrups for marketing within the country as well. Germany, Hungary, Japan, Netherlands and the USA are the main markets. The export of leaves and pods from India is of the value of Rs. 20 million annually. In addition, the erstwhile USSR countries buy sennosides (concentrates) and the current annual export of the sennoside concentrate is around Rs. 20 million.

India exported Senna leaves and their pods to various Asian and European countries. In 1999-2000, India exported its leaves and pods nearly 7466.33 tonnes and earned foreign exchange Rs. 2254.2 lakhs. Nearly 100% supply of this herb has been fulfilled by cultivated land of Tamil Nadu (60%) and Kutch (40%). According to estimation, the Annual production of senna (10000-12000 Mts) *vis-a-vis* their percentage consumption (10%) by Indian Ayurvedic Pharmacies (*Source:* ADMA, Mumbai). The present market price of senna is Rs. 15-20 per kg.

Economics of Cultivation

The market of Senna and their products are highly volatile. The economics (per ha) worked out here are subject to fluctuations, depending upon time and place.

Expenditure (in Rs.)	—	3,500.00
Returns (in Rs.)	—	30,000.00
Net Profit* (in Rs.)	—	**26,500.00**

* Estimation of the net profit is analysed on the basis of grower/collectors price. The grower/collector price is always 25-40% less than wholesale market price due to various stakeholders *viz.* Agent, middle man, Brokers, Commission agents.

SHATAVAR (*Asparagus racemosus* Willd.)

Vernacular Names

San.–Shatavari, Abhiru; *Hin.*–Shatavar, Satamuli; *Tel.*–Satavari, Callagadda; *Tam.*–Kilavari, Satavali; *Mal.*–Shatavari, Satavali; *Kan.*–Majjige-gadde, *Thali periyan*; *Guj.*–Ekalkanto; *Mar.*–Shatavari; *Eng.*–Wild asparagus; *Trade*–Shatavar.

Introduction

Shatavar is a slender climber and belongs to family Liliaceae. It is known to have several medicinal properties. Its roots are borne in a compact bunch, fleshy & spindle shaped, sweet & bitter in taste and silvery white or light ash-coloured externally and white internally. The roots are used as a remedy for tuberculosis, leprosy, epilepsy, dysentery and night blindness. Sometimes, the roasted seeds of shatavar are used as a substitute for coffee in Europe.

Botanical Description

It is climbing undershrub with woody terrate stem and curved or straight spines. Young stem are brittle, smooth, and delicate. Leaves have a minute spiny appearance. These have cladodes curved in groups of 2-6. The flowers are white. Fruits are globular berries, purplish black in colour when ripe, with hard seeds. The roots are 25 cm to 1 m in length tapering at both ends and fascicled at the base.

SHATAVAR (*Asparagus racemosus*)

Medicinal Uses

The tuberous roots are bitter, sweet, cooling, indigestible, appetizer, alterative, stomachic, tonic, aphrodisiac, galactegogue, and astringent to the bowels. It is also useful in the treatment of tumors, inflammations, tuberculosis, leprosy, epilepsy, dysentery and night blindness. It is also used as a uterine sedative.

Dried tuberous roots has ulcer healing resistance or cytoprotection. It has also been identified as one drug to control the symptoms of AIDS. Alcoholic extract of the roots increased the weight of mammary glands in post-partum and estrogen-primedrats.

Geographical Distribution

It is commonly occurs in upper Gangatic plains, Bihar plateau & cultivated in gardens. It is also distributed throughout tropical and sub-tropical India in areas upto 1400 m elevation. It is commercial cultivated in Madhya Pradesh, Uttar Pradesh and Uttaranchal.

Soil and Climate

It thrives better in porous black to rich loam sandy soil, well drainage (pH 6-7) however also tolerant to alkaline conditions. It is a crop of tropics and requires an optimum temperature of 10-40°C with plenty of sunshine and annual rainfall of 50-100 cm. It prefers warm, hot humid climate.

Chemical Constituents

Tuberous roots contains four saponins, *viz.* shatavarin I to IV. Shatavarin is a glycoside of sarsasapogenin. Flower contains quercetin, rutin (2.5% dry basis) and hyperoside.

Improved Varieties

White and Yellow colour types.

AGRO-TECHNIQUE

Land Preparation

Before cultivation, the land should be ploughed twice and the field should be cleaned thoroughly of weeds. About 25 t/ha of cowdung manure is also added. Ridges and furrows are opened at 45 cm apart and planted 15 cm within the row.

Mode of Propagation

It can be propagated by seeds as well as clumps or crowns.

Direct Sowing

For commercial planting it is propagated by seeds. About 3-4 kg seeds are required for one hectare land. The best time for Direct Sowing is May to July.

Nursery Raising

The seeds are sown during May-June in well-prepared raised nursery beds of 5 m × 1m and 20 cm high. Seeds have to be sown in line 3-4 cm apart and covered with a thin layer of fine sand. The beds are lightly watered. The seeds take 15-20 days to germinate. The germination percentage is 70-80%. The nursery duration is for 2-3 month.

Transplanting

Seedlings are transplanted from beds when 5 cm in height, in to polybags. After 2-3 months they should be planted in the main field in 45 cm cubic pits at a spacing of 2 feet. The best time for Transplanting is July-August.

Manures and Fertilizers

It is a newly domesticated crop. A dose FYM @ 20 t/ha is recommended however the recommended fertilizers schedule is not yet been standardized so far.. The fertilizers are applied in the soil near the root-system, after the plants are established.

Irrigation

Shataver in its natural habitat requires optimum moisture content for its normal growth and development. Looking into the plant's water requirement Irrigation was given once in four days interval in summer and ten days interval during winter.

Interculture

Weeding is required at fortnightly interval. However, in the field conditions, weeds *like Cynadon dactylon* and *Parthinium* were noticed.

Pests and Diseases

No major diseases and pest are observed under field conditions, except scale and mites which can be controlled by 0.07% Endosulfan.

Harvesting and Storage

The Crop is ready for harvesting after 14-18 month of planting. The best time is October to December for harvesting. The crop is harvested manually by digging and uprooting the individual plants. The fresh tuber root barks are to be removed by making a long incision in the root. The central canal threat is to removed and then dried in shady place for 3-4 days, yielding 10-12% of dry tubers. The dried tuberous roots are to be stored in gunny bags.

Production and Yield

On an average a yield of 8-10 tonnes per hectare of dried tubers may be obtained under proper management conditions.

Marketing and Trade

The tuberous roots of shatavar and formulations are sold in the market. This herb market is growing day by day due to their demand in the global herbal market. An estimation, the annual production of kalmegh (10,000 Mts) *Vis-a-Vis* their consumption (15%) by Indian ayurvedic pharmaceutical industries in 1999 (*Source*-ADMA, Mumbai). The 65% supply of this herb has been fulfilled by wild resource (forest) of Madhya Pradesh, Uttaranchal and Nepal while rest 35% from cultivated land of Madhya Pradesh. On the other hand, the annual requirement of *shatavar* only in Kerala is nearly 250 Mts in 1995 (*Source*: AFC study on Medicinal Plants farm Project, 1995). In 1999-2000, the total annual consumption of shatavar roots in Natural Remedies (P) Ltd. is nearly 10 tonnes. The current market price of white and yellow colour types shatavar roots is Rs. 25-40 and 120-180 per kg respectively.

Economics of Cultivation

The market of Aloe and their products are highly volatile. The economics (per ha) worked out here are subject to fluctuations, depending upon time and place.

Expenditure (in Rs.)	—	19,500.00
Returns (in Rs.)	—	40,000.00
Net Profit* (in Rs.)	—	20,500.00

* Estimation of the net profit is analysed on the basis of grower/collectors price. The grower/collector price is always 25-40% less than wholesale market price due to various stakeholders *viz*. Agent, middle man, Brokers, Commission agents.

SILYBUM (*Silybum marianum* (L.) Gaertn.)
Syn. *Cardus marianum* L.

Vernacular Names

Kan.–Pavitramullina gida; *Eng.*–Milk thistle, Holy thistle.

Introduction

Holy thistle is an important medicinal plant used by man from ancient times. According to the great Greek physian, Dioscorides, who accompanied the occupying Roman army in Britain, in the first century AD, they used parts of the holy thistle as a remedy for those who are punished by having their body drawn backwards. It is also mentioned in the writings of the venerable Bede (872-735 BC) as a plant which was cure for all ailments and had supernatural properties. It belongs to the family Asteraceae. Because of its spiny nature, this plant can be very easily grown where cattle or other grazing animals are a problem. The plant is also very hardy in nature and facilitates easy and low-risk cultivation.

Botanical Description

An erect annual or biennial, growing up to 120-130 cm tall. Leaves are large, mottled with white, pinnatified into ovate, triangular, sinuate-toothed, spiny lobes. The heads are globular, 6-10 cm broad and concave at the base. The flowers are purple or white, the limb is dilated below and is five fold. Fruit cypsela.

Geographical Distribution

It is distributed in Asia, Africa and Europe. It is found growing wild in Punjab and the North-western Himalayas from Kashmir to Jammu at 6000-8000 ft. altitude. It has been naturalized in North and South America.

Medicinal Uses

In Europe, the herb is used against jaundice, intermittent fevers, dropsy and uterine trouble. As a diet or infusion, the leaves are said to be reliable galactagogues. The seeds are pungent, demulscent, antihepatotoxic (antagonizing liver damage), and antispasmodic.

SILYBUM (*Silybum marianum*)

They are used for the treatment of calculi of the liver and gall bladder and in controlling haemorrhage. The flowering heads are good for diabetics. The extract of seeds is reported to increase the peristalsis of the small intestine and is mildly purgative. It is a good anti-oxidant.

Chemical Constituents

The seeds contain a group of Flavonolignans collectively known as Silymarin , contains silybin, silydianin, silychristin and tocopherol.

Soil and Climate

It is a very hardy plant and is found growing on a variety of soils. However, sandy loam soils which have good drainage are ideal. The crop can be grown successfully even on marginal lands by supplementing the nutrients through fertilizer application. It prefers a moderate type of climate for its successful growth. Places with extreme climates are unsuitable for its cultivation. The plant can be grown up to an altitude of 2400 m from sea-level.

AGRO-TECHNIQUE

Land Preparation

Before cultivation, the land should be ploughed twice and the field should be cleaned thoroughly of weeds. The field is prepared well, divided into convenient sized plots and thoroughly mixed with organic manure. About 25 t/ha of cowdung manure is also added.

Mode of Propagation

It can be propagated by seeds.

Sowing

The crop is raised by using the seeds obtained from the previous season's crop. It is sown during September-October in the plains and March-April in hilly areas. Then the seeds are either sown by broadcasting them in the rows spaced at 60 cm and later thinned to maintain an approximate spacing of 60 cm between the plants within the row, or they can also be sown at 60 × 60 cm spacing using a seed-drill. It enough moisture is not available, the field should be irrigated immediately after sowing.

Manures and Fertilizers

The crop responds well to the application of manure and fertilizers. FYM at the rate of about 8-10 t/ha should be added at the time of Land Preparation. In addition, a dose of 120 kg nitrogen, 100 kg phosphorus and 75 kg potash/ha, is recommended for obtaining a good yield. Among the fertilizers, a half-dose of N and the whole dose of P and K are applied at the time of planting and the remaining half-dose of N is given as a top-dressing after the seedlings are one month old.

Irrigation

It is a sufficiently hardy crop and does not require very frequent Irrigations, except during the early stages of growth, when the crop has to be irrigated at weekly intervals. Subsequently, watering once in a fortnight will suffice. In all, about 10-12 Irrigations during the entire cropping season will enable the crop to grow luxuriantly.

Interculture

Weeds are a problem in this crop only during the early stages of growth. Once the crop picks up, it covers the field with its large foliage and the weeds are automatically suppressed. One or two weeding when the plants are young will help to avoid competition from weeds.

Pests and Diseases

It is practically free from most Pests and Diseases. However, the pest *Tanyameus polliatus* which can be controlled by a spray of 0.03% Methyl Parathion. Among the diseases, root-wilt caused by the Rhizoctonia species can be controlled by drenching the soil in fungicides like Captan or Brassicol (0.1%).

Harvesting and Storage

It is a 4 months duration crop. The crop is prickly and hence it is to be harvested carefully by cutting the individual heads with the help of sharp instrument. Since all the heads do not reach maturity at the same time, harvesting is done 2-3 times. There is also a problem of seed shattering; therefore, the heads should be harvested when the seeds are just turning brown. If the crop is to be harvested mechanically, it should be harvested when the majority of the heads have reached maturity.

The leads are spread in the threshing yard for 3-4 days for drying. While drying, quite a few of them burst and release the seeds, and the remaining are threshed, separated and stored.

Production and Yield

A good crop, on an average, 10 quintals/ha of seeds may be obtained.

STEROID-SOLANUM (*Solanum viorum* Dunal.)

Vernacular Names

Kan.–Kandanka mullu; *Mal.*–Punnarichunda.

Introduction

Among the various plants which are being used as raw materials for the production of steroidal drugs, steroid-bearing solanurn holds an important place due to its quick growth and low initial investment in its commercial cultivation.

Botanical Description

A stout, branched, woody shrub attaining a height of 0.75 to 1.5 m. The stem has spines, the leaves are ovate to lobed with spines on both the surfaces, the flowers are borne on axillary clusters, white; the berries are yellowish when ripe or greenish; the seeds are small, brown in colour and abundant, embedded in a sticky mucilage.

Geographical Distribution

It is widely distributed in the subcontinent, extending from sea-level up to 2000 m and is reported from Khasi, Jaintia and the Naga Hills of Assam and Manipur. It occurs in Sikkim, West Bengal, Orissa, the Upper Gangetic Plains and in the Nilgiris, ascending to an altitude of 1600 m. It is reported from North-east, North-west, southern as well as Central India, and extends into Burma and China. Its commercial cultivation is mainly confined to the Akola-Jalgoan tract of Maharashtra in an area of about 3 000 ha.

Medicinal Uses

These steroidal compounds have anti-inflammatory, anabolic and antifertility properties, due to which they find large-scale use in health and family planning programmes all over the world.

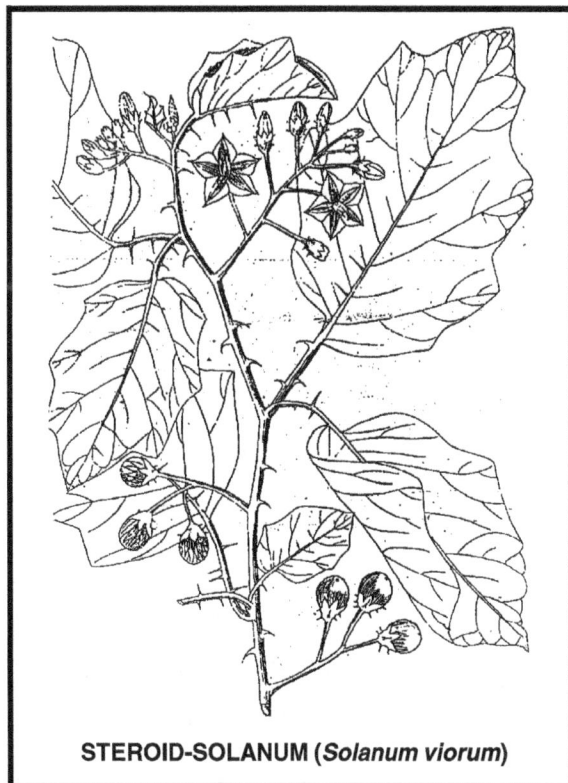

STEROID-SOLANUM (*Solanum viorum*)

Chemical Constituents

It yields a glyco-alkaloid, solasodinc, a nitrogen analogue of dio.Senine. Solasocline through, I 6-clehydru-prcgnenolone (16 DPA) is converted to a group of compounds like testosterone and methyltestosterone and corticosteroids like predinisolone and hydrocortisone.

Soil and Climate

It is a hardy plant and can be cultivated on a wide range of soils under various agroclimatic conditions, but it cannot withstand water- logging. the best soil for its successful cultivation is red lateritic soil with a moderate quantity of organic matter. It is found growing under different kinds of climates throughout the length and breadth of the country, but prefers a moderate climate for its successful growth.

Improved Varieties

GaIxo strain, BARC Strain, Pusa-1, RRL 20-2 , RRL-GL-6, Arka Sanjeevini and Arka Mahima.

AGRO-TECHNIQUE

Land Preparation

The field is thoroughly prepared, about 25 t/ha FYM is incorporated into the soil, it is divided into blocks of convenient sizes, and ridges are opened at a distance of 60 cm.

Mode of Propagation

It can be propagated by seeds.

Nursery Raising

The best sowing time may vary from June to September or October for better germination. Under rainfed conditions, the crop should be sown from mid-June to the first week of July. The seedlings are raised in nursery-beds of 10 m × 1 m size which are convenient for weeding, Irrigation and removing seedlings. To each strip, 5 baskets of FYM and 1 kg of calcium ammonium nitrate are applied. Seeds which are presoaked in water for 24 hours are sown in lines of I to 1.5 cm-deep furrows, 10 cm apart and covered with a thin layer of soil. About 1.0-1.5 kg seeds sown in nursery beds will provide enough seedlings for planting I ha of land. Their germination is completed in 7-10 days.

Transplanting

In nursery bed within 4-5 weeks of, when the seedlings are 10-12 cm high and develop 6 leaves, they are ready Transplanting into the main field. At the time of Transplanting, the seedlings uprooted from the beds are thoroughly cleaned of soil and dipped in 0.1% Bavistin solution for at least 5 minutes to overcome the wilt problem and, thereafter, transplanted at a distance of 45 cm within the row.

Manures and Fertilizers

Green manure before planting has been found to increase the yield by 20%. For better yields, an application of 100:60:40 kg of 100 kg nitrogen, 60 kg phosphorus and 40 kg potash, depending upon the soil condition, is recommended. The entire quantity of phosphorus, potassium and half of the nitrogen are applied at the time of Land Preparation. While the remaining half of the nitrogen is applied when the plants start flowering.

Irrigation

In the absence of sufficient moisture in the soil, the field should be immediately irrigated after

Transplanting. The crop is further irrigated at weekly intervals during the first month and then the interval is increased to once in fortnight, and later as and when required.

Interculture

After 2-3 weeks of Transplanting the first weeding is done, and later when the crop is 2-3 months old. Afterwards, the crop puts on enough canopy to smoother the surface and, hence, no growth of weeds takes place.

Pests and Diseases

Pests

The plant is sufficiently hardy and, therefore, free from any of the serious Pests. However, it is sometimes attacked by leaf-eating caterpillars and wingless hoppers. In case of severe attack, the crop may be sprayed with Enclosulphan (3 ml/l) to control them. The fruit borer (Leucinodes orbonalis) and root-eating grubs are the other pests reported. Ekalux (2 ml/l) can be sprayed to control the fruit borer. Chloropyriphos may used to control grubs.

Diseases

- ☞ Some of the diseases the crop is attacked by are: powdery mildew, collar rot, bacterial blight, leaf-blight and damping-off. Dowdery mildew will be noticed during prolonged dry and warm period, Bavistin (1 g/1) may be sprayed to control this disease.

- ☞ Collar rot or *Fusarium* wilt can be overcome by keeping the field clean and planting the crop in a well drained soil. Dipping the roots of the seedlings in a 0.1% solution of Bavistin for I hour and drenching the seed-beds with 0.25% of copper oxychloride or 0.1% of Bavistin solution can control the disease.

- ☞ Sometimes the plants are attacked by mosaic, caused by three different viruses, which leads to stunted growth and chloratic leaves. Such plants are better removed and destroyed. Leaf-blight is caused by *Pythiurn butleri* and is not. a serious problem in this crop.

- ☞ Bacterial blight is not a serious disease on this crop. However, under severe incidence, a treatment with the solution of 30 g of Streptocyclene and 30 g of Copper sulphate dissolved in 500 1 of water per hectare controls this disease.

Harvesting and Storage

The crop takes about 6 months to be ready for harvesting. The spiny nature of the plant hampers plucking the berries at the right stage of maturity, which is very important. During the first part of the harvesting season, when the fruit is big, on an average, one person with gloves can pluck about 50 kg, of berries, while working 8 hours a day. However, some good workers in the peak season when most of the berries are ready for harvest can pick even 80 kg of berries per day. The figure is reduced to 40 kg towards the closing season when the fruits become smaller in size. The picking operation spreads over 3 months, because the fruits mature at different times. The processing of berries for marketing requires a lot of care. Fresh fruits contain about 80% moisture. The pharmaceutical firms need berries containing about 10% moisture. Hence, the berries must be dried in the sun.

Production and Yield

When the crop is grown by adopting proper cultivation practices, it may yield nearly 8-10 t/ha of fresh berries which, in turn, will give about 2.5 t/ha of dried berries.

Marketing and Trade

The current market price of solanum fruits is Rs. 30-50 per kg.

Economics of Cultivation

The market of Solanum and their products are highly volatile. The economics (per ha) worked out here are subject to fluctuations, depending upon time and place.

Expenditure (in Rs.)	—	10,500.00
Returns (in Rs.)	—	42,000.00
Net Profit* (in Rs.)	—	**31,500.00**

*   Estimation of the net profit is analysed on the basis of grower/collectors price. The grower/collector price is always 25-40% less than wholesale market price due to various stakeholders *viz.* Agent, middle man, Brokers, Commission agents.

SWEET FLAG (*Acorus Calamus* L.)

Vernacular Names

San.–Vacha, ugrandha; *Hin.*–Bach, gorbacc; *Tel.*–Vasa; *Tam.*–Vasampu; *Mal.*–Vayampu; *Kan.*–Baji; *Eng.*–Sweet flag, calamus; *Trade*–Calamus.

Introduction

Sweet flag is herb belonging to the family Aeraceae. The dried root (rhizome) has long been employed in medicinal preparations and for flavouring liquors. It contains a volatile, yellowish-brown oil of peculiar, but pleasant, slightly sweet odour, that can be isolated by steam distillation. The yield and physico-Chemical Constituents and, consequently, the composition of the 'calamus oil' depend upon the source from which the rhizomes are obtained. The European and American oils are similar in composition, consisting of low amounts of 'asarone' which can be compared to the Indian oil. The rhizomes contain a high percentage of starch. Prior to World War II, commercial suppliers of the drug came from USSR, Central Europe, Romania, India and Japan.

Botanical Description

A semi-aquatic perennial herb with a stout midrib, with a creeping and many branched aromatic rhizome. The rhizomes are horizontal and jointed, with a spongy texture. The rhizomes are 1.5-2.5 cm thick and sometimes up to 2 m long. The breadth of the leaf is 2-4 cm and resembles the leaves of an iris. The leaves are bright green, thickened in the middle and the margin is wavy. The spathe is 6-30 cm long and the spadix is slightly curved. The fruit is a berry yellow-green in colour, angular and 1-3 seeded.

Geographical Distribution

It grows wild on the edges of swamps, on the banks of rivers and ponds in North America, Europe and Asia. The countries growing this crop commercially are the USSR, Central Europe, Rumania, India and Japan. In India, the leading states are Karnataka, Kashmir, Manipur, Arunachal Pradesh, Meghalaya and in the foothills of the Himalayas, up to 1800 m. In Karnataka, it is being grown in the

Koratagere Taluk of Tumkur district in an area of 100 acres and the annual production is 400 t of dried rhizomes.

Medicinal Uses

The plant is aromatic on account of the presence of an essential oil, which contains the glucosidic bitter principle, 'Acorin'. It has an expectorant action due to the presence of the essential oil and is commonly used as a remedy for asthma. It is also used as a remedy for chronic diarrhoea. The rhizome is an appetizer and is also useful in the treatment of epilepsy, delirium, hysteria and loss of memory. In Unani literature, the rhizome has been described as having a very bitter taste and is said to be useful as a brain tonic. The powdered rhizomes have been employed in combination with *Rauvolfia serpentina* and other drugs in the treatment of neurosis, insomnia and hysteria. Because of the pleasant aromatic smell, the rhizomes are extensively used as an aromatic agent in agarbathies, dhoops and hawen samagries.

The powdered rhizome and the essential oil can be used as a safe insecticide against flies, mosquitoes, bed-bugs, moths and lice. This property is due to the presence of the 'trans-isomer of asarone'. The leaves and rhizomes are used to control lice. Its essential oil is used in flavouring liquors, beer, gin, vinegar, snuff and various other preparations. The "bach" of commerce, which has

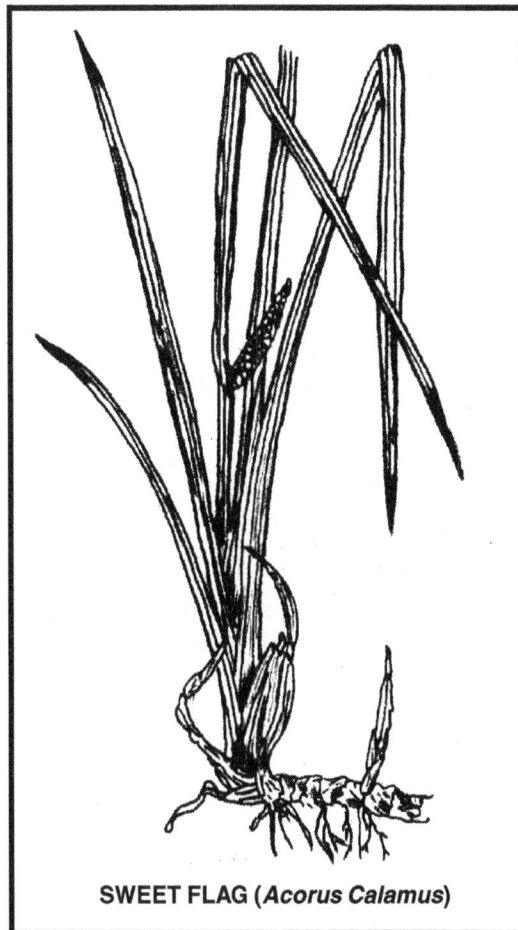

SWEET FLAG (*Acorus Calamus*)

many Medicinal Uses is prepared from the rhizomes. The fresh rhizomes are used in confectionery and also as a substitute for ginger. In moderate doses, the oil produces antispasmodial action on the involuntary muscle tissue, inhibiting the excessive peristaltic movements of the intestine.

Chemical Constituents

It produces an essential oil. The essential oil contains asarone (82%). Major and minor constituents are Calamenol (5.0%), Calamene (4.0%), Calamenone (1.0%), Methyl eugenol (10%), Eugenol (0.3%) etc,

Soil and Climate

Sweet flag is cultivated in almost the same way as rice and may be grown in any part of the country where suitable Irrigation facilities are available. Although it in the wild state grows in water, it may be cultivated in almost any good, but fairly moist soil. It is usually grown in areas where paddy can be grown. It is generally grown in the clayey loam and light alluvial soils of river banks. It is a hardy plant and is thus found growing from tropical to subtropical climates. However, it prefers an area with good and well-distributed rainfall throughout the year.

AGRO-TECHNIQUE

Land Preparation

The field is laid out exactly as for rice, irrigated sufficiently and, after ploughing twice, is watered heavily and ploughed again. The field is then left alone for a few days and puddled. After that, it is leveled with a leveling board.

Mode of Propagation

It can be propagated by rhizome cutting.

Planting

The planting material consists of the live ends or tops of the previous crop. The best planting season is spring (March-April). At harvest, the mature portion of the rhizome is cut off, which is the marketable part and the tender portion formed by the growing end is used for replanting and propagation. The best spacing for sweet flag is 30 cm × 30 cm the rhizome pieces are pressed into the mud at a depth of about 5 cm inside the soil. The rhizomes are planted in such a way that the plant in the second row comes in-between the first row and not opposite to it.

Manures and Fertilizers

The field should be manured with green manure (10 to 12 t) and compost (15 t/ha). Along with this, 100 kg ammonium sulphate, 300 kg super phosphate and 100 kg Muriate of potash should be added, or otherwise 125 kg N, P, and K/ha is recommended to be applied in 3 split doses.

Irrigation

The field must be regularly irrigated. About 5 cm of water is left standing in the field in the beginning. Later, the water level is increased to 10 cm as the plant grows.

Interculture

Careful and regular weeding is required. In all, eight weedings are given and, at each weeding, the growing plants are lightly pressed down into the soil.

Pests and Diseases

This crop is practically free from the attack of any pests or diseases of a serious nature, except mealy bugs, which infest the shoot and roots. The insect can be controlled by spraying the shoot and drenching the roots of grown plants with 0.1% Methyl Parathion or 0.15% Oxydemeton Methyl or 0.2% Quinolphos.

Harvesting and Storage

After about a year, the crop is ready for harvesting. The field is allowed to dry partially, so that sufficient moisture is retained to facilitate the necessary deep digging. The leaves start turning yellow and dry, indicative of maturity. The rhizomes will be at a depth of 60 cm and are about 30-60 cm long. Therefore, harvesting should be done carefully. The rhizomes, after harvest, are cut into short lengths of 5-7.5 cm and all the fibrous roots are removed. The pieces are then washed thoroughly and dried in the sun. The dried material is put into gunny bags and rubbed to free them of leafy scales.

Production and Yield

Under favourable conditions, a yield of 15-18 quintals/ha of dried rhizomes is obtained. The fresh rhizomes yield about 2.95-3.4% of essential oil. The fresh aerial parts yield 0.12% oil.

Marketing and Trade

According to estimation, the annual demand of Acorus roots by the herbal industries and crude drug producers only in Maharashtra is nearly 10-15 tonnes In 1999-2000, the total annual consumption of Acorus roots in Natural Remedies (P) Ltd. is nearly 13 tonnes. The current market price of Acorus roots is Rs. 12-15 per kg.

Economics of Cultivation

The market of Aloe and their products are highly volatile. The economics (per ha) worked out here are subject to fluctuations, depending upon time and place.

Expenditure (in Rs.)	—	3,500.00
Returns (in Rs.)	—	12,000.00
Net Profit* (in Rs.)	—	**8,500.00**

*   Estimation of the net profit is analysed on the basis of grower/collectors price. The grower/collector price is always 25-40% less than wholesale market price due to various stakeholders *viz.* Agent, middle man, Brokers, Commission agents.

Chapter 6

HARVEST TECHNOLOGY AND VALUE-ADDITION OF MEDICINAL PLANTS

The harvesting medicinal herbs requires careful planning to ensure the parts are processed in peak condition and fast enough to retain their active ingredients. Wild plants offer a free and natural source of herbal remedies and give the satisfaction of collecting herbs in the traditional way. Furthermore, active constituents are often more highly concentrated in wild plants as the herb is generally likely to be growing in its preferred environment.

6.1. Collection

Much information is not available on growing medicinal herb. Apart from plants which are cultivated crops like mints, tobacco, digitalis, ginseng, basil, opium, fennel, garlic, caraway, anise etc., there are many medicinal plants at present obtained by collection from wild sources. However, collection of these plants in wild continues to be unscientific, labour-oriented and undependable. Many of the crude drugs are poisonous and dangers of collection by untrained collectors are obvious, because even experts have difficulty in correct identification of some species unless they are in flower. Mistaken identity during collection, particularly of plants with phenotypic similarity, could lead to accidental poisoning.

Medicinal and aromatic plants are so far collected following the traditional methods without adhering to the pharmacopoeial standards laid down for the collection of different drugs and their parts. Thus, no control is kept on the quality of the material. Plants are collected as and when the collectors are free from their farming and other activities or any such time when there is a demand from the market. The collection of underground parts like roots, rhizomes, tubers etc. is done by digging out the entire plant by means of kutki (*Picrorrhiza kurroo*), as may be available from July-August to March-April. In certain cases, the collectors have been busy collecting these herbs during summers *i.e.*, the growing season, without caring for its survival and regeneration. By doing so, on one side, the collection of immature and undeveloped parts/plants results into a product of inferior quality lacking proper active ingredients and on the other side, the life cycle of the plant is broken as it is not allowed to bear flowers, set fruits and shed seeds leading to loss in future regeneration in their natural habitats. As a matter of rule, any plant having underground part, *i.e*. root, rhizome, tuber, corm and bulb as official part must be harvested when these have borne fruits, shed seeds and aerial parts have started withering. This way before uprooting the plant, the seed shed has taken place, which produces plant, crop in the coming season through natural regeneration. Secondly the active Chemical Constituents are also on the higher side at such a dormant stage, when the vegetative growth has ceased and the entire contents are concentrated in the underground parts. Summer collection not only results in breaking the life cycle of these plants, but also in yielding poor quality material.

Drugs like dhoop (*Rhododendron anthopogon*), dioscorea (*Dioscorea deltoidea*), atis (*Aconitum heterophyllum*), salampanja (*Dictylorhiza hatagirea*), valerian (*Valeriana officinale*) etc. fall in this category. Banaksha (*Viola serpens*) and allied spp. are collected by shepherds, graziers and gwalas during March-April along with guchhi (*Morchella esculenta*) mostly with entire plant instead of plucking the flowers collected in winters by ruthlessly cutting the branches, and the fruits of amla (*Embelica officinalis)*, harad (*Terminalia chebula*) and bahera (*Terminalia bellirica*) too are collected haphazardly without waiting for the proper maturity stage and often by cutting the branches thereby reducing the fruit-bearing capacity of the trees during the succeeding years. Though people have acquired near expertise in the art of collection, yet owing to the greed of contractors and ignorance of collectors about the quality, the norms are frequently flourted.

The aerial parts of the growing above growing-stems, leaves, flowers, berries and seeds. The stems are normally cut 5-10 cm above ground shortly after the plant has begun to flower, when it is putting most effort into growth. Perennial may be cut higher above ground to encourage further crops. Leaves are gathered throughout the growing period. Young leaves are considered to be of highest quality so far as content of active principles is concerned.

In case of large flowers, it is picked just after they have opened, usually during the spring or summer. Sometimes only specific parts of the flower are used, such as the petals of marigold (*Calendula officinalis*), while other flowers are used whole. Full bloom flowers or blossoms are not suitable for drug market. The fruits and berries are harvested in early autumn when rope but still firm. If allowed to become over-ripe, they may not dry properly. They can be picked individually or in branches.

Underground parts are storage organs for the plant and accumulate active principles during summer months. Roots and rhizomes of perennials are generally gathered after two to three years of growth. The roots of annuals are generally not collected. This applies to majority the drug plants collected from wild. However, cultivation, the roots or rhizomes are harvested as per the requirements and presence of active constituents *e.g.* turmeric, dioscorea, ginger, thubarb etc.

Tubers, corms and bulbs are collected at the end of flowering or fruiting when all the aerial portions shows signs of senescence. All the underground portions are collected, cleaned and dried in shade or ovens at 30°C. Proper care should be taken to dry the materials *e.g.* gloriosa, squill, garlic, saffron, colchicun etc.

Bark is collected either in spring, when the trees and shrubs begin to bud, or in autumn after they have shed their leaves. The flow of sap is considered to be at maximum at these times and bark readily detaches from the wood *e.g. cinnamon, cinchona, cassia*.

Sap is collected in the spring as it rises, or as it tails in the autumn. Trees such as silver birch (*Betula pendula*) produce huge quantities of sap if tapped, although this reduces the trees vitality. Collect milky juices or latex from plants such as dandelion (*Taraxacum officinale*), by squeezing the stems over a bowl. Wear gloves as latex or sap can be corrosive. The gel from Indian aloe (*Aloe barbedensis*) is scarped out after slicing the teat lengthwise and peeling backs the edges.

6.2. Harvesting

Harvesting is an important operation in cultivation technology, as it reflects upon economic aspect of the crude drugs. An economic point which needs attention over here is the type of drug to be harvested and the pharmacopoeial standards of which it needs to achieve. Harvesting can be done efficiently in every respect by the skilled workers. Selectivity is of advantage in that the drugs other than genuine, but similar in appearance can be rejected at the site of collection. It is, however, a laborious job and may not be economical. In certain cases, it cannot be replaced by any mechanical means, *e.g.* digitalis (*Digitalis purpurea*), vinca (*Catharanthus roseus*) and senna (*Cassia angustifolia*) leaves. The under ground drugs like rhizome, roots, tuber, bulbs, corms etc. are harvested by mechanical devices, such as diggers or lifters. The tubers or roots are thoroughly washed in water to get rid of earthy-matter. Drugs which constitute all aerial parts are harvested by binders for economic reasons. Many a times, flowers, seeds and small fruits are harvested by a special device known as seed stripper. The technique of beating plants with bamboos is used in case of cloves. The cochineal insects are collected from branches of cacti by brushing. The seaweeds producing agar are harvested by long handled forks. Peppermint and spearmint are harvested by normal method with mowers, whereas fennel, corinder and caraway are uprooted and dried. After drying, either they are thrashed or beaten and the fruits are separated by winnowing. Sometimes, reaping machines are also used for their harvesting.

6.3. Drying

Before marking a crude drugs, it is necessary to process it properly, so as to preserve it for a long time and also to acquire better pharmaceutical elegance. This processing includes several operations or treatments, depending upon the source of the crude drugs (animal or plant) and its chemical nature. Drying consists of removal of sufficient moisture content of crude drugs, so as to improve its quality and make it resistant to the growth of microorganisms. Drying inhabits partially enzymatic reactions. Drying also facilities pulverizing or grading of a crude drugs. In certain drugs, some special methods are required to be followed to attain specific standards. *e.g.* fermentation in case of Dalchini (*Cinnamomum* zeylanicum) bark and Indian gentian (*Picrorrhiza kurroo*). The slicing and cutting into smaller pieces is done to enhance drying, as in case of liquorice (*Glycyrrhiza glabra*), squill and calumba. The flowers are dried in shade so as to retain colour and volatile content. Depending upon the type of chemical constitutes, a method of drying can be used for a crude drug. Drying can be of two types: *(i)* Natural (Sun drying), *(ii)* Artificial.

6.3.1. Natural Drying (Sun-drying)

In case of natural drying, it may be either direct sun drying or in the shed. If the natural colour of the drug (e.g. Digitalis, clove, senna) and the volatile constituents of the drugs (peppermint) are to be retained, drying in shed is preferred. If the contents of the drugs are quite stable to the temperature and sunlight, the drugs can be dried directly in sunshine (e.g. gum, acacia, seeds and fruits).

6.3.2. Artificial Drying

Drying by artificial means includes drying the drugs in *(a)* an oven; *i.e.* tray-dryers; *(b)* vacuum dryer, and *(c)* spray dryers.

6.3.2.1. Tray Dryers

The drugs which do not contain volatile oils and are quite stable to heat or which need deactivation of enzymes are dried in tray dryers. In this process, hot air of the desired temperature is circulated through the dryers and this facilities the removal of water contents of the drugs (*e.g.* belladonna roots, cinchona bark, tea leaves and gum are dried by this method).

6.3.2.2. Vacuum Dryers

The drugs which are sensitive to higher temperature are dried by this process e.g. tannic acid and digitalis leaves.

6.3.2.3. Spray Dryer

Few drugs which are highly sensitive to atmospheric conditions and also to temperature of vacuum- drying are dried by spray dryer method. The technique is followed for quick drying of economically important plant or animal constituents, rather than the crude drugs. *e.g.* papaya latex, pectin, tannins, etc.

6.4. Garbling (Dressing)

The next step in preparation of crude for market after drying is garbling. This process is desired when sand, dirt and foreign organic parts of the same plant, not constituting drug are required to be removed. This foreign organic matter (extraneous matter) is removed by several ways and means available and practicable at the site of the preparation of the drugs. If the extraneous matter is permitted in crude drugs, the quality of drug suffers and at times, it dosen't pass pharmacopoeial limits. Excessive stems in case of lobelia and stramonium (*Datura* spp.) need to be removed, while stalks, in case of cloves are to be deleted. Drugs constituting rhizomes need to be separated carefully from the roots and rootlets and also stem base. Pieces of iron must be removed with the magnet in case of castor seeds before crushing and by shifting in case of vinca and senna leaves. Pieces of bark should be removed by peeling as in gum acacia.

6.5. Packing

The morphological and chemical nature of drug, its ultimate use and effects of climate conditions during transportation and storage should be taken into consideration while packing the drugs. Aloe is packed in goat skin. Colophony and balsam of tolu are packed in kerosene tins, while asafetida is stored in well closed containers to prevent loss of volatile oil. Cod liver oil, being sensitive to sunlight, should be stored in such containers, which will not have effect of sunlight, whereas the leaf of the drugs like senna, vinca and other are pressed and baled. The drugs are very sensitive to moisture and also costly at the same time need special attention. *e.g.* digitalis, ergot and squill. The squill becomes flexible, ergot becomes susceptible to the microbial growth, while digitalis losses its potency due to decomposition of glycosides. If brought in contact with excess of moisture during storage. Hence, the chemicals which absorb excessive moisture (desiccating agents) from the drugs are incorporated in the containers. Colophony needs to be packed in big masses to control auto-oxidation. Cinnamom bark, which is available in the form of quills, is packed one side the other quill, so as to facilitate transport and to prevent volatilization of oil from the drug. The crude drug like roots, seeds and others do not need special attention and are packed in gunny bags, while in some cases bags are coated with polythene internally. The weight of certain drugs in lots is also kept constant. *e.g.* Indian opium.

6.6. Storage

Preservation of crude drugs needs sound knowledge of their physical and chemical properties. A good quality of the drugs can be maintained. If they are preserved in well closed and, possibly, in the filled containers. They should be stored in the premises which are water-proof, fire proof and rodent-proof. A number of drugs absorb moisture during their storage and become susceptible to the microbial growth. Some drugs absorb moisture to the extent of 25% of their weight. The moisture, not only increase the bulk of the drug, but also causes impairment in the quality of crude drug. The excessive moisture facilities enzymatic reactions resulting in decomposition of active constituents. *e.g.* ergot, cod liver oil and digitalis. Form or shape of the drug also play very important role in preserving the crude drugs. Colophony in the entire form (big masses) is preserved nicely, but if stored in powder form, it gets oxidized or looses its solubility in petroleum ether. Squill, when stored in powdered form rubbery mass on hygroscopic and forms rubbery mass on prolonged exposure to air. The fixed oil in the powdered ergot becomes rancid on storage. In order to maintain a good quality of powdered ergot, it is required that the drug should be defatted with lipid solvent prior to storage. Lard, the purified internal fat of the abdomen of the hog, is to be preserved against rancidity by adding siam benzoin. Atmospheric oxygen is also destructive to several drugs and hence, they are filled completely in well closed containers, or the air in the containers in replaced by an inert gas like nitrogen; *e.g.* Shark liver oil, papain, etc.

Apart from protection against adverse physical and chemical changes, the preservation against insect or mould attacks is also important. Different types of insects, nematodes, worms, mould and mites infest the crude drugs during storage. Some of the more important pest found in drugs are Coleoptera (*Stegoblium paniceum*) and Calandrum granarium). Lepidoptera (*Ephestia kuehniella* and *Tinea pellionella*), and Archnida or mites (*Tyroglyphus farnae* and *Glyophagus domesticus*). They can be prevented by drying the drug thoroughly before storage and also by giving treatment of fumigants. The common fumigants used for storage of crude drugs are methyl bromide, carbon disulphide and hydrocyanic acid. At times, drugs are given special treatment, such as liming of the ginger and coating of nutmeg. Temperature is also very important factor in preservation of the drugs, as it accelerates several chemical reactions leading to decomposition of the constituents. Hence, most of the drugs need to be preserved at a very low temperature. The costly phyto-pharmaceuticals are required to be preserved at refrigerated temperature in well closed containers. Small quantities of crude drugs could be readily stored in air-tight, moisture proof and light proof containers such as tins, cans, covered metal tins, or amber glass containers. Wooden boxes and paper bags should not be used for storage of crude drugs.

6.7. Value-Addition or Processing

In order to make use of its latent medicinal qualities, a plant must be treated and modified in such a way that its specific curative substances can be enacted. Such transformations form part of what is known as the 'galenic' branch of pharmacy. The active principles contained in every medicinal plant consist of a number of compounds, which individually or in groups, can have a specific action on the organs or our bodies. Different methods have, therefore, been devised to extract these substances, either singly or collectively, depending on the result required. Generally, there are four basic, simple preparations *viz.* decoction, maceration, infusion and juice extraction.

6.7.1. Decoction

Decoction is the method normally used for those medicinal herbs whose active principles are difficult to extract because they are contained in woody parts of the plant, or which require prolonged heating in order to pass into solution. Sometimes, the extraction by decoction involves boiling the whole plant or a part of it in water for a given time and the allowing it to macerate for a further period before filtering. This method is reserved for certain plants such as those with a high mucilaginous content.

6.7.2. Extraction

Extraction is much more suitable method for obtaining the active principles when the parts of the plant being used are soft and fragile, such as leaves bud or flowers. It is preferable, except in certain cases, to use herbs that have been slightly dried, as the reduction of water content will concentrate the principal constituents.

6.7.3. Maceration

This method is used for medicinal plants whose active principles are soluble in cold water for several hours during, which time all the principles that do not need heat to release them (*i.e.* that are not thermolabile) will be released into the solution. The yield of mucilage from certain plants, such as the common mallow (*Malva sylvestris*) and marsh mallow (*Althaea officinalis*), is in fact greater when extraction is carried out by means of cold maceration rather than by infusion.

6.7.4 Dehumidifying

An effective but expensive way to dry herbs is to use a dehumidifier, which literally sucks water out of the plant. The dehumidifier should be placed in a more or less sealed small room in which the herbs are hung in loose bunches or placed on mesh trays. Herbs will dry quickly with this method and, as no heat is used, there is little deterioration or decay.

6.7.5 Freeze-Drying

Freeze-drying retains colour and flavour but is more suited to culinary than to medicinal herbs. Whole sprigs of herbs such as parsley (*Petroselinum crispum*) or sage (*Salvia officinalis*), can be frozen in plastic freezer bags. There is no need to defrost before use as the leaves crumble easily when still frozen. Chickweed (*Stellaria media*) can also be frozen and used topically for itchy and weeping skin conditions. Many plants may be juiced, frozen as ice cubes and thawed as required.

6.8. Preservation

Medicinal plants can be preserved in a number of ways, the most common and simple, being air or oven drying. A warm, dry place, for example an airing cupboard, is ideal. Use plain paper for drying herbs, never printed newspaper. Dried herbs can be stored for man months in a dark glass jar or a brown paper bag.

6.9. Active Constituents

The active ingredients are the main effective compounds of medicinal plants. The quantification of active ingredients is the first step of standardization of crude drugs. A number of chemical test are available for quantification of active ingredients. Some of the main chemical groups (organic compounds) of active ingredients are given below;

(a) Saponins

There are two types of saponins-triterpenoid and steroidal saponins. Many plants containing steroidal saponins have a marked hormonal activity, liquorice (*Glycyrrhiza glabra*), being one of the best known. Triterpenoid saponins, for example those in cowslip root (*Primula veris*), are often strong expectorants, and stay also aid in the absorption of nutrients.

(b) Cardiac Glycosides

It is found in various medicinal plants, such as foxglove (*Digitalis spp.*) contains digitoxin, digoxin and ditoxin. Cardiac glycosides are also significantly diuretic. They help to transfer fluids from the tissues and circulatory system to the urinary tract, thereby lowering blood pressure.

(c) Flavonoids

It possesses anti-inflammatory properties and are especially useful in maintaining healthy circulation. Rutin, a flavonoid found in plants including buckwheat (*Fagopyrum esculentum*), and lemon (*Citrus limon*), strengthens capillary walls.

(d) Mucilage

It is made up of polysaccharides (large sugar molecules) that soak up water, producing a sticky jelly- like mass. Mucilage lines the mucous membranes of the digestive tract, protecting against irritation, acidity and inflammation. This soothing and protective action appears to extend to other areas, including the mucous membranes of the throat lungs, kidneys and urinary tubules. Slippery elm (*Ulmus rubra*), is a typical mucilaginous herb.

(e) Phenols

This group of compounds includes salicylic acid-the natural forerunner of aspirin. Salicylic acid is found in many plants. for example wintergreen (*Gaultheria procumbens*), and white willow (*Salix alba*). Another phenol is thymol - a constituent of thyme (*Thymus vulgaris*). Phenols are antiseptic and reduce inflammation when taken internally yet they have an irritant effect when applied to the shin.

(f) Tannins

Tannins are produced to a greater or lesser degree by all plants. The harsh, astringent taste to tannin - laden bark and leaves makes them unpalatable to insects and grazing animals. Tannins contract the tissues of the body-hence their use to "tan" leather. They draw the tissues closer together and improve their resistance to infection. Oak bark (*Quercus robur*) and black catechu (*Acacia catechu*) are both high in tannins.

(g) Coumarins

Coumarins of different kinds are found in many plant species and have widely divergent actions. The coumarins in melilot (*Melilotus officinalis*), thin the blood, while bergapten, found in celery (*Apium graveolens*), is used as a sunscreen, and khellin, found in visnaga (*Ammi visnaga*), is a powerful smooth muscle relaxant.

(h) Anthraquinones

Anthraquinones have an irritant laxative effect on the large intestine, causing contractions of the intestinal walls and stimulating a bowel movement and also make the stool more liquid. For example; senna (*Cassia angustifolia*), and Indian rhubarb (*Rheum* spp.).

(i) Anthocyanins

These pigments, which give flowers and fruits a blue, purple or red hue, help to keep the blood vessels healthy. For example, Blackberry (*Rubus fruticosus*), and grapes (*Vitis vinifera*).

(j) Glucosilinates

These compound are found exclusively some medicinal plants belong to mustard family, glucosilinates have an irritant effect on the skin, causing inflammation and blistering. Applied as poultices of painful or aching joints, they increase blood flow to the affected area, helping to remove the build-up of waste products (a contributory factor in joint problems). Glucosilinates also help to reduce thyroid function. For example, Radish (*Raphanus sativus*), mustard (*Sinapis alba*).

(k) Volatile Oils

It is extracted from plants to produce essential oils. Some volatile oils contain sequiterpenes, such as azulence, found in chamomile (*Chamomilla* spp.). These constituents have an anti-inflammatory effect. Tea tree (*Melaleuca alternifolia*), for example, is known to contain over 60 different volatile compounds within its volatile oil, many of them being strongly antiseptic.

(l) Cyanogenic Glycosides

Though these glycosides are based on cyanide, a very potent poison, they have a helpful sedative and relaxant effect on the hem and muscles in small doses. Wild cherry bark (*Prunus serotina*), and elder (*Sambucus nigra*), both contain cyanogenic glycosides, which contribute to both plants' ability to suppress and soothe irritant dry coughs.

(m) Vitamins

Some plants contain significant levels of vitamins. Watercress (*Nasturtium officinale*), for example, contains an appreciable quantity of vitamin E, and the tips of dog rose (*Rosa canina*), have particularly high levels of vitamin C. Most other medicinal plants contain at least some vitamins.

(n) Bitters

Bitterness itself stimulates secretions by the salivary glands and digestive organs. Such secretions can dramatically improve the appetite and strengthen the overall function of the digestive system. Many herbs have bitter constituents, wormwood (*Artemisia absinthium*), and chiretta (*Swertia chirata*)

(o) Alkaloids

It is mixed group, alkaloid mostly contains a nitrogen - bearing molecule (-NH2) that makes them particularly pharmacologically active. Some are well-known drugs and have a recognized medical use. Vincristine, for example, derived from periwinkle (*Catharanthus roseus*), is used to treat some types of cancer (especially in breast cancer). Other alkaloids, such as atropine, round in deadly nightshade (*Atropa belladonna*), have a direct effect on the body, reducing spasms, relieving pain and drying up bodily secretions.

(p) Minerals

A number of medicinal plants are rich in minerals. Horsetail (*Equisetum arvense*), for example, has high levels of silica. Dandelion (*Taraxacum officinale*), has large quantities of potassium, and unlike other diuretics which flush this mineral out of the body, it helps to maintain high levels of potassium.

Chapter 7

ADULTERATION AND SUBSTITUTION OF CRUDE (HERBAL) DRUGS

India is a vast country with variety in languages, food, habit, clothing, traditionals and culture besides diversity in climate vegetation and physiotherapy. The indigenous system of medicine have deep roots in our cultural heritage. In ancient times most of the drugs used for therapeutic use were of plant origin and were mentioned in the Ayurvedic literature. But the lack of proper description and voucher specimens of these plants has made it difficult to provide correct identity of the specific drugs described. As a result of this confusion, in many cases, a large number of plant species have now come in wide practices for the same drug in different areas and even in the same locality and *vice versa*.

There are about 600 single drugs of vegetable origin presently used in the preparation of about 1000 popularly used Ayurvedic formulations. The systematic identification of many of these indigenous herbal drugs is still a subject of confusion and controversies, because several different species of herbs are available for one and same medicinal name crude drugs, on one hand, whereas the same plant is being used for different crude drugs on the other hand. It is a challenge for the scientists to identify the controversial drugs of present time so that the real species to be used as a specific drug could be singled out and the controversy may be resolved by laying down the pharmacopoeial standards for the proper identification of such drugs.

The are many examples of botanically different plants linked with one Indian name, *e.g.*, the name of **Ashok** - *Polyalthia longifolia* Benth. & Hook. f. and *Saraca indica* Linn.; **Babuna** - *Corchorus depressus* Linn., *Cotula anthemoides* Linn., *Matricaria chamomilla* Linn., **Badaward** - *Amberboa divaricata* Kuntze, *Cardus nutans* Linn.; **Banda** - *Dendrophthoe falcate* (Linn. f.) Ettingsh, *Viscum album* Linn.: **Mamira** - *Coptis teeta* Wall. *Thalictrum foliolosum* DC.; **Rattanjot**- *Onosma echioides* Linn., *Clausena pentaphylla* (Roxb.) DC., *Anemone obtusiloba* D. Don etc. There are many more such examples to quote.

Similar is the case with the well known group of eight rare Ayurvedic herbs known as **Astavarga**. Studies have revealed that the drug material varies from market to market; *e.g.*, out of the **Astavarga** group the drug **Meda** and **Mahameda** are obtained from the name plant *Polygonatum verticellatum* All., the only difference traced is that **Meda** material comprises of somewhat thicker assorted tubers. **Jivak** and **Rishbhak** are obtained from both *Liparis rostrata* and *Malaxis acuminata* D. Don, which are two different plants. Similarly, **Rasna** and **Jivanti** are obtained from more than one different plant sources.

Adulteration of drugs is so much in prominence that some times 'Anemone' flowers are sold as Viola (**Banafsha**) flowers; *Strychnos potatorum* Linn. f. seeds as *Strychnos nux-vomica* Linn. (Kuchla) seeds; pieces coloured and cut, are sold as 'Saffron' (**Zafran**); ordinary gums in large quantities are mixed with small percentage of 'asafoetida' and passed on to the public as true asafetida (Heenga). *Tinospora cordifolia* (Willd.) Miers extract (Satgilo) is adulterated with arrow-root powder; *Mallotus phillippinesis* Muell-Arg. (**Kamla**) is mixed with fine brick powder. **Silajeet** and **Tabasheer** are prepared synthetically and sold as genuine products. *Acacia arabica* Willd (**Babul**) bark after being used in tanning hides is converted into tea and is sold as such. Almond-oil is generally adulterated with ground nut oil and cotton-seed oil; Honey is adulterated with invert sugar i.e. glucose and fructose, **Musk** is adulterated with dust and dried blood. **Gaozaban** is supposed to be indigenous to India and is mentioned under six species of plants. But it is reported that **Gaozaban** belongs to *Anchusa strigosa* Labill, species not mentioned in the literature and not indigenous to India. The market sample is of *Arnebia benthami* Johnston. The **Chaksu** is reported to be prepared from an adulterant *Dolichos biflorus* Linn. (**Kulthi**) instead of *Cassia absus* Linn.

Some substitutes for the unavailable drugs have also been mentioned in the literature. These are also to be examined on scientific lines. The literature available on drugs is also very old and the names mentioned for some of the drugs have either been changed of the drugs occurring by those name in that period, may have extinguished. As Avicenna in his Canon has designated several drugs *e.g.* Osbed, Ba, Halbeed, Barkeesa shull, Fall, Hast dahanas Indian, but their identity is difficult to ascertain.

It may be noted that *Terminalia arjuna* Wight & Arn. And *Terminalia alata* Heyne ex roth are sold in the market as **Arjun**. *Calotropis gigantean* (Linn.) R. Br. Ex. Ait. And *Calotropis procera* (Ait.) R. Br. Are substituted for each other as **Arka**; the commonest species of *Tinospora cordifolia* is likely to be substituted with *Tinospora sinensis* merr. And *Tinospora crisa* Miers known as **Gilo**; *Derris indica* (Lam.) Bennet is often used as substitute for *Caesalpinia bonduc* (Linn.) Roxb. used as **Karanja**; *Trichosanthes dioica* roxb. and *Trichosanthes cucumerina* Linn. both are used as **Patola**; *Hemidesmus indicus* R. Br., *Ichnocarpus frutescens* R. Br., *Cryptolepis buchananii* Roem. & Schult. And *Decalepis hamiltonii* Wight & Arn. are used as **Sariva** differently.

In view of the prevailing controversy of the single drugs, the available literature has been screened and market samples of the drugs in current were examined. Some of the interesting observations made are given below.

Sl.No.	Botanical Name	Hindi/English Name	Part Used	Substitute	Adulterant
1.	Acacia catechu	Khair/Catch tree	Gum	—	Senecio jacquemontianus
2.	Acacia nilotica	Babul/Indian Gum tree	Gum	Ramalina fraxinea	—
3.	Acacia catechu	Khair/Catch tree	Stem, Bark	—	Rhus paniculata
4.	Arnica montana	Arnica	Flower	—	Tragopogon pratensis
5.	Ephedra geradiana	Asmania	Leaves	—	Ephedra intermedia E. major
6.	Aleurites fordii	Tung oil tree	Karnel oil	Mallotus philippensis	Sapium sebiferum
7.	Aconitum nephallus	Mithavis/Indian aconite	Roots	Aconitum chasmanthum	Gloriosa superba
8.	Aconitum heterophllum	Atis/Indian Atis	Rhizome	—	Aspergus spp)
9.	Acorus calamus	Buch/Sweet flag	Rhizome	Alpinia galanga	—
10.	Atropa belladonna	Angurshefa/Belladona	Leaves	Atropa acuminata	Althaea officinalis
11.	Coptis teeta	Mamira/Gold thread	Rhizome	Thalictrum foliolosum	Thalictrum foliolosum Picrorhiza kurroo
12.	Cassia angustifolia	Sanaya/Senna	Rhizome	Pluchea lanceolata	Cassia obtusa
13.	Curcuma aromatica	Van haldi/Wild Turmeric	Rhizome	Curcuma domestica	—
14.	Coscinium fenestratum	Daru haldi/Turmeric tree	Stem	Jateorhiza palmate	—
15.	Colchicum luteum	Hirunitutiya/Meadow saffron	Corm	Colchicum autumnate	—
16.	Chenopodium hybridum	Wormseed	Leaves	Chenopodium ambrosioides	Datura stramonium
17.	Gentiana lutea	Great yellow gentian	Rhizome	Gentiana kurroo	—
18.	Taxus baccata	Thuner/Himayan Yew	Leaves	—	Caphalotaxus sp.
19.	Tylophora fasiculata		Root & Leaf	Cephaelis ipecacuanha	—
20.	Digitalis purpurea	Tilpushpi/Common Foxglove	Leaves	Verbascum thapsus	Inula racemosa
21.	Dipcadi erythraeum		Bulb	Urginea indica	Urginea indica
22.	Thalictrum javanicum	Pilijari	Roots	Thalictrum foliolosum	—
23.	Nardostachys grandiflora	Jatamansi	Rhizome	Valeriana officinalis	Selinum vaginatum
24.	Picrorhiza Kurroo	Kutki/Picrorhiza	Roots	Gentiana Kurroo	—
25.	Polygonum viviparum	Snake root	Root Stock	Polygonum bistorta	—
26.	Plantago lanceolata	Isabgol/Psyllium	Seed/Coat	—	Plantago psyllium
27.	Premna obtiusifolia	Agetha/Headache tree	Roots	Premna coriacea	—
28.	Prunus cornuta	Sicklewort	Karnel oil	Prunus amygdalus	Telfairia pedata
29.	Potentilla fruticosa	Tormentilla	Dried leaf	Camellia sinensis	—
30.	Polygala senega	Seneca Snakeroot	Roots	Primula veris Primula denticulata	—

Contd...

Sl.No.	Botanical Name	Hindi/English Name	Part Used	Substitute	Adulterant
31.	*Pimpinella anisum*	Saunf/Anise	Grains	—	*Setaria glauca*
32.	*Piper nigrum*	kali Mirch/Black pepper	Fruit	—	*Schinus molle*
33.	*Fumaria vaillantii*	Pilpara	Drug	—	*Rungia pectinata*
34.	*Boerhaavia diffusa*	Punarnava/Hogweed	Herb	—	*Trianthema protulacastrum*
35.	*Mentha piperata*	Piper mint	Leaves	*Micromeria capitellata*	—
36.	*Mentha piperita*	Piper mint	Starch	*Tacca leontopetaloides*	—
37.	*Urginea indica*	Jangli Payag/Indian squill	Bulb	*Scilla hyacinthiana*	—
38.	*Rauvolfia serpentina*	Sarapgandha/ Serpentine root	Roots	*Rauvolfia tetraphylla*	*Rauvolfia densiflora*
39.	*Rheum palmatum*	Rheum/Chinese Rhubarb	Rhizome	*Rheum webbianum R. emodi*	—
40.	*Limnanthemum cristatum*		Dried plant	*Swertia chirayata*	—-
41.	*Viola odorata*	Banafshah/Sweet violet	Roots	*Cephaelis ipecacuanha*	*Viola sylvestris*
42.	*Valeriana jatamansi*	Tagar/Indian valerian	Roots & Rhizome	*Valeriana pyrolaefolia*	—
43.	*Shorea robusta*	Sal	Fatty oil	*Theobroma cocao*	—
44.	*Schinus molle*		Essential oil	*Piper nigrum*	—
45.	*Saussurea costus*	Kuth/Costus	Roots	*Saussurea hypoleuca*	*Senecia jacque montianus*
46.	*Swertia chirayita*	Nirmali/Clearling nut tree	Dried plant	*Swertia decussata*	*Andrographic paniculata*
47.	*Salvia lantana*	Clary sage	Rhizome		*Saussurea lappa*
48.	*Cinchona officinalis*	Quinine/Quinine bark	Root Bark	*Hymenodictyon obovatum Cinchona lencifolia*	*Swietenia mahagoni*
49.	*Centella asiatica*	Mandukparni/Indian pennywort	Herb	*Hydrocotyle javanica*	—
50.	*Sida acuta*		Fibres	*Corchorus capsularis*	—
51.	*Cephaelis ipecacuanha*	Ipecac	Fruit	*Strychnos potatorum*	—
52.	*Centella asiatica*	Mandukparni/Indian pennywort	Roots	—	*Richardia scabra*
53.	*Santalum album*	Chandan/Sandal tree	Wood powder	*Ximenia americana*	—
54.	*Pseidotsuga menziesii*		Dried leaves	*Coffea arabica*	—
55.	*Pseudarthria visicida*	Chapakno	Roots	*Desmodium gengeticum*	—
56.	*Hemidesmus indicus*	Anantamul/Indian sarasaparilla	Dried Roots	*Smilax aspera Ichnocarpus frustescens*	—
57.	*Holarrhena antidysenterica*	Kutaja/Kurchi bark	Stem bark	*Wrightia tomentosa*	*Wrightia tinctoria*

Although scientific approach based on the systematic identification of the single drugs of controversial nature from botanical, pharmacognostical, chemical and pharmacological point of view may be an answer to solve the prevailing controversies resulted due to the above reasons. There is also a need for a multi-disciplinary approach and coordinated efforts of various scientists working on indigenous drugs and an integrated policy for medicinal plants to be framed by the Government.

Chapter 8

MARKET POTENTIAL (EXPORT/ IMPORT) OF CRUDE DRUGS

India has a rich treasure of medicinal plants due to the diversity of agro-climatic conditions spread all over the country from tropical to temperate zones, coastal plains to high altitudes and semi-arid to highly humid evergreen forests, therefore, it is in an advantageous position to produce a number of crude drugs. The medicinal plants grown in India may comprise different plant components or parts such as whole plants (Brahmi, Kalmegh), Floral parts (Cloves, Saffron etc.) or Fruits (Cardamom, Harar, Amla etc.) or Berries (Black pepper) or Seed husk (Isabgol.) or Rhizomes (Atis, Jatamansi) or Roots (Rauvolfia) or Leaves (Tobacco) or Bark (Acacia, Cinchona) or Bulbs (Garlic) or other parts of medicinal plants.

Among the different medicinal plants grown in India, the important among them are Atis, Amla, Jatamansi, Podophyllum, Carium, Harad, Behrda, Neem, Cardamom, Ginger and Turmeric that were originated in this country. Medicinal and aromatic plants such as geranium, patcholi, lemon balm, ruta, linaloe, thymus, sage and salvia were introduced by the invaders. Some Indian spices have potential medicinal value such as black pepper is called as 'The King of Spices', Cardamom as the 'Queen of Spices' and turmeric is once known in the West as 'Indian Saffron'. Most of them are believed to be originated in the Western Ghats of the country.

Medicinal plants possesses appetizers, emollient, cooling, astringent and anti-oxidant properties. Medicinal plants are used in the pharmaceutical industries in the preparation of herbal products as well as value added (semi-process items) and consumer articles like cosmetics, tooth pastes, soaps etc.

The consumption of crude drugs varies from one country to another and is influenced to a large extent by the size of the population and the rate at which it grows. It is also influenced by the disposable income, which in the case of developing countries is a major factor. In the developed countries, medicinal herbs are used in the industrial sector, principally in herbal processing whereas in developing countries, medicinal plants are mainly consumed in traditional systems of medicine (Vaidya's, home remedies).

India exports entire gamut of medicinal plants. The major ones are rauvolfia, costus, ginseng, belladona, plantago, swertia, cassia, tamarind, hemidesmus, catharanthus as well as essential oil, oleoresin, and alkaoids. Senna, ginseng, plantago, santalum are some of India's favoured and preferred medicinal plants in the international markets. In terms of volume, exports have recorded a whooping growth of 54 percent from 38744.43 tonnes in 1991-92 to 59494.17 tonnes in 1999-2000. The shift in emphasis from commodities in whole form to value added products has given a new dimension to the pharmaceutical industry in the country. On the other hand, India has not yet succeeded in certain medicinal plants like Chitarak (*Plumbago zylanica*), Kuth (*Saussuria costus*), Kokum (*Garcinia indica*), Manjistha (*Rubia cordifolia*), Ashok (*Saraca asoca*), Kapur kachari (*Hydichium spicatum*), Akarkara (*Anacyclus pyrethrum*), Rakta chandan (*Pterocarpus santalinus*), Cubeb (Piper cubeba), Guggul (*Commuphora wightii*), Liquorice (*Glycyrrhiza glabra*), Dalchini (*Cinnamomum zeylanica*) etc. and the requirements are met through imports. Indian medicinal plants and their products are mostly exported to East, European countries, Central Asian countries, American zone and to the Middle East countries in large volumes and to the African and Australian zones in small volumes. As a whole, the recent performance of India is improving in it's exports of medicinal plants, still it has not yet recovered in the international herbal market. Once it had the monopoly in it, for this, there appears to be the prevalence of several hurdles in this industry.

The export and Import of crude drugs (from 1991-1992 to 1999-2000) as foreign exchange earner are given below:

Table 1: Export of Ayurvedic, Unani, Homeopathic Medicines and Alkaloids

Year	Ayurvedic & Unani Medicines		Homeopathic Medicines		Alkaloid	
	Quantity (in tonnes)	Amount (in Lakh Rs.)	Quantity (in tonnes)	Amount (in Lakh Rs.)	Quantity (in tonnes)	Amount (in Lakh Rs.)
1991-92	3360.56	3318.48	—	567.19	—	40.68
1992-93	4117.25	5022.43	—	164.56	—	68.17
1993-94	3112.34	5361.13	—	517.34	—	119.02
1994-95	4173.46	9338.94	—	180.02	—	115.01
1995-96	3716.29	9644.97	—	352.51	—	402.95
1996-97	12986.70	15503.58	56.32	128.53	154.80	179.39
1997-98	8939.52	6499.84	51.99	309.51	271.49	1989.65
1998-99	10898.79	7451.59	32.56	37.47	73.18	613.11
1999-2000	10399.20	5474.00	121.96	67.39	122.45	899.04

Table 2: Import of Ayurvedic, Unani, Homeopathic Medicines and Alkaloids

Year	Ayurvedic & Unani Medicines		Homeopathic Medicines		Alkaloid	
	Quantity (in tonnes)	Amount (in Lakh Rs.)	Quantity (in tonnes)	Amount (in Lakh Rs.)	Quantity (in tonnes)	Amount (in Lakh Rs.)
1991-92	1732.50	583.61	—	447.21	—	0.28
1992-93	914.95	442.83	—	427.53	—	—
1993-94	2201.95	268.14	—	581.67	—	1.65
1994-95	1814.29	236.94	—	524.91	—	11.96
1995-96	1287.37	1270.26	—	485.61	—	3.34
1996-97	3640.05	3395.02	126.21	496.96	—	—
1997-98	1637.19	507.04	102.13	572.49	6.41	97.94
1998-99	3761.57	1863.54	171.63	936.42	0.63	53.99
1999-2000	3934.49	3956.77	146.06	799.69	0.07	0.27

Table 3(a): Export of Major Crude Drugs from1991-92 to 1995-96

Crude Drug Items	1991-92		1992-93		1993-94		1994-95		1995-96	
	Quantity (in Ton.)	Value (in Lakh Rs.)	Quantity (in Ton.)	Value (in Lakh Rs.)	Quantity (in Ton.)	Value (in Lakh Rs.)	Quantity (in Ton.)	Value (in Lakh Rs.)	Quantity (in Ton.)	Value (in Lakh Rs.)
Glycyrrhiza glabra (Roots)	–	–	0.31	0.82	1.26	0.66	–	–	0.96	0.70
Alpinia sp. (Rhizomes)	72.5	11.77	75.61	15.56	98.5	22.01	22.44	6.25	83.47	22.28
Curcuma zedoaria (Roots)	46.95	4.63	48.72	7.93	46.69	5.03	33.1	4.63	30	3.74
Rauvolfia serpentina (Roots)	0.33	3.23	–	–	0.11	0.17	–	–	3.93	3.2
Saussurea costus (Roots)	–	–	2.00	0.75	2.50	0.55	–	–	8.00	1.52
Panax ginseng (Roots)	1325.73	386.67	2034.79	838.18	1911.32	774.55	2945.2	1314.94	3072.11	1627.55
Atropa belladonna (Roots & leaves)	–	–	2.00	0.53	–	–	–	–	3.16	0.99
Plantago psyllium (Husk & Roots)	17544.40	6854.74	15347.99	7741.40	15475.44	7546.76	20889.43	11372.11	19120.27	15440.97
Swertia chirayita (Whole plant)	13.69	3.87	0.575	0.61	5.6	1.36	–	–	–	–
Cassia angustifolia (Pods & leaves)	5121.22	1116.29	4476.56	1222.99	3658.28	819.76	400.48	850.99	6279.77	1391.61
Tamarindus indica (Seeds & Powder)	3557.75	141.08	2726.72	129.31	2745.33	210.33	2736.89	174.37	2718.2	189.58
Catharanthus roseus (Roots & leaves)	270.77	91.47	128.2	49.14	150.11	67.73	151.7	43.89	191.92	64.97
Hemidesmus indicus (Roots)	0.24	0.07	12	4.67	11.76	4.57	20.61	13.31	1.1	0.3
Ziziphus mauritiana (Fruits)	8.67	3.22	61.5	13.53	86	2.69	21.16	8.12	29.45	13.44
Vitis vinifera (Water)	1062.91	59.73	–	181.37	–	20.64	–	69.7	–	82.07
Ricinus communis (Oil)	–	3330.10	–	1514.95	–	3816.23	–	1125.22	–	19.64
Humulus lupulus (Dry leaves)	–	–	–	–	–	–	0.89	0.08	19.00	1.37
Piper longum (Fruits)	4.180	1.140	15.720	1.470	88.660	41.360	45.660	30.830	38.530	17.020
Piper nigrum (Garbled/Ungarbled/ Dehydrated/Crushed)	20460.47	7428.68	23540.60	7674.65	461056.10	18654.51	36851.42	23428.81	25888.56	19273.00
Cinnamomum zeylanicum (Bark)	23.930	9.760	0.031	0.094	5.350	0.580	1.010	2.290	2.480	0.580
Syzygium aromaticum (Buds)	0.035	0.090	0.000	0.000	0.01	0.02	35.78	11.07	0.29	0.56
Myristica fragrans (Fruits)	0.3	0.68	2.07	1.99	11.41	5.23	5.02	5.69	5.88	6.82
Other Medicinal Products	6009.67	1026.45	8253.17	1449.89	5470.18	1018.67	4452.73	753.98	–	–

Table 3(b): Export of Major Crude Drugs from 1996-97 to 1999-2000

Crude Drug Items	1996-97		1997-98		1998-99		1999-2000	
	Quantity (in Ton.)	Value (in Lakh Rs.)	Quantity (in Ton.)	Value (in Lakh Rs.)	Quantity (in Ton.)	Value (in Lakh Rs.)	Quantity (in Ton.)	Value (in Lakh Rs.)
Glycyrrhiza glabra (Roots)	0.01	0.02	9.51	8.35	3.42	9.22	70.39	81.93
Alpinia sp. (Rhizomes)	64.3	12.8	29.00	6.59	75.42	29.62	201.53	74.94
Curcuma zedoaria (Roots)	31.7	5.02	37.23	6.38	43.17	7.42	36.15	8.04
Rauvolfia serpentina (Roots)	—	—	NA	NA	0.19	0.17	9.03	5.70
Saussurea costus (Roots)	—	—	NA	NA	NA	NA	—	—
Panax ginseng (Roots)	2768.38	2074.05	3787.40	2399.29	4470.56	2995.34	1379.46	1015.14
Atropa belladonna (Roots & leaves)	99.35	21.31	5.83	1.29	44.27	18.52	22.74	8.19
Plantago psyllium (Husk & Roots)	17842.63	13697.52	20634.25	15884.62	14782.76	13742.48	15295.31	10815.18
Swertia chirayita (Whole plant)	45.03	33.1	16.24	4.65	1.62	0.50	50.29	37.74
Cassia angustifolia (Pods & leaves)	5948.61	1380.82	5011.84	1377.31	5180.58	2070.82	7466.33	2254.20
Tamarindus indica (Seeds & Powder)	2415.76	222.16	1174.36	138.95	1179.03	166.71	2763.08	423.98
Catharanthus roseus (Roots & leaves)	24.57	7.85	—	—	277.71	104.42	541.54	213.19
Hemidesmus indicus (Roots)	11.69	3.07	23.14	4.19	32.02	27.22	14.71	6.22
Ziziphus mauritiana (Fruits)	38	30.01	—	—	81.90	58.65	35.90	15.25
Vitis vinifera (Water)	104.85	59.21	61.04	34.75	84.708	61.16	66.91	49.63
Ricinus communis (Oil)	11.89	12.38	124.79	51.92	270.29	97.26	136.50	78.49
Humulus lupulus (Dry leaves)	0.02	0.04	—	—	—	—	20.00	3.41
Piper longum (Fruits)	348.780	204.930	611.00	354.80	272.66	395.47	319.23	660.22
Piper nigrum (Garbled/Ungarbled/ Dehydrated/Crushed)	46904.56	40865.36	34632.32	483035.27	32798.39	59055.42	33908.33	66816.66
Cinnamomum zeylanicum (Bark)	9.930	8.180	111.53	62.24	3.31	2.69	—	—
Syzygium aromaticum (Buds)	83.09	24.04	—	—	1.25	5.56	310.73	384.77
Myristica fragrans (Fruits)	5.25	4.61	31.40	5.69	2.81	6.09	130.58	83.99
Other Medicinal Products	—	—	3062.69	1430.32	4736.34	2333.74	2734.60	1136.40

Source: Monthly statistics of the foreign trade of India Vol. I (Export) (1991-2000).

These hurdles are lower productivity, non-availability of high yielding disease resistant quality planting materials, improper marketing activities, price fluctuations, increasing cost of production, increasing level of domestic consumption, competition from the new entries, stringent sanitary and phyto sanitary conditions, demand for organic herbs, Pests and Diseases etc. So as to regain the status in the international herbal market and to improve Indian position both in terms of volume and value these problems have to be tackled with the help of a planned strategy.

Table 4(a): Import of Major Crude Drugs from 1991-92 to 1995-96

Crude Drug Items	1991-92		1992-93		1993-94		1994-95		1995-96	
	Quantity (in Ton.)	Value (in Lakh Rs.)	Quantity (in Ton.)	Value (in Lakh Rs.)	Quantity (in Ton.)	Value (in Lakh Rs.)	Quantity (in Ton.)	Value (in Lakh Rs.)	Quantity (in Ton.)	Value (in Lakh Rs.)
Glycyrrhiza glabra (Roots)	650.46	63.88	488.02	35.78	650.85	48.48	1285.92	97.50	581.15	49.03
Alpinia sp. (Rhizomes)	5.00	0.64	44.18	3.48	—	—	70.82	6.78	55.3	6.34
Curcuma zedoaria (Rhizomes)	—	—	—	—	—	—	—	—	—	—
Rauvolfia serpentina (Roots)	—	—	—	—	—	—	—	—	—	—
Sassurea costus (Roots)	—	—	—	—	—	—	14.07	2.81	243.00	14.69
Panax ginseng (Rhizome)	66.07	23.36	29.46	26.72	15.55	2.73	61.82	13.28	324.30	24.10
Plantago psyllium (Roots & Husk)	—	—	—	—	—	0.55	0.82	—	—	—
Swertia chirayita (Whole plant)	24.23	2.51	93.04	12.23	172.15	25.85	152.18	22.68	58.22	14.61
Cassia angustifolia (Leaf & Pod)	—	—	15.39	4.96	—	—	—	—	—	—
Catharanthus roseus (Roots)	2.79	0.23	—	—	—	—	—	—	—	—
Hemidesmus indicus (Roots)	—	—	1.83	0.55	2.99	0.56	—	—	3.50	1.08
Ziziphus mauritiana (Fruits)	5.47	0.27	2.0	0.13	66.08	6.29	41.57	3.58	25.18	1.71
Humulus lupulus (Dried leaves)	14.42	36.89	57.25	41.24	56.47	208.07	118.49	263.85	45.68	114.50
Rosa water	—	—	—	—	0.03	1.09	—	—	—	—
Resinoids	37.44	102.23	35.86	141.83	47.02	176.23	50.90	206.29	41.36	158.26
Other Medicinal Plants	1063.55	142.46	1301.85	179.02	1175.88	171.25	907.63	170.19	—	—

Table 4(b): Import of Major Crude Drugs from 1996-97 to 1999-2000

Crude Drug Items	1996-97		1997-98		1998-99		1999-2000	
	Quantity (in Ton.)	Value (in Lakh Rs.)	Quantity (in Ton.)	Value (in Lakh Rs.)	Quantity (in Ton.)	Value (in Lakh Rs.)	Quantity (in Ton.)	Value (in Lakh Rs.)
Glycyrrhiza glabra (Roots)	1622.96	167.12	594.68	70.73	1077.74	157.34	1106.63	178.45
Alpinia sp. (Rhizomes)	92.81	14.31	55.18	10.19	48.49	9.76	65.00	24.29
Curcuma zedoaria (Rhizomes)	—	—	—	—	—	—	9.00	1.64
Rauvolfia serpentina (Roots)	—	—	27.96	5.14	—	—	25.80	4.86
Sassurea costus (Roots)	245.00	18.94	194.15	26.60	147.10	31.78	266.31	43.16
Panax ginseng (Rhizome)	62.85	32.19	24.47	10.07	46.79	36.11	15.27	13.07
Plantago psyllium (Roots & Husk)	—	—	3.28	5.72	1.73	2.31	—	—
Swertia chirayita (Whole plant)	52.10	12.29	271.63	22.82	47.49	11.75	53.87	15.41
Cassia angustifolia (Leaf & Pod)	—	—	—	—	29.39	46.03	—	—
Catharanthus roseus (Roots)	4.50	0.24	—	—	0.38	8.40	—	—
Hemidesmus indicus (Roots)	1.03	0.28	8.53	1.80	2.90	0.63	—	—
Ziziphus mauritiana (Fruits)	8.2	0.68	187.44	34.36	97.60	24.97	—	—
Humulus lupulus (Dried leaves)	61.64	277.41	137.97	192.54	159.18	248.28	106.72	211.15
Rosa water	0.01	0.66	0.12	3.03	0.006	0.04	—	—
Resinoids	60.18	255.41	50.76	263.39	97.73	278.35	—	—
Other Medicinal Plants	—	—	588.53	203.15	956.85	617.97	1017.12	659.23

Source: Monthly statistics of the foreign trade of India Vol. II (Import) (1991-2000).

The world demand for medicinal plants and its products are ever increasing, further more, herbs are building blocks to a series of value added derivatives - such as alkaloids, food supplements (single herbal drugs), herbal oils, oleoresins, crude drugs, ground herbs, herbal powder, freeze dried herbs, etc. We have the expertise and world class facilities to manufacture all these products which now dominate the international market. So the future prospects for the growth of Indian herbal industry lies in focusing on value-added products, consistent drive on quality and ensuring competitiveness through enhanced productivity. In this regard, the National Medicinal Plants Board and State Medicinal Plants Board with other organizations engaged in research, development and export promotion are also working positively over the years and are trying to maintain India's image in the international market as the only country in the world capable of supplying almost all medicinal plants.

Chapter 9

STANDARDISATION AND QUALITY CONTROL OF MEDICINAL PLANTS

Plant materials are used throughout developed and developing countries as home remedies, over-the-counter drug products and raw materials for the pharmaceutical industry, and represent a substantial proportion of the global drug market. With ever-growing commercialization of herbal drugs, there is an urgent need to develop some kind of standards and quality control parameters for the raw materials as well as the medicines comprising single or multiple plant materials. The botanical raw drugs before reaching pharmaceutical line pass through various stages, all of which can influence the nature and quality of secondary metabolites present, which in turn have a direct bearing on the efficacy of drugs.

A question arises as to what kind of quality control and standardization measures serve any useful purpose. There are mainly three aspects for ensuring, quality control and standardization of plant drugs:

- ☞ Botanical identity and pharmacognostical examination of raw materials.
- ☞ Protection of therapeutic potentials of the raw materials.
- ☞ Biological assay

9.1. Botanical Identity and Pharmacognostical Examination of Raw Materials

Before collection of a medicinal plant, it should be correctly identified and one should know which part is required to be collected. For whole herb it is easy to identify by following taxonomical characters with the help of floras and matching them with authentic specimens preserved in various herbaria. But the isolated plant parts, such as roots, rhizome, tuber, bulbs, leaves and seed procured from the market are to be identified through pharmacognostic standards. It is, therefore, necessary to maintain such standards for each herbal drug used by the industry.

9.2 Protection of Therapeutic Potentials of the Raw Materials

A medicinal plant from field to the drug preparation stage has to face various riddles particularly under storage. The collected plant materials are stored under unhygienic condition which make these prone to microbial infestation which in turn deteriorate active principles of the crude sample. Some of the problems and measures to control the quality of drugs have been studied and discussed in the present text.

9.2.1. Collection and Drying

Freshly collected chopped raw material has high moisture content (55-90%); therefore, it has to be subjected to drying. Drying prevents moulding, the enzymatic action, microorganism infestation and chemical changes. It fixes the constituents and facilitates grinding and milling as well as converting the raw drug into a more convenient form for industrial handling.

9.2.2. Storage and Preservation

Proper storage and preservation are important factors in maintaining a high degree of quality of raw materials. If these are not properly stored, they may absorb moisture which not only causes increase in weight of raw material, but also reduces the percentage of active principles. It also favours enzymatic action and hastens microbial destruction of the material.

(a) Microbial Association with Crude Drug Samples Under Storage

The plant samples collected from forests are stored in traditional warehouse where they are usually packed in gunny bags or spread as such on ground and have to face fluctuation, environment and diverse range of microbabes.

(b) Biodeterioration of Active Ingredients

Fungi are capable of infecting a wide varieties of substrates due to their uniquitous nature. During storage, the fungal organisms thrive on drug plants by utilizing various components including the active ingredients.

9.2.3. Biological Assay

The advantages of these methods are to be utilized for standardization of the final products, i.e. the medicines in respect of the following aspects:

- To establish bio-availability parameters, chemical and biological
- Toxicity studies, acute and chronic
- Dose determination
- Clinical trials

The future of the medicinal plants industry in Indian holds great promise since modern infrastructure and expertise are available. It is expected that in the coming years various methodologies would be adopted to bring revolutionary progress which include genetic engineering, better crop management and improved processing technologies so that India could compete in the international market in a big way.

The credibility of the traditional systems of medicine, which have stood the test of time, will certainly go up and enhance their efficacy, reputation as well as acceptability, if quality control parameters are developed and strictly followed.

Chapter 10

LEGISLATION AND POLICY OF MEDICINAL PLANTS

A National Policy has to admit the needs of various user groups of medicinal plants and therefore one of the basic factors that should influence the design of the policy framework is the users' profile. In the case of medicinal plans it is important to note that there exist non-commercial users who constitute a much larger population than commercial users.

The policy framework has to include temporal consideration *viz.* long-term availability of medicinal plants, as well as the more immediate needs of the user groups. It needs to consider financial resources and were needed incentives for encouraging conservation actions.

The policy since it is formulated for the purpose of guiding implementation, has to review the need for strengthening and where necessary the building of new institutions in both the government and non-government sectors. It also needs to examine the need for national level co-ordination with appropriate representation of user groups.

Finally, there is legal matter specific to the subject of medicinal plants for *e.g.* regulation of collection of plants from the wild and the new issue of property rights over traditional knowledge and Indian biological resources. A national policy framework needs to take these legal matters into consideration.

10.1. National Medicinal Plants Board

The medicinal plants sector at present is not well organized and needs special attention. Although different Ministries and Departments in the Government sector and NGOs and individuals in the private sector are making efforts in different directions, yet there is a need to co-ordination and implementation of policies relating to medicinal plants both at the National and State levels is necessary to facilitate inter-ministry, inter-state and institutional collaboration and to avoid duplication of efforts. Therefore, a need for the establishment of a national level nodal body was felt to formulate policies for the medicinal plants sector and develop the potential of this sector through schemes and projects that encourages investment in this sector.

As such, the Medicinal Plants Board was set up under a Government Resolution notified on 24th November, 2000 under the Chairmanship of Union Minister for Health & Family Welfare. The objective of establishing a Board was to establish an agency which would be responsible for co-ordination of all matters related to medicinal plants, including drawing up policies and strategies for conservation, proper harvesting, cost effective cultivation, research and development , processing, marketing of raw material in order to protect, sustain respective, department, organizations but the Board would coordinate and provide a direction and an important to the activities. The Board will undertake the following activities:

- ☞ Promote encouragement for cultivation of selected medicinal plants backed by buy-back arrangements.

- ☞ Encourage Sates and UTs to registering raw drug traders and cultivators so that source of supply of medicinal plants is monitored as a measure to promote quality control, safety and efficacy of drugs.

- ☞ Facilities measures, which enhance efficiency, cost effectiveness and storage of medicinal plants.

- ☞ The NMPB has identified thirty-one (31) species *viz.* Amla (*Embilca offinalis*), Ashok (*Saraca asoca*), Ashwagandha (*Withania somnifera*), Atis (*Aconitum heterophyllum*), Bael (*Aegle marmelos*), Bhui amlaki (*Phyllanthus amarus*), Brahmi (*Bacopa monnieri*), Chandan (*Santalum album*), Chirata (*Swertia chirata*), Giloe (*Tinospora cordifolia*), Gudmar (*Gymnema sylvestre*), Guggal (*Commuphora wightii*), Isabgol (*Plantago ovata*), Jatamansi (*Nardostachys jatamansi*), kalihari (*Gloriosa superba*), Kelmegh (*Andrographis paniculata*), Kokum (*Garcicina indica*), Kuth (*Saussurea costus*), Kutki (*Picrorrhiza kurroa*), Makoy (*Solanum nigrum*), Mulethi (*Glycyrrhiza glabra*), Safed Musli (*Chlorophytum borivillianum*), Coleus (*Coleus barbatus*), Pipli (*Piper longum*), Rasaut (*Berbaris aristata*), Sarpgandha (*Rauvolfia serpentina*), Senna (*Cassia angustifolia*), Shatawar (*Asparagus racemosus*), Tulsi (*Ocimum sanctum*), Vai Vidang (*Embelia ribes*), Vatsnabh (*Aconitum ferox*), which are in high demand both in domestic and international markets are to be brought into cultivation status as these constitute a bulk of the ingredients used in the preparation of ISM&H and herbal products. Such list will naturally undergo changes from time to time.

- ☞ Undertaken general and specialized surveys of the national and international market for medicinal plants and products for identifying niche areas.

- ☞ Motivate and encourage State/Uts to set up State Medicinal Plants Board/Vanaspati Van Societies who can give a focus and direction to medicinal plants related activities.

- ☞ Support manufactures/NGOs and representative individuals for participation in international fairs, seminars and meeting with a view to create awareness and explore the international market for plants based herbal products.

- ☞ Support R & D studies in the areas of post harvest management including increasing shelf-life, introducing better storage techniques and agro-techniques, enhance bio-availability to be taken up through CSIR, NBRI, CIMAP, ICFRE, RRLs, DBT, Horticulture and Forest Departments.

☞ Launch efforts to create mass awareness about the importance of medicinal plants in all strata of society, rural and urban.

Objectives of National Medicinal Plants Board

The National Medicinal plants Board (NMPB) has tried to fulfilled their objectives with the help of State Medicinal Plants Board or State Node Agency, are given below:

☞ To identify and estimate the natural plant wealth in various state including trees, roots, stems, bark, leaves, flowers, seeds, fruits, resins and other useful parts of medicinal and cosmetic use.

☞ To harness the plant wealth to promote industries to produce phyto-chemicals, essential oils, medicines and cosmetics use.

☞ To identify the commercially viable species in each agro-climatic zone and to standardize the agro-practices for cultivation and preservation of these medicinal plants.

☞ To transfer the agro-technology to the farmers and to encourage them to take up the cultivation of medicinal plants particularly in forests, wastelands and orchards as an additional generating activity.

☞ To reduce the pressure on forest to conserve the biodiversity and protect the natural environment.

☞ To strengthen the capacity for undertaking research and development in evolving new herbal products and technique for their manufacture by utilizing the existing capacity within and outside in an optimal manner.

☞ To standardize intermediate and final products using the medicinal plants grown or to be grown in the various states and to introduce better technology in the exciting pharmaceutical industries.

☞ To provide professional and technical assistance to the public, private and joint sector enterprises to manufacture high quality products at comparative prices for the sale in the domestic and international market.

☞ To ensure adequate supply of quality medicines in Indian Systems of Medicines and Homeopathy for use in Government hospitals and dispensaries and to sell the surplus in open market.

10.2. Goals of National Policy

The goals of national medicinal plants conservation policy should include:

☞ To ensure long-term conservation *in-situ*, of viable breeding populations of medicinal plants diversity in a range of natural habitats corresponding to their natural distribution.

☞ To encourage and support a range of *ex-situ* conservation measures as a supplementary and complementary measure to the above.

☞ To encourage medicinal plant requirements of industry to be met sustainably through policy instruments that create a favourable economic environment for the large-scale cultivation of medicinal plants.

- To ensure access of medicinal plants to village communities for their primary health care needs and address issues relating to participation and empowerment of village communities in conserving and managing forest and non-forest resources.

- To regulate, improve and where necessary ban medicinal plants collection from the wild.

- To carry out appropriate legislation and/or modify existing ones to regulate internal and external trade in medicinal plants and to ensure adequate staff training of the agencies involved in regulation.

- To ensure community benefits ansing from commercial utilization of indigenous medical knowledge of plant resources as provided for in the international "Convention on Biodiversity"

- To develop a cadre of taxonomists, para-taxonomists and forest scientists to support and sustain rapid, prioritized baseline surveys for medicinal plants conservation status and to monitor changes on a long-term basis.

- To identify, review and clearly mandate appropriate institutions, governmental and nongovernmental to implement the national policy; and to strengthen/build adequate institutional structures and capacity in Government and Non-Government Organizations for research, training and conservation action which are essential to initiate and sustain effective policy implementation.

- To encourage and facilitate time-bound 'projects' through governmental and nongovernmental institutions, to further the goals of the policy.

- To create widespread public awareness on the importance and need for medicinal plants conservation.

- To encourage local processing and enterprise development for value addition at the local levels where plants are collected for the market.

- To continually monitor, evaluate and analyze the impact of this Medicinal Plants Conservation Policy and for this purpose to create and national-level co-ordination structure.

10.3. Future Action Plan

Forests have been exploited, butchered and assaulted for a variety of economic gains for many centuries. Their production capacity is at the lowest ever Many valuable plant species have disappeared and quite a large number are on the fringe margin. It is, therefore, a compelling obligation on the society that this valuable treasure responsible for our health and longevity is protected, replenished and properly managed. Cultivation on large scale is the most viable factor that can substantially supplement this erosion of genetic diversity.

- Creation of protectorates/biosphere reserves to conserve the genetic stock of endangered species.

- Large scale cultivation of rare and endangered species used in Indian Systems of medicines and Homoeopathic Pharmacopoeias to be initiated through the use of appropriate technology and its transfer to farmer's fields. Transplant and tissue culture techniques should be adopted for such species which are either shy seed producers or their propagation through seeds is not easy.

- Establishment of demonstration plots for herbal drugs at institutional level and a chain of herbal gardens in all states rich in medicinal plants followed by establishment of Herbal

Farms and a well stocked central herbal garden. Integration of medicinal and aromatic plants in social/agro-forestry and cropping patterns of our farmers will provide more economic gains per unit area per unit time.

☞ As a result of thorough survey, an inventory of medicinal plants including the plants used in local folklore, has to be prepared for judicial exploitation and sustainable management of our resources both for the present and for posterity.

☞ Establishment of nodal agencies both at the centre and the states for coordination of cultivation, collection, extraction and utilization of the herbal resources by those involved in the vocation.

☞ Generation of trained and skilled manpower to handle all aspects of medicinal plants.

☞ Public awareness programmes, specifically conservation consciousness need to be intensified among school and college students through the Introduction of topics on medicinal plants in their curricula. To start with, this can be initiated by including most common plants in the form of botanical names, local/common Vernacular Names, brief description and broad medicinal uses.

Chapter 11

SUSTAINABLE CONSERVATION AND DEVELOPMENT STRATEGIES FOR MEDICINAL PLANTS

Several national and international agencies have formulated appropriate policies and strategies for the conservation of medicinal plants. The world conservation strategy (IUCN, UNEP & WWF, 1980) defines conservations as "the management of human use of the biodiversity so that it may yield the greatest sustainable benefit to present generation while maintaining its definitions invokes two complementary components "conservation" and "sustainability". The primary goals of biodiversity conservation as envisaged in the World Conservation Strategy can be summarized as follows:

- ▷ Maintenance of essential ecological processes and life support systems on which human survival and economic activities depend,
- ▷ Preservation of species and genetic diversity and
- ▷ Sustainable use of species and ecosystems which support millions of rural communities as well as major industries.

Medicinal plants are potential renewable natural resources. Therefore, the conservation and sustainable utilization of medicinal plants must necessarily involve a long term, integrated, scientifically oriented action programme. This should involve the pertinent aspects of protection, preservation, maintenance, exploitation, conservation and sustainable utilization. A holistic and systematic approach envisaging interaction between social, economic and ecological systems will be a more desirable one. The most widely accepted scientific technologies of biodiversity conservation are the *in-situ* and *ex-situ* methods.

11.1. Conservation of Biological/Genetic Diversity

Conservation has been defined as the management of resources of the biosphere for the benefit of all life including human, so that sustainable benefit may be derived by the present with potentials to meet the needs and aspirations of the future generations. Biological diversity refers to the richness of species i.e. number of species in a community of living systems while genetic diversity is a heritable diversity within and between the species of a genus. Basic rationale behind the conservation of biological diversity is the ensure sustainability in the utilization of species and ecosystems by maintaining essential ecological processes in the life-support system. Indian region is the center of diversity of 152 economic species including medicinal species like cardamoms, Nutmeg, Dioscorea, Pepper, Ginger, Rhododendrons, Jasmines, Betels leaf etc.

The concept of biosphere reserve is a method wherein specific ecosystems are conserved and managed for posterity. The germplasm collection in the herbal gardens/drugs farms of valuable medicinal species can be a major tool in conserving genetic/ biological diversity. It will provide the basic material for future cultivation of desired plants.

11.1.1. *In-Situ* Conservation

In has been established that the best and cost effective way of protecting the existing biological and genetic diversity is the 'in-situ' or on the site conservation wherein a wild species or stock of a biological community is protected and preserved in its natural habitat. The prospect of such a 'ecocentric', rather than a species centered approach is that it should prevent species from becoming endangered by human activities and reduce the need for human intervening to prevent premature extinctions. To achieve this *in-situ* conservation the central government has initiated several measures recently involvement appropriate NGOs.

11.1.1.1. Medicinal Plants Conservation Areas (MPCAs)

Between 1993 and 1997 FRLHT (Foundation for Revitalization of Local Health Traditions) Bangalore (Karnataka), in collaboration with State Forest Departments has established a coordinated network of 30 *in-situ* medicinal plants conservation areas located with in the protected areas of Kerala. Tamil Nadu, Karnataka with involvement of local communities un conservation, inventorization, threat assessment and Nursery Raising programme.

The concept has received further boost through "Task force on Conservation and Sustainable use of Medicinal plants" set up by GOI (1999) in the Planning Commission which has recommended the establishment of 200 MPCAs, each of about 500 ha in medicinal plants rich Protected Areas (PAs) of the country to cover different forest types. The key activities of the MPCAs includes:

- ꝑ Systematic survey and documentation of intra-specific variation; generation of data on distribution patter, macro- and micro-habitats, association and cultural inputs of each species.

- ꝑ Development of suitable species recovery programme of endangered and enrichment planting programme for economically valuable species.

- ꝑ Building up and strengthening community institution for long term management of the sites.

↦ Training of wildlife staff and other for *in-situ* conservation of medicinal plants. The involvement of Botanical Survey of India (BSI), University of Botany Department alongwith NGOs would ensure success of such initiatives.

11.1.1.2. Sacred Groves

These are patches of forest dedicated to local deity whose entire biodiversity along with other natural resources are conserved by the worshippers, the forest village communities. Revitalisation of sacred grove concept alongwith phytoletry thus would not only help in *in-situ* conservation of forest biodiversity but also add to the greenery of towns and cities and in *ex-situ* conservation of threatened taxa. It would serve as a best example of people's participation in conservation of biodiversity alongwith medicinal plants.

11.1.1.3. Conservation of Ethnomedicinal Plants

Of the 8000 species known to have medicinal value, the vast majority, (about 5000 to 5500 species) form components of tribal or folklore medicine which is traditionally passed on orally through generations. There is an urgent need for systematic documentation of this knowledge to safeguard it from biopiracy in this age of patenting regime. This documentation study needs to be used in the process of planning and programming of conservation of rich medicinal plants diversity of the country.

11.1.2. *Ex-Situ* Conservation

Conservation of medicinal plants can be accomplished by the *ex-situ* i.e. outside natural habitat by cultivating and maintaining plants in biotic gardens, parks, other suitable sites, and through long term preservation of plant propagules in gene banks (seed bank, pollen bank, DNA libraries, etc.) and in plant tissue culture repositories and by cryopreservation).

11.1.2.1 Ethno-medicinal Plant Gardens

A valuable *ex-situ* conservation measure would be the creating of a network of regional and sub-regional ethno-medicinal plans gardens, which should contain accessions of all the medicinal plants known to the various ethnic communities in different regions of India. The chain of gardens will act as regional repositories of our cultural and ethno medicinal history and embody and living traditions of our society's knowledge of medicinal plants.

11.1.2.2. Gene Banks

While it is known that the largest proportion of local biodiversity in all our eco-system is used for medicinal purposes, very little is known about their conservation status in the wild. What is likely is that a large number of medicinal plant species are under various degrees of threat. The 'Precautionary principle' would suggest that an immediate and countrywide exercise be taken up to deposit seeds of wild medicinal plants with a first priority to known Rare, Endangered, Threatened (RET), species and endemic species. A simultaneous effort should also be launched to evaluate the genetic variation of economic plants with a view to promoting their domestication and breeding.

11.1.2.3. Nursery or Demonstration Plots

The most urgent task in order to ensure immediate availability of plants and planting materials to various user groups like farmers, the professional Indian System of Medicines (SM) community, plants breeders, industry and to conservation organizations, is to promote and nation-wide network of medicinal plant nurseries which will multiply all the regional-specific plants that are used in the current practice of traditional medicine. These nurseries should become the primary sources of supply of plants and seed material that can be subsequently multiplied by various users. The forest departments, agricultural extension agencies, village panchayats, Non-Governmental Organizations (NGOs) and private enterprises should be encouraged to establish these nurseries or grow medicinal plants in existing ones. In the initial years interest-free short-term loans or loans at different rates of interest can be offered by banking institutions to this nursery network.

In view of its importance and urgency, Government can promote national co-ordination of this program through a national nodal institution, which may be called NATIONAL MEDICINAL PLANTS DEVELOPMENT BOARD.

11.2. Cultivation or Domestication

The medicinal and aromatic plants should be domesticated and brought under cultivation to maintain constant supply of quality materials and thus reduce the pressure on the wild populations. The process of domestication of a species involves characterizations of its reproductive biology to decide on the method of propagation-seed or propagule and definition of the area in which to cultivate based on the Soil and Climate characteristic of the plant's natural habitat. The genetic resources of the species are screened for its adaptability and yield and quality of the material that will make the end product, to identify suitable genotypes for cultivation. Soil type to be used, Irrigation and fertilizer amounts and application schedule and sowing and harvesting times are standardized. Disease and pest problems of standing crop and harvesting material are solved to keep the product safe for consumption. Initiation of plants breed programme will ensure that in future the crop will be high yielding, resistant to pest and domestic quality.

Cultivation and processing of material harvested from medicinal and neutraceutical (Supplements) plants, already identified useful for restoration and protection of human health, can provide much needed avenues of self-employment to educated unemployed in villages and small towns. Prospection of plant species for new drugs and other products and genetic engineering for certain crop plants for the production of raw material for the secondary metabolites for pharmaceutical industry are expected to further open such employment opportunities. To make the above developments possible, it is of immediate paramount importance to inventories evaluate and conserve plant species in nature and to *ex-situ* conserve the genetic resources domesticate and cultivate species whose medicinal and industrial importance is already established.

Cultivation of medicinal plants, however, is inversely linked to prevalence of easy and cheap collection from the wild, leak of regulation in trade, cornering of the profits from wild collection by a vast network of traders and middlemen and absence of industry's interest in providing buy-back guarantees to growers. Cultivation of medicinal plants is also difficult due to lack of standardized agronomic practices for not species and unavailability of sources of quality planting materials.

Policy measures to promote cultivation of medicinal plants, therefore, need to facilitate industry's role by way of tax incentives, etc. for investments in cultivation and agricultural research and simultaneously regulate indiscriminate, destructive collection of medicinal plants from the wild, while improving their conservation status *in situ*. In the context of medicinal plants, there is a special case for encouraging organic systems and polyculture models instead of the conventional mono-culture models in agriculture and agroforestry.

There is also a special case for encouraging in an 'organized' way, small, marginal farmers and tribals to grow medicinal plants in their household gardens, bunds, wastelands because this can encourage economic participation of the rural poor in the growth of the herbal industry.

11.3. Involvement of Primary Stakeholders

The very tribal communities who are responsible for sustenance of ethnomedicines are engaged in collection and sale of medicinal plants alongwith other Non-Timber Forest Produce (NTFPs) to agents of big dealers at nominal prices for meeting out their subsistence requirement. According to one estimate collection of medicinal plants and plant products through forest contributes at least 35 lakh mandays/year to the national economy as a source of both full and part employment and poorest of the poor are the involved communities. The forest dwellers are therefore best bets to involve in conservation and cultivation of medicinal plants in various ecogeographical regions of the country. This is necessary as the ownership of MFP including medicinal plants has now been conferred on *gramsabhas* in the fifth scheduled areas of the country.

Attempts have been made to involve these primary stakeholders in conservation, cultivation and trade of medicinal plants in different regions of the country through various schemes sponsored by Ministry of Environment and Forest, Ministry of Health and Family Welfare, Ministry of Rural Development and the State Governments. Ministry of Health and Family Welfare has initiated a scheme for improving awareness and availability of medicinal plants and remedies for ISM for Reproductive Child Health (RCH) Programme. Identified NGOs are assisted for Nursery Raising and to educate rural population about the use of locally available medicinal plants for preventive health and for curative purposes.

11.4. Model for Cultivation of Medicinal Plants

There is possibility that the agrotechniques developed so for medicinal herbs may not suit the purpose as they are mainly geared to achieve healthy growth and higher dry biomass. Such organs need not be necessary rich in bio-active principles as most of them are secondary metabolites and produced under trees conditions. A strong case therefore exists to evolve natural organic farming systems for cultivation of medicinal plants. A multi-tiered model with vertical zones involving trees, shrubs, herbs and climbers would be more suitable and profitable rather than monocultures as generally practiced in agricultural crops.

A proposal to establish 200 "Vanaspati Vans" on large areas of 3000 to 5000 ha through scheme of ISM should aim at establishing multi-tiered natural organic farming of medicinal plants in degraded forest and fringe forest areas under different eco-geographical zones of the country on "Care and Share" principles of JFM. Important medicinal plants of each tier for that zone should be included in this type of farming. These should be " Vans" in real sense of the term closely simulating natural forests of that zone. Apart from intensive production of medicinal plants these "Vanaspati Vans" will produce quality herbal products an degenerated productive employment to about 50 lakhs people specially women who are skilled in herbal production, collection and utilization.

11.5. Sustainable Harvesting

This is a very important aspect in sustaining the resource base either from wild or cultivated plants. Therefore, optimal methods of harvesting practices are necessary to be adopted or worked out in the absence of any foolproof system. This will cater to meet the following aspects also.

11.6. Community Based Enterprises

Considering the low value of many medicinal plant species (in raw drug form), a strong case exists for promotion of community-level enterprise for value addition to medicinal plants through simple, on site techniques like drying, cleaning, crushing, powdering, grading, packaging, etc. This will also increase the stake of village communities in conservation. Isolated micro enterprises on their own may find it difficult to manufacture standardized products and market them. Therefore, a rare to promote a macro organization to network and provide financial, technical and marketing support to rural enterprises exists.

11.7. Research

In order to help conservation of medicinal plants, the thrust of research should focus on immediate field-related problems. Priority areas in medicinal plants research need to include *inventorisation* and distribution mapping; threat categorization based on IUCN guidelines,-, conservation biology, propagation of rare species for reintroduction into their natural habitats; agro-technology of economically and clinically important species along with effective and quick technology transfer.

Research on medicinal plants based on Indian System of Medicines (ISM) theories like *desh vichar* traditional guidelines on suitable collection times and habitats and advice based upon indigenous texts like *Vrksh-ayurveda* and *krishi shastra* need to be promoted so that parameters for quality of plants can be evolved based on Indian System of Medicines (ISM) principles.

11.8. Training

In-service training programs for staff of various government and non governmental agencies like forests, wildlife, botanic gardens, schools and colleges in taxonomy and conservation biology need to be developed and supported in a large way. Training of community-based para-taxonomists: is also relevant to encourage community involvement in conservation.

To encourage the use and sustainable economic benefit from medicinal plants so that local communities can develop a stake in their conservation, suitable training courses in sustainable harvest (specially for tribal collectors), value addition and marketing of medicinal plants, needs to be promoted.

One of the major user groups of medicinal plants is the community of practitioners of traditional medicine, both classical and folk. Unfortunately, this group is least involved with the conservation of medicinal plants. With adequate orientation and training, this large group could make a significant contribution to the conservation of medicinal plants and its sustainable use.

11.9. Documentation and Dissemination of Information (Computerized Database)

This is the greatest drawback as the entrepreneur are not able to procure adequate information on various aspects like propagation and harvesting methods, pricing, marketing, policies, research. etc. at one place, from database or documented literature. A large number of persons/organizations have been working in isolation in this direction in India and abroad without recognizing any nodal agency WHO could be contacted. The entrepreneurs are, therefore, moving from place to place from pillars to posts to gather, information for any particular medicinal plant regarding their appropriate utilization, procurement of raw material, their policies, methods of cultivation and value addition for adequate returns, etc. The following areas are required to be covered for dissemination knowledge.

- ☞ Geographical Distribution and resource base.
- ☞ Methods for sustaining the resource base including packages for cultivation and value addition for short term and long term enterprises.
- ☞ Market status of medicinal and aromatic plants for formalizing and organization markets.
- ☞ Policies: Domestic policy, conventions, rules and regulations for harvesting/extraction marketing, industries and trade (both internal and external).

> ⮞ Pricing patterns/regimes for equitable distribution of profits in marketing channels from harvest to enterprise main markets.

> ⮞ Social and economic dimensions.

> ⮞ Coordination of research initiation and results of trials.

> ⮞ Identification of national and international level auth.

11.10. Procedures/Steps to Boost Trade in Medicinal Plants

The general procedures/steps of NTFP (include medicinal and aromatic plants) trade in Nepal are indicated below. Other countries may also adopt same measures.

Certificate of Income Tax Registration

Traders of NTFP have to register their business with the office of income Tax and obtain the certificate of tax clearance.

Collection Permit

A collection permit either from the DFO (in government managed forests) or from a PUG (in the case of Community Forests) is required for harvesting NTFPS. An application should state the type of NTFPs, the area, the quantity and the purpose of collection.

Royalty Payment

According to the Article 8(3) of Forest Regulation 1995, royalty should be paid before getting the collection permit. But Article 11(3) of The Regulation mentions that the authorized officer shall collect the fee or royalty while issuing the release order. If the NTFPs are collected from a Forest Use Groups (FUG), the fee or the royalty should be paid to the FUG that provides the collection permit.

Checking and Weighing

After harvesting the products, the respective forest office (DFO/Range Post) or FUG shall check and weigh to tally the amount of NTFPS collected against the licensed amount.

Release Order or Transit Permit

For the sale, transport and distribution of herbs collected in the district, a release order must be obtained from the DFO. The release order is valid for fifteen days but can be extended for seven more days at a time.

Local Taxes

With the provision of the new Local Governance Act that came into effect in May 1999, the local development bodies have a right to impose tax on any product including processed and unprocessed NTFPs at any level of transaction and trade. Many District Development Communities and Municipalities have started taxation on trade of NTFP from the districts.

Checking and Endorsement

The traders are required to show both Collection Permits and Release Order at Forest Check-post/Beat located en route and they shall transport only after having been endorsed by the check-post/Beat.

Export Recommendation

DFO recommends the concerned Customs Office for granting the permission to export the products.

Product Certification arid Export Permission for Some NTFPs

Department of Plant Resources (DPR) provides product test and certification services for processed as well as unprocessed NTFPS. To export the processed NTFPs that are prohibited to export in crude, a product certification and export permission must be obtained from the DPR. The Department tests the products, seals the containers and grants the export permission.

Export Recommendation for Processed NTFPs to India

Only the processing and manufacturing industries can export processed NTFPs to India. For this, the company should apply with Federation of Nepalese Chambers of Commerce and Industry (FNCCI) for an export recommendation letter. This FNCCI forwards the letter to the Department of Industry. The technical committee of Department of Industry makes necessary inquiries and provides an Export Recommendation to the processing company for the products processed by the company in Nepal.

Certificate of Origin

Certificate of origin must be obtained from the branches of FNCC or from the Nepal Chamber of Commerce (NCC) to export NTFPs. To export processed NTFPs to India the Certificate of origin is provided to the processing companies that have the export recommendation letter for the products from FNCCI.

Certificate of General System of Preference

If exporters want to have the privilege of exemption of import duties abroad, they must obtain a Certificate of General System of Preference from Trade Promotion Centre (TPC) and Customs Office of Nepal.

Export Permission and Duty

Export permission is granted by customs office en route after taking the export duties equivalent to 0.5 per cent of the Free on Board (FOB) value of processed and unprocessed NTFP.

Import Permission and Duty

The customs office of importing country grants Import Permission. Import Duty on Cost, Insurance and Freight (CIF) value must be paid to the Customs Office. For processed or unprocessed NTFP, the import duty can be exempted from Indian Customs Office if the certificate of origin is shown. And for the developed country, Certificate of origin and certificate of general system of preference should be shown to exempt the import duties if the country has given such privilege to Nepalese products. When importing the forest products from foreign country, the importers have to submit an application alongwith the customs declaration form and the authentic evidence from the concerned country

Government policies in some cases are supportive for the trade of some NTFPs. Almost all NTFP based manufacturing enterprises can entertain income tax holiday for an initial five-year period, which can be extended up to 10 years. To assist the trade and processing of NTFPS, the export companies are given the tax exemption facility for the export value of the products.

However, to boost the NTFP trade, the ban and taxation provisions and trade and export procedures of NTFP have to be reconsidered. Market support systems and a mechanism for the fair distribution of profits among the NTFP stakeholders have to be promoted.

LIST OF ENDANGERED MEDICINAL PLANTS

Sl.No.	Botanical Name	Family	English Name	Part Used	Habit & Habitat	IUCN Status	Active Constituents	Threats*	Current Regional Distribution
1.	Aconitum ballourii Stapf.	Ranunculaceae	—	Roots	Herb/Sub-alpine to Alpine Himalayas	CR/N	Pseudoaconitine (0.4-0.5%)	1, 2, 3, 4 & 5	Uttaranchal to Nepal Himalayas (3,000-4,500 m altitude)
2.	Aconitum deinorrhizum Stapf.	Ranunculaceae	Safed Bish	Roots	Herb/Alpine grassy slopes	CR	Pseudoaconitine (0.51%)	2, 4, 1 & 5	J&K to H.P. Himalayas (3,000-3,500 m altitude)
3.	Aconitum falconeri Stapf.	Ranunculaceae	Meetha Tellia	Roots	Herb/Sub-Alpine to Alpine meadows	CR	Pseudoaconitine (0.47%)	1, 2, 3 & 5	Garhwal Himalayas of Uttranchal (3,300-4,000 m altitude)
4.	Aconitum ferox Wallich ex Ser.	Ranunculaceae	Indian Aconite	Roots	Herb/Moist forest Temperate to Sub-alpine Himalayas	CR/R	Total alkaloids (0.63-4.7%)	2, 4, 1 & 5	Western Indian Himalayas (2,100-3,800 m altitude)
5.	Aconitum heterophyllum Wall. ex Royle	Ranunculaceae	Atis roots	Roots	Herb/Sub-alpine to Alpine regions in grassy meadows	CR/N	Atisine (0.4%), Aconitic acid	2, 4, 3, 1 & 5	J & K to Garhwal regions of Uttaranchal Himalayas (3,000-4,200 m altitude)
6.	Aconitum violaceum Jacq. ex Stapf.	Ranunculaceae	Patis	Roots	Herb/Alpine Himalayas	CR/N	Pseudoaconitine (1%)	1, 2, 3, 4 & 5	Indian Northwestern Himalayas (3,000-5,100 m altitude)
7.	Acorus calamus Linn.	Araceae	Sweet Flag	Rhizome	Herb/Swampy & marshy open places	VU/R	Essential oil (1.8%)	2, 1 & 5	W. Bengal, Sikkim & Meghalaya (2,000-3,000 m altitude)

Contd...

Table Contd...

Sl.No.	Botanical Name	Family	English Name	Part Used	Habit & Habitat	IUCN Status	Active Constituents	Threats*	Current Regional Distribution
8.	Angelica glauca Edgew.	Apiaceae	Chora	Roots	Herb/Shady, moist slopes of temperate Himalayas	CR/R	Essential oil (0.4-1.3%)	1, 4 & 5	J & K to Kumaon region of Uttaranchal Himalayas (2,500-3,500 m altitude)
9.	Aquilaria malaccensis Lam. Syn. A. agallocha Roxb.	Thymelaeaceae	Agarwood	Wood	Tree/Along with foothills & river valleys	CR/N	Essential oil, Agrol (0.08%)	2, 1, 7, 6 & 5	North South India (100-1,000 m altitude)
10.	Arnebia benethamii (Wall ex G. Don) John. Syn. Macrotomia benthamii (Wall) DC.	Boraginaceae	Gojaban	Leaves, Flower	Perennial herb/Open slopes, stony or rocky places, forest of birch & rhododendron	CR/N	Arnebinus (0.37%), essential oil	4, 5 & 1	Kashmir to Uttaranchal (3,500-4,500 m altitude)
11.	Atropa acuminata Royle ex Lindl.	Solanaceae	Indian atropa	Leaves, roots	Perennial herb/In open slope	CR/N	Total alkaloids (0.81%)	4, 2 & 5	Kashmir to Himachal Pradesh (2,000-3,500 m altitude)
12.	Baliospermum montanum Muell.-Arg. Syn. Jatropha Montana Willd.	Euphorbiaceae	Danti	Seeds, Roots	Perennial herb/Forest openings, edges & wastelands	LRnt/R	β-Sitosterol, resin & starch	4, 2, 1 & 5	Central India (upto 300 m altitude)
13.	Berberis aristata DC. Syn. B. sikkimensis (Schneid.) Arhndt.	Berberidaceae	Indian berberry	Bark of root	Spiny shrub/Open temperate forest of Himalaya	EN/R	Berbarine $(C_{20}OH_{18}O_4N)$	1, 2, 3 & 5	Indian North Western Himalayas (1,800-3,500 m altitude)
14.	Berberis chitria Lindl.	Berberidaceae	—	—	Spiny shrub/Sub-tropical to temperate forest	EN/N	Berbarine (5%, $C_{20}OH_{18}O_4N)$	1, 4, & 3	Kashmir to Kumaon regions of Uttaranchal (1,500-3,000 m altitude)
15.	Berberis kashmirana Ahrendt.	Berberidaceae	—	Bark of root	Shrub/Sub-alpine forest of Himalaya	CR	Berbarine $(C_{20}OH_{18}O_4N)$	2, 4, 1 & 5	J & K Himalayas (3,500-4,000 m altitude)
16.	Berberis lycium Royle.	Berberidaceae	Daruhaldi	Bark of root	Spiny shrub/In exposed arid places, sub-tropical to temperate	EN	Berbarine $(C_{20}OH_{18}O_4N)$	2, 1 & 5	J & K to Uttaranchal Himalayas (1,200-2,100 m altitude)

Contd...

Table Contd...

Sl.No. Botanical Name	Family	English Name	Part Used	Habit & Habitat	IUCN Status	Active Constituents	Threats*	Current Regional Distribution
17. *Berberis peliolaris* Wall ex G. Don.	Berberidaceae	—	Bark of root	Large shrub/Exposed slopes in temperate forest of Himalaya	CR	Berberine $(C_{20}OH_{18}O_4N)$	1, 2, 3, 4 & 5	Yamuna Valley, Tehri Garhwal Himalayas (2,400-3,000 m altitude)
18. *Bergenia ciliata* (Haw.) Sternb.	Saxifragaceae	Pasanbhed	Rhizome	Herb/Shady & moist rocks	VU/N	Bergenin (0.75%), gallic acid	1, 3, 4 & 5	Kashmir to Kumaon regions of Uttaranchal (1,200-2,500 m altitude)
19. *Bunium persicum* Boiss Fedtcsh. Syn. *Carum bulbocastanum* acut. Non Koch.	Apiaceae	Black caraway	Fruit & Seeds	Perennial herb/On moist, open slopes.	EN/N	Essential oil (2.5%)	4, 8 & 9	Kashmir to Garhwal regions of Uttaranchal (2,000-3,500 m altitude)
20. *Butea monosperma* (Witt.) Maheshwari Syn. *B. lrondosa* Koe. ex Roxb.	Papilionaceae	Flame of the forest	Leaves, root & bark	Small tree/Dry deciduous forest	DD/R	Butin, Leucocinidin	4, 2, 1 & 5	Central India (upto 300 m altitude)
21. *Celastrus paniculata* Willd. Syn. *C. Montana* Wight & Am.	Celastraceae	Black oil plant	Seeds, roots & bark	Woody liana/Sub-montane forest up to terai region	LRnt/R	Pristimerin	4, 2, 1 & 9	Central India (300-500 m altitude)
22. *Cinnamomum tamala* (Ham.) Nees & Eberm.	Lauraceae	Indian cassia	Leaves	Small tree/Sub-tropical regions	LRnt/R	Essential oil (0.3-0.6%)	2, 4 & 5	Indian North Western Himalayas (upto 1,000 m altitude)
23. *Clerodendrum colebrokianum*	Verbenaceae	—	Leaves & root	Shrub/ Subtropical forest undergrowth along forest margins	VU/N		2, 3, 1 & 5	Northeastern India (800-1,500 m altitude)
24. *Clerodendrum serratum* (L.) Moon	Verbenaceae	Bhargi	Leaves & root	Woody shrub/Tropical moist deciduous forest	VU/R	Sapogenin	2, 4, 1 & 5	Central India (400-500 m altitude)
25. *Coptis teeta* Wall.	Ranunculaceae	Gold thread	Rhizome	Herb/Moist, tropical, shady forest slopes	CR	Renin (0.5-2.7%) Berbatin (7.1-8.6%)	4, 2 & 5	Mishmi hills (2,500-3,000 m altitude)

Contd...

Table Contd...

Sl.No.	Botanical Name	Family	English Name	Part Used	Habit & Habitat	IUCN Status	Active Constituents	Threats*	Current Regional Distribution
26.	Cordia rothii Roem & Schultz.	Ehretiaceae	—	Fruit & bark	Medium tree/Mixed dry deciduous forests & scrub forests	LRnt/R	Tanin	10, 2 & 5	Throughout Central India (upto 300 m altitude)
27.	Costus lacerus	Zingiberaceae	—	Bark	Shrub/Moist & marshy locations	DD/N	Not Known	—	Northwestern India (500-1,000 m altitude)
28.	Crateriostigma plantagineum Hochst.	Scrophulariaceae	—		Perennial herb/Exposed, gravelly & shallow soils	CR/R	—	1, 2, 5 & 6	Western parts of M.P. (100-200 m altitude)
29.	Curculigo orchioides Gaertn. Syn. C. malabarica Wt.	Amaryllidaceae	Black Musli	Rhizome	Herb/Dry deciduous forest undergrowth in moist shady areas	VU/R	—	2, 4 & 5	Central, peninsular & western India. (500-700 m altitude)
30.	Curcuma angustifolia Roxb.	Zingiberaceae	East Indian Arrow	Rhizome	Herb/Dry deciduous forest	LRnt/R	—	4, 2, 3 & 5	Central India (500-800 m altitude)
31.	Curcuma caesia Roxb.	Zingiberaceae	Black Zedoary	Rhizome	Herb/Moist deciduous forest	CR/R	—	1, 4, & 5	Central India (500-800 m altitude)
32.	Dactylorhiza hatagirea D.Don Syn. Orchis latifolia auct. Non L.	Orchidaceae	Palm herbal medicine	Rhizome	Herb/Subalpine & Alpine Meadows	CR/R	Starch, Mucilage, volatile oil	4, 1 & 5	Kashmir to Kumaon region of Uttaranchal (2,500-4,000 m altitude)
33.	Delphinium denudatum Wall. ex Hook. f. & Thoms.	Ranunculaceae	Nirbisi	Roots	Perennial herb/Open grasslands & margins of fields	CR/N	Ulphiuline	2	Kashmir to Kumaon region of Uttaranchal (2,500-3,000 m altitude)
34.	Dioscorea deltoidea Wall. ex Kunth.	Dioscoreaceae	Dioscorea	Rhizome	Herb/Open places	CR/R	Diosgenin (4.8-8.0%)	2, 4, 1 & 5	Indian Northwestern Himalayas. (1,500-2,500 m altitude)
35.	Drymia indica (Roxb.) Jessop. Syn. Urginea indica (Kunth.) Roxb.	Liliaceae	—	—	Herb/Rocky places along streams	VU/R	—	1, 3, 5 & 6	Central India (500-700 m altitude)
36.	Evolvulus alsinoides Linn. Syn. E. hirtus Lam.	Convolvulaceae	Sankhpuspi	Leaves & bark	Herb/Moist open fields	LRnt/R	—	2, 4, 1 & 5	Central India (upto 200 m altitude)

Contd...

Table Contd...

Sl.No.	Botanical Name	Family	English Name	Part Used	Habit & Habitat	IUCN Status	Active Constituents	Threats*	Current Regional Distribution
37.	Fritillaria roylei Hook.	Ariacaeae	—	Bulb	Perennial herb/ Open alpine meadows	CR/N	Pimin ($C_{26}H_{43}O_3N$)	4, 2, 1 & 5	Kashmir to Kumaon region of Uttaranchal (3,000-3,500 m altitude)
38.	Gastrochilus longillora	Zingiberaceae	—	Rhizome	Perennial stemless herb/On tree trunks in moist, broad-leaved subtropical forest	CR/N	—	—	Meghalaya, Mizoram, Arunachal Pradesh (1,000-1,500 m altitude)
39.	Gentiana kurroo Royle.	Gentianaceae	Indian Gantian	Rhizome	Herb/In Quercus forests, on bare hills & edges of rocks	CR/N	Gentianin	2, 4, 5 & 1	Kashmir to Kumaon region of Uttaranchal (2,000-4,000 m altitude)
40.	Gloriosa superba Linn.	Liliaceae	Glory lily	Roots	Herb/Along edges of moist tropical forest	EN/R	Gloriosine ($C_{22}H_{25}O_6N$), superbine, colchicine (1.3%)	4, 2, 1, 3 & 5	Tropical hills of Central India (upto 2,350 m altitude)
41.	Gymnema sylvestre (Retz) R. Br. Syn. Asclepias germinata Roxb.	Asclepiadaceae	Gurmar	Leaves	Climber/Deciduous forests	VU/R	Gymnemic acid	2, 1, 3 & 5	Central India (200-300 m altitude)
42.	Hedychium coronarium Koering	Zingiberaceae	Common Ginger lily	Rhizome	Perennial herb/ Moist tropical forest near stream & canals	EN/R	Essential oil	4, 1 & 5	Madhya Pradesh & Chhatisgarh (700-1,000 m altitude)
43.	Hedychium spicatum Ham. Ex Sm.	Zingiberaceae	Spiked Ginger	Rhizome	Perennial herb/ Moist & shady places	VU/N	Essential oil (4%)	4, 5 & 2	Himachal Pradesh to Kumaon region of Uttaranchal (1,500-2,500 m altitude)
44.	Heracleum candicans Wall. ex DC. Syn. Heracleum lanatum Mictux.	Apiaceae	Padara	Fruit	Perennial herb/ Along water courses, amongst stones & open slopes	EN/R	Xanthotoxin, coumarins	4, 5 & 1	Kashmir to Kumaon region of Uttaranchal (1,500-2,500 m altitude)
45.	Hydnocarpus kurzii	Flacourtiaceae	True Chaulmoogna	Seeds	Tree/Tropical evergreen forests	EN/N	—	1, 3 & 5	Northeastern India (200-800 m altitude)
46.	Ilex khasiana	Aquifoliaceae	—	Leaves & bark	Tree/Sun-tropical mixed evergreen forests	CR	lixine ($C_{53}H_{60}$, 2%)	2	Khasi hills (1,00-1,500 m altitude)

Contd...

Table Contd...

Sl.No. Botanical Name	Family	English Name	Part Used	Habit & Habitat	IUCN Status	Active Constituents	Threats*	Current Regional Distribution
47. Inula racemosa Hook. f.	Asteraceae	Poshkar	Roots	Perennial herb/ Forest clearings & shrubberies	CR/N	Inulin, essential oil	1, 3 ,5 & 4	Kashmir to Kumaon region of Uttaranchal (2,000-3,200 m altitude)
48. Jurinea dolomiaea Boiss. Syn. J. macrocephala (Royle) Clark.	Asteraceae	Doop	Roots	Perennial herb/ Grassland & open meadows in sub-alpine & alpine areas	LRnt	—	5, 11 & 2	Kashmir to Kumaon region of Uttaranchal (3,000-4,200 m altitude)
49. Lavatera cashmeriana Cambess.	Malvaceae	—	Leaves	Perennial herb/ open grasslands, stony substrates	EN	—	4 & 5	Kashmir Himalaya (2,000-2,500 m altitude)
50. Luvunga scandens	Rutaceae	Sugandh Kokila	Fruits	Shrub/Temperate, sacred forest	CR/N	Essential oil	2 & 3	Meghalaya (upto 1,500 m altitude)
51. Meconopsis aculeate	Papaveraceae	Blue poppy	Roots	Herb/Rocky in alpine areas in between boundaries	EN/R	—	4, 2, 1 & 5	Indian Northeastern Himalayas (3,000-3,500 m altitude)
52. Nardostachys jatamansi. DC. Syn. Nardostachys grandiflora DC	Valerianaceae	Spikenard	Rhizome	Perennial herb/ In rock crevices & moist shady places	CR/R	Essential oil	2, 4, 1 & 5	Uttaranchal (2,500-4,000 M altitude)
53. Nepenthes khasiana	Nepenthaceae	Pitcher Plant	Leaves	Herb/Along road cutting & streams	CR	—	4, 2 & 5	Meghalaya (upto 500 m altitude)
54. Operculina turpethum (L.) Silva. Syn. Ipomea turpethum R. Br.	Convolvulaceae	Nisoadh	Whole plant	Twining herb/Dry deciduous forest & fallow lands, wasteland	VU/R	Glucidine resin (10%)	1, 2, 4 & 5	Central India (upto 1,000 m altitude)
55. Paeonia emodi Wall. ex Royle	Paeoniaceae	Himalayan Poney	Roots	Perennial herb/ Fringes of temperate forests	VU/N	Essential oil	1, 2, 4 & 5	Uttaranchal Himalayas (1,800-2,500 m altitude)
56. Panax pseudo ginseng Linn.	Araliaceae	Asiatic Ginseng	Roots	Herb/Shady areas with an undergrowth of rhododendron forests along stream	CR/N	Panaqulon, Saponin (1%)	1, 2 & 5	Sikkim, Assam & Arunachal Pradesh (2,000-4,000 m altitude)
57. Picrorhiza kurroa Royle ex Benth.	Scrophulariaceae	Kutki	Rhizome	Herb/On slopes & meadows of alpine Himalayas	EN/N	Picrorhizine, Kutkin (3.4%)	4, 2, 11 & 5	Indian Himalayas (2,500-4,000 m altitude)

Contd...

Table Contd...

Sl.No.	Botanical Name	Family	English Name	Part Used	Habit & Habitat	IUCN Status	Active Constituents	Threats*	Current Regional Distribution
58.	*Podophyllum hexandrum* Royle. Syn. *P. emodi* wall ex Royle.	Podophyllaceae	Indian Podophyllum	Rhizome	Herb/Sub-alpine & alpine meadows	CR/N	Podophyllin	3, 4, 2, 1 & 5	Indian Himalayas (3,000-4,500 m altitude)
59.	*Polygonatum verticillatum* (L.) All.	Liliaceae	Meeta Dudhiya	Rhizome	Perennial herb/ Stony substrate, open places	EN/R	—	4, 5, 2 & 1	Kashmir to Kumaon region of Uttaranchal (2,000-3,500 m altitude)
60.	*Prezwalskia tangutica*	Solanaceae	Ma-niao-Pao	Roots	Perennial herb/ Alpine moraine areas	CR/N	Hyocine	11	Northern Sikkim (about 5,000 m altitude)
61.	*Rauvolfia serpentina* Benth.ex. Kurz. Syn. *Ophioxylon serpentina* L.	Apocynaceae	Rauvolfia	Roots	Herb/Moist to dry deciduous forest	EN/R	Reserpine	2, 4, 3, 1 & 5	Central India (100-300 m altitude)
62.	*Rheum austrate* D. Don Syn. *R. emodi* Wall. ex Meissn.	Polygonaceae	Indian Rhubard	Rhizome	Perennial herb/ Moist open grassy slopes & meadows	VU/R	Emodin, Chrysophanic acid	2, 4, 1 & 5	Kashmir to Kumaon region of Uttaranchal (3,000-4,000 m altitude)
63.	*Rheum nobile*	Polygonaceae	—	Rhizome	Perennial herb/ Open hilly slopes & screes	EN/N	Anthuqunon	2, 1 & 5	Sikkim, Assam & Arunachal Pradesh (3,500-4,500 m altitude)
64.	*Rhododendron anthopogon*	Ericaceae	Dhoop	Leaves	Shrub/Temperate to sub-alpine regions	VU/N	Anthroqunon	2 & 3	Sikkim, Assam, Darjeeling & Arunachal Pradesh (3,000-4,500 m altitude)
65.	*Rhus semialata*	Anacardiaceae	Dry galls	Fruit & Leaves	Tree/Subtropical semievergreen forest	VU/N	Tanin (50-80%)	2, 1 & 5	Northeastern India & foothills of Darjeeling (800-1,500 m altitude)
66.	*Saussurea costus* (Falc.) Lipsch. Syn. *S. lappa* (Decne.) Sch-Bip.	Asteraceae	Kuth	Roots	Perennial herb/ Open temperate to sub-alpine meadows	CR/N	Essential oil (1.5%), Saussurin	1, 2, 4 & 5	Kashmir to Garhwal region of Uttaranchal (2,000-3,000 m altitude)
67.	*Saussurea gossypiphora* D.Don	Asteraceae	Scared saussurea	Roots	Perennial herb/ Open places of alpine Himalayas	CR/R	—	4, 2 & 3	Kashmir to Kumaon region of Uttaranchal (3,500-5,700 m altitude)

Contd...

Table Contd...

Sl.No.	Botanical Name	Family	English Name	Part Used	Habit & Habitat	IUCN Status	Active Constituents	Threats*	Current Regional Distribution
68.	Saussurea obvallata (DC.) Edgew.	Asteraceae	Bramkamal	Roots	Perennial herb/Alpine zones-near glaciers	EN/R	—	4, 2, 1 & 5	Indian Northwestern Himalayas (3,000-5,100 m altitude)
69.	Saussurea simpsoniana (Field & Gard.)	Asteraceae	Doop	Roots	Woody herb/Glacier zones of alpine Himalayas	EN/R	—	4, 2, 1 & 5	J&K, H.P. Uttaranchal (4,300-5,000 m altitude)
70.	Swertia angustifolia Buch.-Ham. ex D.Don. Syn. Ophelia angustifolia (Buch.-Ham. ex D.Don.) G. Don	Gentianaceae	Hilly Chirayata	Whole plant	Herb/Marshy areas frequently occurring in moist forests	EN/R	—	2, 4 & 5	M.P. & Chhatisgarh (400-700 m altitude)
71.	Swertia chirayita Roxb. Ex Flem.	Gentianaceae	Chiretta	Whole plant	Herb/Moist shady places	CR/R	Ophelic acid, chiratin	4	H.P. & Uttaranchal (1,500-3,000 m altitude)
72.	Taxus wallichiana Zuec. Syn. T. baccata Hook. f.	Taxaceae	Common Yew	Leaves & bark	Tree/In temperate mixed forests; occurring along with rhododendron, betula etc.	CR/R	Taxine ($C_{35}H_{49}O_{35-10}N$, 1.3%)	1 & 5	Sikkim, Assam & Arunachal Pradesh (1,500-3,000 m altitude)
73.	Thalictrum foliolosum DC.	Ranunculaceae	Yellow Zari	Rhizome	Perennial herb/Open hill slopes and forests	VU/R	Bebrine (0.35%), Thalictrine (0.03%)	1, 2, 4 & 5	Uttaranchal (1,300-3,400 m altitude)
74.	Tylophora indica (Burm.f.) Merriel Syn. Tylophora asthmatica (L.f.) Wight & Am.	Asclepiadaceae	Emetic Swallow Wort	Leaves	Climber/Dry deciduous forests in the hills	VU/R	Tylophorine ($C_{24}H_{27}O_4N$)	1, 3 & 5	Central India (up to 1,000 m altitude)
75.	Valeriana jatamansi Syn. V. wallichii	Valerianaceae	Indian Valerian	Rhizome	Herb/Temperate Himalayas	CR/R	Volatile oil (0.5-2.12%)	2, 1, 4 & 5	Indian northeastern Himalaya (1,200-1,800 m altitude)

Source: BCPP Report, 1998

IUCN Status (Categories)
CR = Critically endangered ; EN = Endangered ; VU = Vulnerable ; LR = Lower risk ; nt = near threatened : LR-lc = Least concern ; LR-cd- Conservation dependent: DD = Data deficient; NE = Not evaluated.

Threats*
1. Harvest for Medicine, 2. Loss of Habitat, 3. Human interference, 4. Overexploitation , 5. Trade , 6. Harvest for wood, 7. Edaphic factor, 8. Harvest for food, 9. Trade for parts, 10. Trampling, 11. Not Known.

GLOSSARY OF TECHNICAL MEDICAL TERMS

Abortifacient: Drug or agent inducing expulsion of a non-viable fetus.

Abscess: Localized collection of plus formed as a reaction to pyegenic organisms.

Acne: A skin condition common in adolescence, in which blackheads (comedones) are associated with a popular and pustular eruption of the pilosebaceous follicles.

Acute: Short and severe, not long drawn out or chronic.

Affection: Any disease or pathlogical condition.

Ague: Malarial fever with successive stages of fiver and chills.

Alopecia: Baldness which can be congential, premature or senile.

Alterative: Able to restore health.

Amenorrhoea: Absence of the menses.

Amoebiasis: Infection of large intestine by the protozoon *Entamoeba histolytica*, which it causes ulceration by invasion of the mucosa.

Anaemia: A disorder due to a deficiency in the number of red blood cells or of their haemoglobin content, or of both.

Analgesic: A drug which relieves pain.

Anasarca: Serous infilteration of the cellular tissues and serous cavities; generalized edema.

Anodyne: A remedy which relieves pain.

Anthelmintic: Any remedy for the destruction or elimination of intestinal worms.

Antidote: A remedy which counteracts or neutralizes the action of a person.

Antifebrile: Any agent which reduces or allays fever.

Anti-inflammatory: Any agent which prevents inflammation.

Antiperiodic: Any agent which prevents the periodic returns of a diseases.

Antipyretic: Any agent which allays or reduce fever.

Antiscoebutic: Any agent which prevents or cure scurvy.

Antispasmodic: Any measure used to relieve spasm occurring in muscle.

Aperients: Drugs which stimulate evacuation of the bowel.

Aphrodisiac: An agent which stimulates sexual excitement.

Appetizer: A small amount of food or drink taken to stimulate the appetite.

Arthrorsis: An articulation or joint.

Asthma: Paraxysmal attack of difficulty in breeding (disease of lungs causing difficulty in breathing).

Astringent: An agent which contracts organic tissue.

Beri-beri: A deficiency diseases caused by lack of aneurine (Vitamin B1).

Bilharziasis: Infestation of the human body by schistosoma (Blood flukes) from drinking, or bathing in infected water.

Bitters: Substrates, the extracts of which are used as stomachics.

Boil: A furncle. A acute inflammatory condition, surrounding a hair follicle; caused by the *Staphylococcus aureus*.

Bowel: An intestine.

Cardialgia: Pain in the heart.

Carminative: Having the power to relieve flatulence and associated colic.

Cataract: An opacity of the crystalline lens or its capsule. Eye complaint of old age.

Catarrh: Inflammation of a mucous membrane with constant flow of mucus.

Cathartic: Purgative.

Cephalagia: Pain in the head.

Cholagogue: A drug which causes and increased flow of bile into the intestine.

Chronic: Lingering, lasting, opposed to acute.

Colic: Severe pain resulting from periodic spasm in an abdominal organ.

Conception: The act of becoming pregnant by the impregnation of the ovum by the spermatozoon.

Consumption: A wasting away of the tissues of the body, in tuberculosis of the lungs.

Contagion: Communication of diseases from body to body.

Contraceptive: An agent used to prevent conception.

Convulsions: Involuntary contractions of muscles.

Corn: A cone-shaped, overgrowth and hardening of epidermis.

Cramp: A painful involuntary contraction of a muscle.

Cystitis: Inflammation of the urinary bladder.

Demulcent: A slippery, mucilaginous fluid which allays irritation and soothes inflammation, especially of mucous membranes.

Dentifrice: Paste or powder, for use in cleaning the teeth.

Depressant: Able to reduce nervous or functional activity.

Diaphoretic: Increased secretion of urine.

Diuretic: An agent which increases the flow of urine.

Dyspepsia: Indigestion.

Dystrophy: Defective nutrition.

Dysuria: Difficult or painful micturtion.

Ecbolic: Any agent which stimulates contraction of the gravid uterus and hastens expulsion of its contents.

Emaciation: Excessive leanness, or wasting of body tissue.

Emetic: Medication which stimulates vomiting.

Emmenagogue: Drug used to bring on menstruation.

Emollient: An agent which softens and sooths skin or mucous membrane.

Epilepsy: The 'fit' is caused by an abnormal electrical discharge that disturbs cerebration and usually results in loss of consciousness.

Expectorant: A drug which promotes or increase expectoration.

Febrifuge: Any drug or agent for reducing fever.

Flatulence: Gastric and intestinal distension with gas.

Galactagogue: An agent inducing or increasing the flow of milk.

Gargle: The act of washing the throat.

Gastralgia: Pain in the stomach.

Germicide: An agent which kills germs.

Gonorrhea: An infectious disease of venereal origin in adults.

Haemoptysis: Blood in the urine.

Haemorrhoids: The condition resulting from infestation with worms.

Hemicrania: Unilateral headache as in migrain.

Hemicrania: Paralysis of one side of the body.

Hypnotic: A drug which produces asleep resembling natural sleep.

Hypotension: Abnormally low blood pressure.

Hypothermia: Below normal body temperature.

Hysteria: A mental disorder characterized by emotional outbursts, and often physical symptoms such as paralysis.

Inflammation: The reaction of living tissues to injury, infection, or irritation; characterized by pain, swelling, redness and heat.

Influenza: An acute viral infection of the nasopharynx and respiratory tract.

Insanity: Relatively permanent disorder of the mind.

Jaundice: A condition characterized by a raised bilirubin level in the blood.

Laxative: A mild aperients.

Leucorrhoea: A sticky, whitish vaginal discharge.

Leukaemia: Blood disease in which the white cells are abnormal in type or number.

Malignant: Virulent and dangerous.

Menorrhagia: An excessive regular menstrual flow.

Metorrhagia: Uterine bleeding between the menstrual periods.

Narcotic: A drug which produces abnormally deep sleep.

Nausea: A feeling of sickness without actual vomiting.

Neuralgia: Pain in the distribution of a nerve.

Oedema: Dropsy. Abnormal infiltration of tissues with fluid.

Ophthalmia: Inflammation of the testicle.

Otitis: Inflammation of ear.

Ozaena: Atrophic condition of the external auditory meatus.

Poultice: A local moist and often heated application for the skin used to improve the circulation, treat inflamed areas etc.

Refrigerant: Increasing and decreasing at periodic intervals.

Rubefacients: Substances which, when applied to the skin, cause redness.

Sedative: An agent which disease caused by lack of vitamin C.

Sialagogue: A agent which induces profound sleep.

Stomachics: Agents which increase the appetite.

Tonic: A medical preparation that improves the functioning of the body or increases feeling of well-being.

Tympanitis: Inflammation of the tympanum.

Ulcer: An open sore in a body surface.

Urticaria: An allergic skin eruption characterized by multiple, circumscribed, smooth, raised, pinkish, itchy weal's, developing very suddenly, usually lasting a few days, and leaving no visible trace.

Vermifuge: An agent that expels intestinal worms.

Wart: Any firm abnormal elevation of the skin caused by a virus.

BIBLIOGRAPHY

Abbas M.; Nigarn, K.B. and Abbas, M. (1994) Effect of plant density and cultivars of ashwagandha (*Withania somnifera* Dunal) on its productivity. *Res. Dev. Rep.,* Vol. 11(1-2), 26-28.

Action, N. and D.L. Klayman. (1985) Artemisinin, a new sesquiterpene endoperoxide from *Artemisia annua. Planta Med.,* Vol. 51, 441-443.

Ahmed, G.U. (1981) Studies on rusts from Nagaland and Arunachal Pradesh. *Indian Phytopath* Vol. 34, 240-41.

Ambad, S.N. and Kadarn, U.S. (1998) Effect of different planting dates on the yield of pyrethrum flower. *J. Maharashtra Agric. Universities,* Vol. 23(l), 66-67.

Anon (1945-1976) Wealth of India, Raw Material, Vol. I-XI, Publication and Information Directorate, C.S.I.R., New Delhi.

Anon (1948) Report on the committee on Indigenous system of medicine, Vol. 1, Ministry of Health , Govt. of India, New Delhi.

Anon (1986) The useful plants of India. Publication and Information Directorate, C.SI.R., New Delhi.

Anon (1991-2001) Monthly Statistics of the foreign Trade of India: Export/Import Vol. I & II Directorate General of Commercial & Statistics, Ministry of Commerce, GOI. Kolkata.

Anon (1998) Market trends in production, price, export, import etc., *Jour. Med. Aro. Pl. Sci.,* Vol. 20,106-112.

Anon (1999) Cites listed Medicinal Plants of India. WWF-India, New Delhi.

Anon (2000) Report of the Task force on conservation & sustainable use of medicinal plants. Planning Commission (Govt. of India), New Delhi.

Anon. (1997a) All India Coordinated Research Project on Medicinal and Aromatic Plants. Progress of work for the period 1st Jan.1986 to 31st Dec. 1996, submitted to QRT on AICRP on M & A.P. Vol. 1. Anand. pp.7 & 4-6.

Anon. (1997a), Progress work for the period Jan. 1986 to Dec. 1996, submitted to QRT on AICRP on medicinal and aromatic plants. Vol. 1 (Delhi): 9, 41-47.

Anon. (1997b) All India Co-ordinated Research Project on Medicinal and Aromatic Plants. Progress of work for the period 1. 1. 1986 to 31.12.1996, submitted to QRT on AICRP on M&AP. Vol. 1. Akola. p. 15.

Anon. (1997c) All India Co-ordinated Research Project on Medicinal and Aromatic Plants. Progress of work for the period 1. 1. 1986 to 31.12 1996, submitted to QRT on AICRP on M&AP. Vol. 1. Faizabad. p. 68.

Anon. (1997c) Progress of work for the period Jan. 1986 to Dec. 1996, submitted to QRT on AICRP on Medicinal and aromatic plants. Vol. 1 (Udaipur): 64.

Anon. (1997c) Progress work for the period Jan. 1986 to Dec. 1996, submitted to QRT on AICRP on medicinal and aromatic plants. Vol. 2 (Mandsaur): 43

Anon., (1997a), Progress of work for the period Jan. 1986 to Dec. 1996, submitted to QRT on AICRP on Medicinal and aromatic plants. Vol. 2 (Indore): 103-104.

Anon., (1997a), Progress of work for the period Jan. 1986 to Dec. 1996. Submitted to QRT on AICRP on medicinal and aromatic plants. Vol. 1 (Akola): 15-16.

Anon., (1997b), Progress of work for the period Jan. 1986 to Dec. 1996, submitted to QRT on AICRP on Medicinal and aromatic plants. Vol. 2 (Mandsaur): 58.

Anon., (1997b), Progress of work for the period Jan. 1986 to Dec. 1996. Submitted to QRT on AICRP on medicinal and aromatic plants. Vol. 2 (Indore): 59, 94-95.

Anon., (1997d) Progress work for the period Jan. 1986 to Dec. 1996, submitted to QRT on AICRP on medicinal and aromatic plants. Vol. I (Udaipur), 40-55.

Anonymous (1950) *Coleus forskohlii* In: Wealth in India- Raw materials, Vol II, CSIR, New Delhi pp. 308.

Anonymous (1982) Market for Selected Medicinal Plants and the Derivatives. International Trade Centre UNCTAD/GATT, Geneva., pp-286.

Anon (1991-2000) Monthly statistics of the foreign Trade of India, Vol I (Export) & Vol II (Import) March Issue, Directorate General of Commercial Intelligence and Statistics , Kolkata.

Anonymous (1998) Medicinal plants significant Trade Study (CITES)-India Country Report, TRAFFIC-India, New Delhi, India.

Arias-Castro, C.; Scragg, A.H. & Rodriguez, M. (1993) The effect of cultural condition on the accumulation of formation by suspension cultures of *Glycyrrhiza glabra. Plant Cell Tissue & Org. Culture* Vol 34, 63-70.

Arora, R.K. (1993) Himalayan plants resources: Diversity and Conservation. In; Himalayan Bio-diversity Conservation Strategies (Ed. U. Dhar). Gyanodaya Prakashan, Nainital, 39-44.

Atal, C. K. & Kapoor B. M. (1989) Cultivation and Utilization of Medicinal Plants. RRL, Jammu-Tawi, J & K.

Atkinson, E. T. (1980) Flora of the Himalayas (with special reference to Kumaon, Garhwal and parts of Tibet; Cosmo Publication, New Delhi.

Atul, C. K. & Kapur, B. M. (1977) Cultivation and Utilization of medicinal and Aromatic plants. Regional Research laboratory, Jammu-Tawi.

Awashti, R.K & Rana, R.S. (2001) Large Cardamom: Production and Forest Conservation in Sikkim. WWF-INDIA. New Delhi.

Badhwar, R. L.; Rao, P. S. & Sethi, H. (1964) Some useful plants. FRI Dehradun (UA)

Bahl, J.R. Tyagi (1988) Colchicine induced autotetraploidy in *Coleus forskohlii* (Wild) Briq. *Nucleus* Vol. 31, 176-180.

Bajaj, Y.P.S. (1990) *In vitro* Production of haploids and their use in cell genetics and plant breeding. In Biotechnology in Agriculture and forestry 12 (ed.), Medicinal and Aromatic Plants 111, Springer-Verlag, Berlin.

Bajaj,Y.P.S. & Simola, L.K. (1991) *Atropa belladona* L. *In vitro* culture, regeneration of plants, cryopreservation and the production of tropane alkaloids. In Biotechnology in Agriculture and forestry 15, Medicinal and Aromatic Plants 111, Springer- Verlag, Berlin.

Banerjee, R. C. (1994). Standardization and grading of Minor forest Products. Agricultural Marketing. Apr-Jun issue; pp 24-27.

Baranov, A. (1966) Recent advances in our knowledge of the morphology, cultivation and uses of ginseng (*Panax ginseng*). *Econ. Bot.*, Vol. 20: 403-406.

Baricevic, D; Urnek, A.; Kreft, S; Maticic, B. & Zupancic, A., (1999) Effect of water stress and nitrogen fertilization on the content of hyscyamine and scopolamine in the roots of deadly nightshade (*Atropa belladonna*). *Environ. Exptl. Bot.*, Vol. 42(10), 17-24.

Baruah, J. N.; Mathur, R. K.; Jain, S. M. & Kataky, J. C. S. (1991) "Sugandhit Tapobhumi". Attar Association of India, Kannauj (U.P.)

BCCP (1998) Biodiversity Conservation Prioritization Project (BCCP) India, Zoo Outreach Organisation/ BSG, India.

Bhadwar, R.L.; Kanira, G.V. and Ramaswamy, S. (1956) *R. serpentina* - Method of propagation and their effect on root production. *Indian J. Pharmaceuticals* Vol. 18(5), 170-175.

Bhakuni, D S. & Jain S. (1995) Chemistry of cultivated medicinal plants. In: Advance, in Horticulture, Vol. 11. Medicinal and Aromatic Plants (Eds. Chadha, K L. & Gupta. R.), *Malhotra Publishing House*, New Delhi. pp. 55- 56.

Bhattacharya, B. D. & Barua, H. K. (1952) *Science & Culture*, Vol. 18, 240-41.

Bhattacherjee, S. K. (2001) Hand book of Medicinal Plants (3rd Ed). *Pionter Publishers*, Jaipur, India.

Bhattarai, N.K. (1989) Traditional phytotherapy among the Sherpas of Helambu, Central Nepal. *Jour. Ethenopharmacology*, Vol. 27 (1&2), 45-54.

Bhattee, S. S. & Beniwal, B. S. (1988) Coptis Teeta-an important and valuable medicinal plant of Arunachal Pradesh and its cultivation. *Indian Forester*, May issues, 251-260.

Bir, S S. & M.I.S. Saggoo. (1985) Cytological studies on members of family Labiatac from Kodaikanal and adjoining areas (South India). *Proc. Indian Acad. Sci. (Plant Sci.)* Vol. 94 (46), 619-626

Bir, S. S. & M.I.S. Saggoo. (1982) Cytology of some members of Labiatae from Central India, *Proc. Nat.l Aca. Sci., India. Section B (Biol. Sci.)*, Vol. 52(l), 107-112.

Bisht, P.S. & Kediyal, VK. (1995) Effect of chemical treatments on germination and survival of the seedlings in *Atropa belladonna* Linn. *Indian J. Forestry*, Vol. 18(3), 208-210.

Biwas, K (1956) Common Medicinal Plants of Darjeeling and Sikkim Himalayas, Govt. Printing Press, West Bengal.

Bohra, N.K. & Sankhla, P.S. (1997) Senna - a cash crop for arid regions. *Vaniki Sandesh*, Vol. 21(3),19-23.

Booij, R. and Meurs, E.J.J. (1994) Flowering in celery (*Apium graveolans* L. var. rapaceum (Mill.) DC): effects of photoperiod. *Scientia Horticulturae*, Vol. 58(4), 271 - 282.

Bose, C.K. & Kapur, B.M. (1982) The nature of Agar formation. Sci & Cult. Vol. 4(2), 89-91.

Brandis, D (1874) Forest flora of North West and Central India. Allen & Co. London.

Bryson, C.T. & C.T. Groom. (1991) Herbicide inputs for new agronomic crops, annual wormwood (Artemisia annua). Weed Technol. 5: 117-124.

Carlson, A.W. (1986) Ginseng: America's botanical drug connection to the orient. Econ. Bot., 40: 233-249.

Ceylan, A; Vomel, A.; Yurtseven, M. & Kaya, N. (1980) The effect of nitrogen fertilization and time of cutting on the yield and alkaloid content of *Atropa belladonna* in Menemen. *Lzmer. Ege Universitisi Sireat Fakultesi Dergisi*, Vol. 17(l), 43-57.

Chadha, K.L. & Gupta, R. (1995) Advances in Horticulture, Medicinal and Aromatic plants, Vol.11, Malhotra publishing house, New Delhi.

Chander, H. and Ahmed, S.M., (1982) Extractives of medicinal plants as pulse protectants against *Callosobruchus chinesis*, L. infestation. *J Food Sci. Tech. - India*. Vol. 19(2), 50-52.

Charles, D.J. Simon, J.E., K.Y. & P. Heinstein. (1990) Germplasm variation in arteminsinin content of Artemisia annua using an alternative method of artemisinin analysis from crude plant extracts. *J. nat. Prod.*,157-160.

Chatterjee, S & Dey, S (1997) A perliminery survey of the status of Taxus baccata in Training District of Arunachal Pradesh. The Indian Forester. Vol. 123 (8), 746-754.

Chauhan, N. S. (1999) Medicinal and Aromatic plants of Himachal Pradesh. *Indus Publishing Company*, New Delhi.

China Coop. (1982) Research Group on quinghaosu and its derivatives as antimalarials. *J. Trad. Chin. Med.*, Vol.2, pp 3.

Chopra, R. N., Nayar, S. L. & Chopra, I. C. (1974). Supplement to the glossary of Indian Medicinal Plants, C.S.I.R. Publication, New Delhi.

Chopra, R.N.. Nayyar, S.L. and Chopra, I.C., (1956) Glossary of Indian medicinal plants. *CSIR. New Delhi*.

Chowdhary, H. J. & Wadhava, B. M. (1984). Flora of Himachal Pradesh Analysis. Vol. 1-3, Botanical Survey of India, Kolkatta (W.B).

Cramer. L.H. (1978) A revision of Coleus (Labiatae) in Sri Lanka (Ceylon) - new texa. *Kew Bull* Vol. 32, 551-561.

Curtis, J. T. & Cotton, G (1956) Plant ecology work book. Laboratory field reference manual. Burger's Publishing Co., Minnesota, USA.

Dahama, A. K. (2002) Organic Farming for Sustainable Agriculture. *Agrobios (India)*, Jodhpur.

Dahatoude, B.N.; Joshi, B.G. & Vitkare, D.G. (1983) Studies on response of nitrogen fertilization on the root yield of aswagandha (*Withania somnifera*). *PK V Res. J.*, Vol. 7(l), 7-8.

Dastur, J. F. (1951). Useful Plants of India and Pakistan, T. Sons & co. Ltd., Mumbai.

de Souza. N.J. and V. Shah (1988) Forskolin - An adenylate cyclase activating drug from Indian herb. In: Economic and Medicinal Plant Research. Vol. 2. Academic Press Ltd., New York

de Souza. N.J.; A.N. Dohadwaha and R.M. Rupp (1986) Forskolin - Its chemical, biological and medicinal potential. *Hoechst India Ltd.*, Mumbai.

Deb, B. D. (1981) The Flora of Tripura State, Vol. 1, 238-239.

Dey, A.C. (1964). Indian Medicinal Plants Used in Ayurvedic Preparations, Bishen Singh Mohendra Pal Singh, Dehradun.

Dhiman, A.K. (2003). Sacred Plants & their Medicinal Uses. *Daya Publishing House*, New Delhi.

Dhua, R.S. (1993) Basella In: vegetable Crops. (Ed. Bose. T.K., Som. M.G. and Kabir. J.) *Naya Prakash*. Kolkata, pp. 797-801.

Drury, C. H. (1978 Reprint) The Useful Plants of India, International Book Distributors, Dehradun.

Dyduch, J. (1995) Temperature as a vernalizing factor in celery production for seeds. *Zeszvy Problemowe Postepow Nauk Rolnicych*, Vol. 419, 29-34.

Dyduch, J. (1996) Studies on optimum density of celery (*Apium graueolens* L. var. dulce Pers.) grown for seeds. *Annales Universitatis Mariae Curie Sklodowsku Sectio EEE Horticultura*, Vol. 4, 29-36.

Edwards, D.M. & Bowen, M. R. (1993) Focus on Juri buti. FRSC, Kthmandu, Nepal.

Elhag, H.M. ; El- Domiaty, M.M.; El Feraly, F.S.; Mossa, J.S. & El-Olemy,M.M. (1992). Selection and micropropagation of high artimisinin producing clones of *Artimisia annua* L. *Phytotherapy Res*. Vol. 6, 20-24.

Elhag, H.M., E1-Domaity, F.S., E1-Feraly, F.S. Mossa, J.S. & M.M. E1-Olemy. 1992. Selection and micropropagation of high artemisinin producing clones of *Artemisia annua* L. *Phytotherapy Res.*, 6:20-24.

Faroogi, A. A.; Khan, M. M. & Vasundhara, M (1999) Production Technology of Medicinal and Aromatic Crops. Natural Remedies Pvt. Ltd. Banglore.

Farooqi, A.A & & Sreeramu, B.S. (2001) Cultivation of Medicinal & Aromatic Crops. *University Press*, Banglore.

Farooqi, A.H.A.; Kumar, R; Sharma, S; Sushil Kumar & Kumar. S. (1999) Effect of plant growth regulators on flowering behaviour of pyrethrum (*Chrysanthemum cinerariaefolium*) in north Indian plains. *J. Med. Aromatic Pl. Sci.*, Vol. 21(3), 681-685.

Formanowiczowa, H. & Kozlowski, J. (1980) Biology of germination of medicinal plant seeds. Part XIIA. Seeds of species of solanaceae. *Herba Polonica*, Vol. 26(l), 21 -38.

Fransworth, N.R. & Morris, R.W. (1976) Higher plants: the sleeping giant of drug development. *Am. J. Pharm. Edu.* Vol. 148, pp. 46-52.

FRIS Project Paper 4. Forest Research Information System Project, Ministry of Forest and Soil Conservation, Kathmandu, Nepal. Pp. 27.

FRLHT (1996) "Conserving a National Resource: Need for a National Policy and National Programme on Medicinal plant conservation". Unpublished background paper for National Consultation on formulating a medicinal plant conservation policy in India. Med plant Conservatory Society, Bangalore.

Gangopadhyay. S. & Isswar, S.C. (1972) A new record of charcoal rot disease on *Datural stramonium*. *Indian Phytopath*. Vol. 25, 155-156.

Ganguly, D. and Pandotra, VR. (1962) Some of the commonly occurring diseases of important medicinal and aromatic plants in Jammu and Kashmir. *Indian phytopath* Vol. 15, 50-54.

Garmasy E, A; Gendy E .A.; Gamasy El, K.M. and Deib, S.A. (1980) Studies oil some cultural treatments on the growth and active ingredient contents of *Atrapa belladonna* plants. 2. Effect of spraying with urea. *Annals of Agricultural Sciences-Ain Shams University*, Vol. 25(1-2), 313-322.

Gaur, B L (1987) Production technology for opium poppy in Rajasthan. *Indian Hor*. Vol. 31(4), 21-24.

Gaur, B.L. & Rathroe, M.S. (1991) Varietal response of opium poppy (*Papavar somniferum*) to nitrogen fertilization on vertisols. *Indian J. Agron.*, Vol. 36(l), 100-101.

Gaur, B.L. & Sharma, D D. (1987) Effect of date of sowing and spacing on opium poppy. *Indian J. Agron.*, 32(2):134-135.

Gautam, P.L.; Singh, B.M. & Sharma, N. (1998) Bioprospecting and conservation of Non-Timber Forest Products *vis a vis* Medicinal and Aromatic Plants, Some approaches to forest Management. *Forest Usufructs*, Vol.1 (1&2) pp 143-150.

Gopalaswanuengar, K.S (1970) Complete gardening in India, G. Kasturi Rangan, Bangalore. p. 565.

Graham, S.A. (1966) The ginseng of Araliaceae in southern United States. Jour. Am. Arb., 47: 132-134.

Grieve, M. (1989) A Modern Herbal, Vol I& II. *Hafner Publishing Co.*, New York.

Gruenwald, J. (1998) The emerging role of herbal medicine in health care in Europe. In 32nd DIA annual meeting. The challenge of world wide pharmaceutical development on an era of regulatory change; accelerated approval with quality and contained Cost. San Diego. California, USA 9-13 June1996, *Drug information journal* 32, 151-153.

Grushnitzky, I.V. (1956) Instructions for accelerated germination of ginseng seeds. Jour. Bot. USSR., 41: 1621-1623 (in Russian).

Gupta, R. and Pareek, S.K. (1995) Senna, In: Advances in horticulture. Vol. 11 -Medicinal and aromatic plants (Chadha, K.L. and Gupta, R., Eds.). *Malhotra Publishing house*, New Delhi, India, 325-336.

Gupta, R. K. (1968) Cultivation of *Digitis lanata*. In Chamba Hills, Himachal Pradesh. Indian Forester, Vol. 93(1), 33-40.

Gupta, R., (1973) Belladonna-a new cash crop for temperate region. *Indian Farming*, Vol. 23(9), 9-11.

Gupta. R.; Modi, J.M. & Mehta, K.G. (1977) Studies on cultivation of senna (*Cassio angustifolia*) in north Gujarat. *South Indian Hort*. Vol. 25(l), 26-29.

Gusgiao, Li, Guo, X., Jim, R., Ilang, S., Jain, H. & Z.Li. (1982) Clinical studies on treatment of cerebal malaria with quinghaosu and its derivatives. J.Trad.Chin. Med., 2: 125-130.

Heaton, J B. and Dullahide, S.R. (1994) Control of celery leaf curl disease caused by *Colletotrichum acutatum.*, *Aust. Pl. Path.* Vol. 22(4), 152-155,

Hegde, D.M. (1988 b) Response of periwinkle (*Catharanthus roseus* (L.) G. Don) to nitrogen, phosphorus and potassium fertilization. *Agri. Res. J.* (Kerala), Vol. 26(2), 227-233.

Hegde, D.M., (1988 a) Growth and productivity of periwinkle (*Catharanthus roseus* (L.) G. Don) in relation to plant type and method of propagation. *Agri. Res. J.* (Kerala). Vol. 26(l), 50-58.

Hegde, L. & R. Krishnan (1991) Varietal response to chromosomal doubling in *Coleus forskohlii* Briq. *Pro. Agril. Res.* (Sri Lanka) Vol. 3, 77-89.

Hegde, L. & R. Krishnan. (1998) Variability in *Coleus forskohlii* Briq. - an endangered medicinal plant. *Plant Genetic Resources News Letter*, Vol.113, 50-52.

Hegde, L. & R. Krishnan (1994) Growth analysis in autotetraploid *Coleus forskohlii. Indian J. Agril Sci.* Vol. 64(6), 405-408.

Hegde, L. & R. Krishnan (1998) Genetic divergence and character interrelationship in forskolin yielding *Coleus forskohlii* Briq. In: Gautam, P L, R. Raina, Umesh Srivastav, S. P Raychaudhuri and B. B. Singh. (1996) Proceedings of UHF-IUFRO International Workshop on Prospects of Medicinal Plants, Nov. 15-18, University of Horticulture and Forestry, Nauni. H.P., pp. 243-248.

Hegde, L. & R. Krishnan. (1992) Morphological variations in *Coleus forskohlii* Briq. Collections. *Proc. Conf. Cytol. Genet.* Vol. 3, 11-16.

Hegde, L. & R. Krishnan. (1994) Intervarietal hybridisation for genetic upgrading of *Coleus forskohlii* Briq., *Proc. Conf. Cytol. Genet.* Vol. 4, 79-82.

Hegde, L. (1992) Studies on germplasm evaluation, induced autotetraploidy and hybridisation in *Coleus forskohlii* (Willd.) Briq. (Syn. *C. barbatus* Benth) Ph.D. Thesis, UAS. Bangalore.

Hemmerly, T.E. (1977) Ginseng far in Lawrence country, Tennessee. *Econ. Bot.*, Vol. 31, 160-162.

Hendry, A. & Daulay, H.S. (1992) Relative performance of isabgol varieties. *Madras agri. J.*, Vol. 79 (10), 585-587.

Hu, Shiu Ying. (1976) The genus Panax (ginseng) in Chinese medicine. *Econ. Bot.*, Vol. 30, 11-28.

Hussain, A (1992) Dictionary of Medicinal Plants. *CIMAP* Lucknow, India.

Hussain, A. and Janardhana, K.K. (1965) Stolen rot of Japanese mint. *Curt. Sci.* Vol. 34, 156-157.

Io, F.U., Sun, X. and C. congun. (1980) Analysis of chromosome morphology and C-banding in ginseng. *Scient. Agric.*, Vol. 5, 31-35.

IUCN, WHO, & WWF (1993) Guidelines on the Conservation of Medicinal Plants. International Union for the Conservation of Nature and Natural Resources, Gland, Switerland.

Jain, N.K., (1985) Status report of diseases in Madhya Pradesh. Proceedings of Fifth All India Workshop on Medicinal and Aromatic Plants, Bangalore, 22-25 December 1995.

Jain, P.M. (1990) Effect of split application of nitrogen on opium poppy. *Indian J. Agron.*, Vol. 35(3), 240-242

Jain, S. K. (1968) Medicinal Plants. *National Book Trust*, New Delhi.

Janardhanan, K.K (2002. Diseases of Major Medicinal Plants. *Daya Publishing House*, New Delhi.

Janardhanan, K.K. and Husain, A. (1972) A new Alternaria blight of *Datura innaria* Mill. *Indian Phytopath.* Vol. 25, 461-463,

Jaryal, G.S., (1998) Suitable medicinal and Aromatic plants for commercial cultivation. *Udhmita Vikas Kendra, Jahagirabad,* Bhopal (M.P.) pp 216.

Jo, J., Blazich, F.A. & T.R. Konseler (1988) Postharvest seed maturation of American ginseng: Stratification temperatures and delay of stratification. *Hort. Sci.,* Vol. 23, 995-997.

Kahar, L.S.; Tomar, S.S.; Pathan, M.A. & Nigam, K.B. (1991) Effects of sowing dates and variety on root yield of ashwagandha (*Withania somnifera*). *Indian J. Agric. Sci.,* Vol. 61(7), 495.

Kandaswamy, D., Oblisami. G.. Mohandass, S. & Santhanakrishnan, P., (1986) Influence of VA-Mycorhizal inoculation on the growth of pyrethrum in the nursery. *Pyrethrum Post,* Vol. 16(3), 81-83.

Kanjilal, U.N.; Kanjilal & R.K.;Das, A (1940) The Flora of Assam. Vol. 4, 112.

Karki, M. (1998) A review of the medicinal plants Sector in India. IDRC, Canada.

Kattimani, K.N.; Reddy, Y.N. & Rao, B.R. (1999) Effect of pre sowing seed treatment on germination, seedling emergence, seedling vigour and root yield of ashwagandha (*Withania somnifera* Dunal). *Seed Sci. Tech.,* Vol. 27(2), 483-488.

Kaul, M.K. (1996) Strategies for sustainable use of threatened medicinal plant resources in Western Himalaya 1n : Ethnobiology in Human Welfare, (Ed. S.K. Jain) *Deep Publisher,* New Delhi, 284-286.

Kaul, M.K. (1999) Medicinal plants of Kashmir and Ladakh. *Indus Publishing Company,* New Delhi.

Kaushik, P. & A. K. Dhawan (2000) Medicinal Plants & Raw Drugs of India. *Beshan Singh, Shiva off press,* Dehradun.

Kharwara, P.C.; Awasthi, O.P. & Singh, C.M. (1988) Effect of sowing dates, nitrogen and phosphorus levels on yield and quality of opium poppy. *Indian J. Agro.* Vol. 33(2), 159-163.

Khatva, D.C.; Chakrabarty, D.K. & Sen, C. (1981) Some new diseases of vegetable ornamental and plantation crops. *Indian Phytopath.* Vol. 34, 231-232.

Kirtikar, K. R. & Basu, B. D. (1975) Indian Medicinal Plants. Vol. I-IV. Allahabad, India.

Klayman, D.L. (1985) Quinghaosu (artemisinin): An antimalarial drug from Chinna. Science, 228: 1049-1055.

Konno, C., sugiyama, K., Kano, M., Takahashi, M. & H. Hikino. (1984) Isolation and hypoglycaemic activity of panaxons A, B, C, D and E glycans of panax ginseng roots. *Planta med.,* Vol. 50, 434-436.

Konsler, T.R. (1988) Ginseng - Production guide for North Carolina. *Farm Bull.* North Carolina State Univ. P. 15.

Krishna, S. & Badhwar, R. L. (1952) Aromatic Plants of India. Part XV, CSIR, New Delhi.

Krishnar, R. (1995) Periwinkle. In: Adances in Horticulture Vol. 11. Medicinal and aromatic plants. (Chadha, K L. & Gupta, R., Eds.). *Malhotra Publishing House,* New Delhi, India. pp 409-429.

Kuipers, S. E. (1997) Trade in Medicinal plants. In: Medicinal plants for forest Conservation and Health care. NWFP Series No. 11. FAO, Rome.

Kulmi, G.S. (1998) Effect of herbicides on weed flora, growth and yield of isabgol (*Plantago ovata*). *Crop Res.*, Vol. 6(l), 28-32.

Kumar, N.; Arumugam, R. & Kandaswarry, O.S. (1982) The effect of NPK on flower production of pyrethrum (*Chrysanthemum cinerariaefolium*). *South Indian Hort.*, Vol. 30(2), 99-103.

Kumari, V. K.; K. Hima Bindu; A.H. Reena; Laxminarayan H. & Vasantha K. T. (2000) Performance of hybrid lines of *Coleus forskohlii* for tuber and forskolin yield. *Proc. National Seminar on Hi-tech Horticulture*, June 26-28, pp.90.

Lacy, M.L.; Berger, R.D.; Gilbertson. R.L. & Little, E.L., (1996) Current challenges in controlling diseases of celery. *Pl. Disease*, Vol. 80(l), 1084-1091.

llangovan, R.; Subbiah, R. & Natarajan. S. (1989) Flowering and yield in senna (*Cassia anguistifolia* Vahl.) as influenced by spacing, nitrogen and phosphorus. *South Indian Horr.*, Vol. 37(2), 103-107.

Maheshwari *et al* (1988) Studied Economics of Intercropping of *Rauvolfia serpentina* with other crops such as soya bean, onion, garlic, etc.

Maheshwari, P. and Singh, U. (1981). Dictionary of Economic plants in India. ICAR, New Delhi.

Maheshwari, S.K.; Dhahtonde, B.N. & Gangrade, S.K. (1988) Differential response of intercropping of *Rauvolfia* with other crops. Progress in Ecology, Symp. Proc. (*Today and Tomorrow Printers and Publishers*, New Delhi) 10 : pp 121-126.

Maheshwari. S.K; Trivedi, K.C.; Asthana, O.P & Gangrade, S.K (1984) Rauvolfia - A remunerative crop for Western Madhya Pradesh. *Indian Hort.* Vol. 28(4), 43-44.

Maitra S.; Jana, B.K.; Delmath S.; Maitra, S. and Dednath. S. (1998) Response of plant nutrients on growth and alkaloid content of ashwagandha (*Withania somnifera* Dunal). *J. Interacdemicia.* Vol. 2(4), 243-246.

Malla, S.B., Shakya, P.R.; Rajbhandari, K.R.; Bhattarai, N.K.; Subedi, M.N (1995). Minor Forest Products of Nepal: General status and trade.

Mallik, S.C., (1993) Celery In: Vegetable crops, (Ed. Bose, T.K., Som, M.G. and Kabir, J.) *Naya Prokash*, Calcutta, India.

Manoharachary, C., (1975) New host records of some interesting fungi from India. Sci. Cult. 41 : 237-38.

Mathur, R.S.; Singh, D & Singh. B.K. (1964) *Cercospora jamaicensis* Chopp. on *Datura stramonium* L. A new record for India. *Curr Sci.* Vol. 33, 378-379.

May, R.M. (1988). How many Species are there on earth? *Science*, Vol. 20, September, 16.

Menghini, A. and Venanzi, G. (1978) The effect of growth regulators on the germination of seeds of various medicinal plants.*Annali della Facolta di Agraria, Perugia*, Vol. 32(2), 771—783.

Menon, A. K. (1960) Indian Essential oil (A review). C.SI.R., New Delhi.

Mishra, R (1968) Ecology work book. Oxford and IBH Publishers, New Delhi.

Mohandass, S. & Sampath, V. (1985) Influence of temperature on flowering in pyrethrum. *South Indian Hort.*, Vol. 33(6), 404-407.

Mohandass, S. & Sampath, V., (1983) Influence of gibberalic acid on pyrethrum seed germination. *Pyrethrum Post*, Vol. 15(3), 85-86.

Mohandass, S. (1986) Effect of some growth regulants on growth and flowering behavior in pyrethrum (*Chrysantheinum cinerariaefolium*). *Madras Agric. J.*, Vol. 73(5), 281-283.

Mohandass, S.; Sampath, V & Gupta, R. (1986) Response of ambient temperature on production of flowers in pyrethrum at Kodaikanal (India). *Acta Horticulturae*, Vol. 188, 163-168.

Mulge, R.; Madalageri, M.B. & Mukesh, C. (1998) Seed germination and seedling growth as influenced by seed treatments in basella (*Basella rubra*). *South Indian Hart.*, Vol. 46(3-4), 148-151.

Muller-Dombois, D & Ellenberg, H. (1974) Aims and methods of vegetation ecology. John Wiley, New York.

Munjal, R.L.; Lall. G. & Chona, B.L. (1959) Some Cercospora species from India-11. *Indian Phytopath.* Vol. 12, 85-89.

Muthumanickam, D. & Balakrishnamurthy, G. (1999) Optimum stage of harvest of extraction of total withanolides in ashwagandha (*Withania somnifera* Dunal). *J. Spices Aromatic Crops*. Vol. 8(l), 95-96.

Nanaiah, K.M. (1993) Studies on hybridization, chromosomal doubling, grafting and leaf anatomy in *Coleus forskohlii* Briq., Ph.D. Thesis, UAS, Bangalore.

Natural Medicine Marketing (1996) Market Reports-Traditional Chinese Medicines. The Chinese Market and International oppurtunities, London, UK.

Nayar, K.K; Ananthkrishnan, T.N. and David, BY. (1981) General and Applied Entomology. *Tata Mcgraw Hill Publishing Company Limited*, New Delhi. pp 276.

Nigam, K.B. (1984) Ashwagandha cultivation. *Indian Hart*, Vol. 28(4), 39-41.

Nigam, K.B.; Patidar, H.; Kandalkar, V S. & Pathan. M.A. (1991) Perforamance of WS 20: a new variety of ashwagandha (*Winthania somnifera* Dunal). *Indian J. Agric. Sci.*, Vol. 61, 81581-81582.

Nigam, K.B.; Rawat, G.S. and Prasad, B. (1984) Effect of methods of sowing, plant density and fertility level on ashwagandha (*Withania somnifera* Dunal). *South Indian Hort.*, Vol. 32(6), 356-359.

NIIR (2001) Herb Cultivation & medicinal Uses. *National Institute of Industrial Research*, New Delhi

NIIR (2002) Herb Cultivation & Processing. *National Institute of Industrial Research*, New Delhi.

Pal, M.; Badola. K.C. and Bhandari, H.C.S. (1995) Vegetative propagation of *Rauvolfa serpentina* by root branch cuttings. *Indian J. forestry*. Vol. 18(l), 18-20.

Panday, G.B. (1987) Performance of seed treatment on seedling blight of ashwagandha. Proceedings of All India Workshop on Medicinal and Aromatic Plants, Udaipur. 2-5 November 1987. pp.334.

Pandita, P N., (1983) Effects of different temperature regimes on pyrethrum seed germination. *Pyrethrum Post*, Vol. 15(3), 76-77.

Pandita, P.N. & Blurt, B. K. (1984) Evaluation of some pyrethrum strains for flower yield and pyrethrin content. *Herba Hungarica*, Vol. 23(1-2), 89-94.

Pandita, P.N. & Sharma. S D. (1990) Pyrethrin content and dry-flower yield of some strains of Dalmatian pyrethrum (*Tanaceturn cinerariaefolium*). *Indian J. Agric. Sci.*, Vol. 60(10), 693.

Pandita, P.N. & Bhat, B.K.(1978) Effect of pollination on flower yield and pyrethrins content in pyrethrum. *Indian J. Gen. Pl. Breeding*, Vol. 38(l), 138-141.

Pandita, P.N. (1986) Effect of storage on seed germination in pyrethrum (*Chrysanthernum cinerariaefolium*). *Pyrethrum Post*, Vol. 16(3), 91-92.

Pandita, P.N.; Blurt, B K; Sharma. S.D. & Chisti, A.M. (1978) Role of flower maturity on the yield of pyrethrins in pyrethrurn (*Chrysanthemum cinerariaefolium*) *Indian J. Pharmaceutical Sci.*, Vol. 40(5), 163-164.

Pareek, S.K.; Maheshwari, M.L. & Gupta, R. (1985) Cultivation of periwinkle in north India. *Indian Hort.*, Vol. 30(3), 9-12.

Pareek, S.K.; Srivastava, V.K; Maheshwari, M.L. and Gupta, R. (1995) Performance of advanced opium poppy (*Papaver somniferum*) selections for latex, morphine and seed yields. *Indian J. Agri. Sci.* Vol. 65 (4), 498-502.

Pareek, S.K; Maheshwari, M.L. & Gupta, R. (1991) Chemical weed control & water management in Periwinkle, Vetiver and Palmarosa oil grass (S P Raychoudhuri, Ed.). *Today and Tomorrow Printers and Publishers*, New Delhi, pp. 167-170.

Pareek, S.K; Srivastava, V.K & Gupta. R. (1989) Effect of source and mode of nitrogen application on senna (*Cassia angustifolia* Vahl). *Trop. Agri., Clyinidad.* Vol. 660, 69-72.

Paspatis, E.A., (1995) Effects of gibberellic acid (GA.) application and nitrogen fertilization on yield and quality of celery. *Annales de l'Institut Phytopathologique Benaki*, Vol. 17(2), 131145.

Patidar, H., Nigam, K.B. & Kandalkar, V S. (1993) Selections for high yield of ashwagandha. *Curr Res.University of Agricultural Sciences Bangalore.* Vol. 22 (3-5), 50-51.

Patil, S.A. (1989) Cytological, genetical and physiological studies in *Coleus forskohlii* (Poir) Briq. M. Phil. Thesis, Shivaji Univ., Kolhapur, Maharastra.

Patra, N.K.; Chauhan, S.P; Ram, R.S. & Kandpal, A. (1996) High alkaloid yielding big capsulated dominant mutant in opium poppy. *Mutation breeding Newsletter.* Vol. 42, 10-11.

Paturde, J.T; Wankhade, S.G.; Deo, D. & Khode, P.P. (2000) Safed musli- an important medicinal plant. Indian Farming, June issues, 8-9.

Paurohit, A.N. (2000) Harvesting Herbs-2000. *Bishen Singh Mahendra Pal Singh*, Dehradun.

Petrishek, I.A; Lovkova, M. Y; Grinkevich, N.L; Oriova, L.P. & Poludennyi, L.V. (1984) Effects of cobalt and copper on the accumulation of alkaloids in *Atropa belladona*. *Biol. Bull. Acod. Sci. USSR*, Vol. 10(6), 509-516.

Postal, S. (1994) Carrying Capacity: Earth's Bottom Line. State of the World, World Watch Institute, Washington, D.C.

Prajapati, N.D.; Purohit, S.S.; Sharma A.K. & Kumar, T (2003) A Hand book of medicinal Plants. *Agrobios (India),* Jodhpur.

Prakash. (1994) Intervarietal hybridisation and induced autotetraploidy in *Coleus forskohlii* Briq. Ph.D. Thesis, UAS, Bangalore.

Proctor, J.T.A. & M.J. Tsujita (1986) Air and root zone temperature effect on the growth and yield of American ginseng. J. Hort. Sci., 61: 129-134.

Proctor, J.T.A. & W.G. Bailey (1987) Ginseng industry, botany and culture. *Hort. Rev.*, Vol. 9, 187-236.

Proctor, J.T.A., Lee, J.C. & Sung-Sik Lee. (1990) Ginseng production in Korea. *Hort. Sci.*, Vol. 25, 746-754.

Proctor, J.T.A., Wang, T.S. and W.G. Bailey (1988) East meets west: Cultivation of American ginseng in China. *Hort. Sci.*, Vol. 23, 968-973.

Pullaiah, T. (2002) Medicinal Plants of India by (2 Vols.). *Regency Publications*, New Delhi.

Rai, L & Sharma, E (1994) Medicinal Plants of Sikkim Himalaya -Status, Usage & Potential. Bishen Singh Mahendra Pal Singh, Dehradun.

Ramakrishnan, T.S. (1954) Leaf spot disease of turmeric (*Curcuma longa*) caused by *Colletotrichum capsici* (Syd.) Butter and Bisby. *Indian Phytopath.*, Vol. 7, 111- 117.

Ramp, D.D. (1986) Biological extinction in Earth History. *Sciences*, Vol.28.

Rao, B.R.R. & Singh, S P, (1982). Spacing and nitrogen studies in pyrethrum (*Chrysanthemum cinerariaefolium*). *J. Agric. Sci. -UK.* Vol. 99(2), 457-459.

Rao, B.R.R.; Bhattacharya, A.K. & Kaul, P N. (1993) Effect of genotype and time of harvesting on yield and quality of periwinkle (*Catharanthus roseus*). *Indian J. Agron.* Vol. 38(2), 344-345

Rao, B.R.R.; Singh, S.P. & Rao. E.VS.P. (1983) Nitrogen, phosphorus and potash fertilizer studies in pyrethrurn (*Chrysanthemum cinerariaefolium*). *J. Agric. Sci. - UK.* Vol. 100(2), 509-5 11.

Rao, R.R., Singh, S.R & Rao, B R. (1982) Effect of fertilization on the flower yield and pyretrhins content in pyrethrum . Proceedings officinal Seminar on Medicinal and Aromatic Plants.

Rao, VG. (1967) A new leaf spot disease of *Datura metel* from India. *Indian. Phytopath.* Vol. 59, 46-48.

Ratnayaka, H; Meurer-Grimes, B & Kincaid, D. (1998) Increasing sermoside yields in Tinnelvelly senna (*Cassia angustifolia*). *Pl. Medica,* Vol. 64(5), 438-442.

Rawat, T. (1997) Growing market of herbal medicines. *Jour. NTFP.* Vol. 4 (1/2), pp 87-88.

Riley, H.P. & V.J. Hoff (1961) Chromosome studies in some South African Dicotyledons. *Canadian J. Genet,* Vol. 3(3), 260-271.

Roth, R.J. & N. Action (1987) Isolation of arteanuic acid from Artemisia annua. Planta med., 53:501-507.

Runnels, H.A. & J.D. Wilson (1933) Controlof Alternaria blight of ginseng with Bordeaux mixture and injuries accompanying its use. Ohio Agr. Expt. Station, *World Bot.* Vol. 522. pp. 15.

Sadgopal & Verma, B. S. (1952) *Indian Forester.* Vol. 78, 26-33.

Sakson, N. (1980) Effect of irrigation and mineral fertilization on yield and alkaloid content of belladonna. International Society for Horticultural Science: Second international symposium on spices and medicinal plants and 27th scientific meeting of GA. *Acta Horticulture,* pp. 253-260.

Sakson, N. (1981) Effect of different substrate moisture levels and harvesting methods on cropping of *Atropa belladona. Herba Polonica,* Vol. 27(3), 221-228.

Samuel, J. C.; Sivaraman, K & Singh, H. P. (2001) Medicinal and Aromatic plants. South Asian. Agri Business & Horticulture. Vol. 1 (4): 49-54.

Sarin, Y.K. (1996) Illustrated manual of Herbal Drugs, CSIR, ICMR, New Delhi, India.

Sastrv, K.P; Kumar, D; Saleem. S.M. & Radhakrishnan, K., (1997) Effect of phosphorus on growth of pyrethrum plants (*Chrysonthenium cinericefolium*) and on flower yields. *Pyrethrum Post,* Vol. 19(4), 132-138.

Sastry, K.P. & Singh, S.P. (1990) The effect of phosphor-us and potassium application on the flower yields of pyrethrum. *Pyrethrum Post*, Vol. 17(4), 130-132.

Sastry, K.P.; D. Kumar; Saleem, S.M.; Krishnan, K.R.; Sushil K. D. & Kumar. S. (1999) Effective phosphorus fertilization for increased flower yield in pyrethrum, (*Chrysanthemum cinerariaefolium*), under low pH soil conditions. *J. Med. Aromatic Pl. Sci,.* Vol. 21 (1), 4-7.

Sastry, K.P; Dinesh Kumar; Saleem, S.M.; Singh, S.P. & Kumar, D.(1989) Effect of different spacing on the growth and flower yield of pyrethrum (*Chrlysanthernum cinerariaefoliurn*).. *Pyrethrum Post*, Vol. 11(13), 98-100.

Sastry, K.R; Singh, S.P; Dinesh K.; Saleem, S.M. & Kumar. D. (1988) Effect of soil pH on the growth and yield of pyrethrum (*Chrysanthemum cinerarmefolium*) Vis. *Pyrethrurn Post*, Vol. 17, 24-25.

Sattar, A.; Alam, M.; Janardhana, K.K. & Hussain, A. (1981) A new leaf spot disease of Japanese mint caused by *Corcospora cassicola*. *Indian Phytopath*. Vol. 34, pp 404.

Shah, R.R & Dalal, K.C. (1980) *In vitro* multiplication of *Glycyrrhiza*. *Curr. Sci*. Vol. 49: pp 69.

Shah, S. C. and Gupta, L.K. (1979) Response of some presowing treatments on the germination of Atraria belladonna Linn. seeds. *Indian J. Agric. Res.*, Vol. 13(l), 53-54.

Shah, S.C.; Gupta, L.K.; Harish S. & Singh. H. (1994) Effect of growth regulators oil growth yield and pyrethrin content in pyrethrum (*Chrysanthemum cinerariaefolium* (Trev.) Bocc.). *Recent Hart.*, Vol. l(l), 84-87

Shah, V & B.S. Kalakot (1996) Development of *Coleus forskohlii* as medicinal crop. In: Domestication and Commercialisation of non-timber forest products in Agro-forestry systems. Non-timber forest products Series 9. FAO publications, pp. 212-217.

Shah, V. (1996) *Coleus forskohlii* (Willd.) Briq. - An overview In: Supplement to cultivation and utilisation of medicinal plants. (Ed. S.S. Handa and M.K. Kaul) pp. 385-412.

Shah, V; S.Y. Bhat; B.S. Bajwa; H. Dornauer & N.J. de Souza (1980) The occurrence of forskolin in the labiatae. *Planta Med.* 39, 183-185.

Shanta Kumari. P. & Nair, M.C. (1981) Some new host record for *Colletotrichum gloeosporioides* in India. *Indian Phytopath*. Vol. 34, 402-403.

Sharma, A. K (2002) A Handbook of Organic farming. *Agrobios (India)*, Jodhpur.

Sharma, A. K (2002) Biofertilizers for sustainable agriculture *Agrobios (India)*, Jodhpur.

Sharma, J.R; Lal. R.K.; Gupta. A.P.; Misra, H.O.; Pant, V; Singh, N.K. & Pandey, V. (1999) Development of non-narcotic (opiumless and alkaloid free) opium poppy (*Papaver somniferum*). *Pl. Breeding*. Vol. 118(5), 449-452.

Sharma, N. & Chandel, K.P.S. (1992) Low temperature storage of *Rauvolfia serpentina*, An endangered, endemic medicinal plant. *Plant Cell Reports* Vol. 11, 200-203.

Sharma, O.L. & Napalia, V. (1997) Efficacy of selected herbicides in opium poppy. *Madras Agri. J.* Vol.4, 706-707.

Sharma, O.T & Mahmud, K.A. (1951) A leaf rot of *Mentha uiridis* L. caused bv *Rhizoctonia solani* Kuehn. *Agric. College*, Nagpur, 26:23.

Sharma, P (1996) Opportunities in and consultants to the sustainable use of Non Timber Forest resources in the Himalayas. ICIMOD,. Vol. 25, 9-11.

Sharma, R (1999) *Artemisia maritima. MFP News*, Vol. 9, No. 3 p. 13-14.

Sharma, R (1999) *Colchicum lutem. MFP News*, Vol. 9, No. 2 p. 13.

Sharma, R (1999) *Gentiana kurroo. MFP News*, Vol. 9, No. 4 p. 14.

Sharma, R (1999) *Taxus baccata. MFP News*, Vol. 9, No. 1 p. 17-18.

Sharma, R (2000) *Curcuma caesia. MFP News*, Vol. 10, No. 4 p. 14

Sharma, R (2000) *Gloriosa superba. MFP News*, Vol. 10, No. 3 p. 10-11.

Sharma, R (2000) *Saussurea costus. MFP News*, Vol. 10, No. 2 p. 17-18

Sharma, R (2000) *Valeriana jatamansi. MFP News*, Vol. 10, No. 1 p. 14.

Sharma, R (2001) *Poenia emodi. MFP News*, Vol. 11, No. 1 p. 14

Sharma, R (2001) *Tylophora indica. MFP News*, Vol. 11, No. 2 p. 10

Sharma, R (2002) Aushadhiya Evam Sugandhiya Paudhuon ki Krishi Taknik. *Daya Publishing House*, New Delhi.

Sharma, R (2003) Dictionary of Ayurveda. *Daya Publishing House*, New Delhi.

Sharma, R (2003) Medicinal Plants of India - An Encyclopeadia. *Daya Publishing House*, New Delhi.

Shastri, K.S.M. (1969) Investigation and control of some important diseases of medicinal plants. *Indian Phytopath.* Vol. 22, 140-142.

Shibata, S. (1977) Saponins with biological and pharmacological activity. In: New Natural Products and Plant Drugs with Pharmacological, Biological or Therapeutic Activity. (Eds. H. Wagner and P. Wolff). *Springer Verlag, Berlin.* p. 177-196.

Shibata, S., Tanaka, O., Shoji, J. & H. Sato. (1985) Chemistry and pharmacology of Panax. Economic and medicinal Plant Research. Vol. I. (Eds: H. Wagner, H. Hikino, & N.R. Farnsworth) *Acad. Press. Lond.*, 218-274.

Shiva, A & Sharma, R (2000) Annonated bibliography on Non-timber Forest Products (NTFP) : Related articles. COMFORPT, Dehradun.

Shoemaker, P.B. & T.R. Konsler. (1932) Fungicide evaluation for Alternaria blight of ginseng. 1981. Fungal nematicide tests. Am. Phytopath. Soc., Vol. 73, pp. 54.

Shoemaker, Paul B. & T.R. Konsler (1981) fungicide evaluation for ginseng blight (*Alternaria panax* Webel). Proceed. of Fifth National Ginseng Congress.

Singh D.V & Mathur, R.S. (1964) *Cercospora daturicila* (Speg.) Vassiljevsky on *Datura stranionium* L. *Sci. Cult.* Vol. 30, 604-605.

Singh J. & Chand R. (1995) Belladonna In: Advances in Horticulture. Vol. 11. Medicinal and Aromatic Plants (Eds, Chadha. K.L. and Gupta, R.), *Malhotra Publishing House*, New Delhi. pp. 283-284.

Singh S.; Sushil Kumar; Singh, S. & Kumar, S. (1998) *Withania somnifera*: The India ginseng ashwagandha.

Singh, A., Kaul, V.K. Mahajan, V.P.Singh, A., Misra, L.N., Thakur, R.S. & A.Husain. (1986) Introduction or Artemisia annua in India and isolation of artemisinin, a primary antimalarial drug. *Indian J. Pharm. Sci.*, Vol. 48, 137-138.

Singh, A., Vishwakarms. R.A. & A. Husain. (1988) Evaluation of *Artemisia annua* strains for higher artemisinin production. *Planta med.*, Vol. 54: 475-476.

Singh, A.K.; Bisen, S S. & Biswas, S.C. (1996) Potentiality of coal and copper mine overburdens for growing medicinal plants. *Environ. Ecol.*, Vol. 14(2), 415-421.

Singh, A.K; Bisen, S.S.; Singh, R.B. & Biswas, S.C. (1998) Effectiveness of compost towards increasing productivity of some medicinal plants in skeletal soil. *Adv. For. Res.* Vol.18, 64-83.

Singh, G.; Singh. R.N. & Bisht, I.S. (1985) Control of leaf spot disease of belladonna caused by *Ascohyta atropae. Indian Phytopath.*, Vol. 38(1), 183-184.

Singh, S.P.; Sharma, J.R.; Rao, B.R.R. & Sharma, S.K. (1988) Genetic improvement of pyrethrum: 111. Choice of improved varieties and suitable ecological niches. *Pyrethruin Post*, Vol. 17(l), 12-16.

Singh, S.P; Shukla. S. & Khanna (1995) Opium poppy In: Advances in horticulture vol. 11 'Medicinal' and aromatic plants (K.L. Chadha & Gupta, R., Eds.). *Malhotra Publishing House*, New Delhi, India. pp. 535-574.

Singh, V.K; Govil, J.N. & Singh, G (2002) Recent Progress in Medicinal Plants - Ethnomedicine and Pharmacognosy. *SCI Tech Publishing* LLC, USA.

Singh, V.K; Govil, J.N. & Singh, G (2002) Recent Progress in Medicinal Plants - Crop improvement, production technology, trade and commerce. *SCI Tech Publishing* LLC, USA.

Singh. M.; Chaudburi, P K. & Sharma. R.P., (1995) Constituents of the leaves of *Cassia angustifolia. Fitoterapia*, 66(3), pp. 284

Singh. S.T; Sharma, J.R; Misra, H.O.; Lal. R.K; Gupta..M..M. & Tajuddin, (1997) Development of new variety sona of senna (*Cassia angustifolic*). *J. Med. Aromatic Pl. Sci.*, Vol. 19(2), pp. 446.

Sinsh et. al (2003) Herbal Medicine of Manipur (Coloured Encyclopeadia). *Daya Publishing House*, New Delhi.

Snedecor, G.W and Cochran, W.G. (1972) Statistical Methods. Lowa State University Press, USA.

Soldati, F. and O. Tanaka. (1984) Panax ginseng: Relation between age of plant and content of ginsenosides. *Planta Med.*, Vol. 50, 351-352.

Souza, M.V. and A. Ali. (1984) Ginseng: Control of grass weeds. *Res. Rep. Eastern Sect. Exp. Comm. Weeds.* p. 495.

Srivastava, A.K. & Srivastava, R. (1971) *Mentha piperata* L. A new host of *Alternaria tenuis* Auct. *Indian Fing.* Vol. 97, pp 34.

Srivastava, G. N. (1994) Medicinal and Aromatic Plants - A potential source of forest produce. Agricultural Marketing. Apr-Jun issue; pp 50-54.

Srivastava, J.; Lambert J. & Vietmeyer, N. (1995) Medicinal Plants; A growing role in development Washington D.C., USA, Agricultural and Natural Resources Department, The World Bank.

Stenton, D. (1893) American Ginseng. *Kew Bull.* pp. 70-75.

Subrat, N.; Iyar, M & Prasad, R (2002) Ayurvedic medicine Industry: Current Status and sustainability. ETS Publishers , New Delhi.

Suchorska, K. and Ruminska, A (1980) Effect of pre-sowing treatment with GA:, on plant development and drug value of *Datura innoxia* and *Atrapa belladonna*. International Society for Horticultural Science: Second international symposium on spices and medicinal plants and 27th scientific meeting of GA. *Acta Horticulture*, pp. 221-226.

Tamar, S.S; Abbas. M.; T. Singh; Nigam, K.B; & Singh, I. (1994) Effect of phosphate solubilizing bacteria and phosphate in opium poppy (*Papaver somniferum*). *Indian J. Agro.,* Vol. 39(4), 713-714.

Tanaka, O. and Kasai, (1984) Saponins of ginseng and related compounds. In: progress in the Chemistry of Organic Natural Products. Eds. W. Herz, H. Grisbach. G.W. Kirby and C.H. Tammi. Springer Verlag, New York. Pp. 1-65.

Tandon, J.S; M. M. Dhar; S. Ramakurnar, and Venkatesan (1977) Structure of coleonol, a biologically active diterpene from *Coleus forskohlii. Indian J. Chem.* Vol. 15 B, 880-883.

Taylor, D. (1997) Herbal Medicine. *Environmental Health Perspectives* Vol. 105, No.11

Toima. N.M.; Boselah, N.A.E.; Youseel, A.K. & Amme. I.S. (1991) Seed germination of *Atropa belladonna* Linn. *Bull. Fac. Agric. - University of Cairo*, Vol. 42(1), 31-38.

Tomar, S S.; Nigam, K.B.; Pachori, R.S. and Kahar. L.S. (1992) Irrigation schedule based on irrigation water: Cumulative pan evaporation in opium poppy (*Papaver somniferum*). *Indian J. Agric. Sci.,* Vol. 62(5), 313-315.

Tomar, S.S.; Abbas. M. & Nigam, K.N. (1993) Effect of sulphur on growth and yield of opium poppy (*Papaver somniferum*). *Indian J. Agro.,* Vol. 38(2), 346-347.

Tomoda, M., Shimada, K., Konno, C., Sugiyama, K. & H. Hikino. (1984) Partial structure of Panaxan A; a hypoglycaemic glycane of Panax ginseng roots. *Planta Med.,* Vol. 50: 436-438.

Trease, G.E. and W.C. Evans. (1983) Pharmacognosy. Baillier Tindall, London. P. 812.

Tripathi, A.K; Shukla, Y.N.; S. Kumara & Kumar, S. (1996) Ashwagandha (*Withania somnifera* Duna] (Solanaceae): A status report. *J Med. Aromatic Pl. Sci.* Vol. 18(l), 46-62.

Tripathi, YC. (1999) *Cassia angustifolia*- a versatile medicinal crop. *Int. Res Crops J.,* Vol. 10, (2), 121-129.

Trivedi, K.C. (1995) Sarpagandha In: Advances in Horticulture Vol. 11 -Medicinal and aromatic plants i Chadha, K.L. and Gupta, R., Eds.). *Malhotra Publishing House,* New Delhi, India. pp. 453-466.

Trivedi, T P, (1987), Toxicity of pyrethrins against major stored grain insect pests. Pl. Prol. Bull. - India, 39(3):35.

Tyagi, B.R. & R. Dubey (1990) Panchytene chromosome morphology of Artemisia annua L.Cytologia, 55:45-50.

Tyler, V. E. (1996) "Pharmacognosy" what's that? You spell it How? Econ. Bpt. Vol. 50, 3-9.

Uniyal, M.R. (1989). Medicinal flora of Garhwal Himalaya. *Shree Baidyanath Ayurveic Bhawan Pvt. Ltd.* Nagpur.

Valdes,L.J; S.G. Mislankarand & A.G. Paul (1987) *Coleus barbatus (C. forskohlii,)*- The potential new drug forskolin (Coleonol). *Econ. Bot.* Vol. 4, 474-483.

Vasanth, S.Gopal, R.H. & Bhima. (1990) Plant antimalarial agents. *J. Scient. Ind. Res.* 49:68-77.

Vishwakarma, R.A; B. R. Tyagi; B. Ahmed & A. Hussain (1988) Variation in forskolin content in the roots of *Coleus forskohlii. Planta Med.* Vol. 54(5), 471-472.

Vyas, S.T & Nein, S., (1999) Effect of shade on the growth of *Cassia angustifolia. Indian Forester*, Vol. 125(4), 407-410.

Warrier, P.K. *et. al* (1996) Indian Medicinal Plants, *Orient Longman,* Hyderbad, India.

Watt, G. (1889) Economic products of India. Vol. 1, 279-281.

WHO (2003) Quality Control methods for medicinal plants materials.

Williams, L., and J.A. duke. (1978) Growing Ginseng. *U.S. Dept. of Agriculture, Farmers' Bull.* Vol. 2201. p. 8.

Williams, L.O. (1957) Ginseng. *Econ. Bot.*, Vol. 11, 344-348.

Winter, H.F. (1963) *Economic Bot.*, Vol. 17, 195-199.

Yanagihara, H.; R. Sakata; Y. Shoyama & Murakami (1996) Rapid analysis of small samples containing forskolin using mono clonal antibodies. *Planta Med.* Vol. 62, 169-172.

Youngken, H.W (2003) Natural Drugs. *Biotech Publishing Company*, New Delhi.

Zilkey, B.F. and B.B. Capell. (1984) Annual and perennial grass and broad leaf weed control in ginseng. *Res. Rept. Eastern Sect. Exp. Comm. Weeds.* p. 496.

Zilkey, B.F. and B.B. Capell. (1985) Annual and perennial grass and broad leaf weed control in ginseng. *Res. Rep. Eastern Sect. Exp. Comm. Weeds.* p. 444.

INDEX OF SCIENTIFIC BOTANICAL NAMES

INDEX OF COMMON VERNACULAR NAMES

Hin.–Hindi; *Eng..*–English; *Mar.*–Marathi; *Mal.*–Malyalam; *Kan.*–Kannad; *Punj.*–Punjabi; *Kash.*–Kashmiri; *Him.*–Himachali; *Garh.*–Garhwali; *Kaon.*–Kumaon; *Tam.*–Tamil; *Tel.*–Telgu; *San.*–Sanskrit; *Guj.*–Gujrati; *Ori.*–Orissi; *Una.*–Unani.

A

Abhaya (San.)-53
Abhini (Tel.)-145
Abini (Tam.)-145
Achuvagandi (Tam.)-31
Adavi lavanga patri (Tel.)-114
Adavi nabhi (Tel.)-92
Adaviamudamu (Tel.)-71
Afim (Hin., Kan)-145
Afium (Mal.)-145
Afyun (Hin.)-145
Agarwood (Eng.)-227
Agnisikha (San.)-92
Ahifen (San.)-145

Ainskati (Ori.)-149
Alalai (Kan.)-53
Alkushi (Beng.)-68
Allale (Kan.)-53
Amalak (Hin.)-27
Amalakamu (Tel.)-27
Amalaki (San.)-27
Amali (Guj.)-27
Amalpori (Mal.)-170
Amlaki (Beng.)-27
Amlika (Hin.)-27
Amrita (San.)-53
Amukkiram (Mal.)-31
Amvala (Hin.)-27

Anatamul (Hin.)-203
Angurshafa (Hin.)-41
Anjubar (Unani)-50
Aonla (Hin)-27
Appaiccevakam (Tam.)-71
Aralaikai (Kan.)-53
Arili (Kan.)-53
Arjun (Hin.)-201
Arnica (Eng.)-202
Aruveppu (Mal.)-140
Aryaveppu (Mal.)-140
Asgandh (Hin.)-31
Asgandha (Mah.)-31
Ashavgandha (Guj.)-31